Edexcel

higher

GCSE Modular Mathematics

4-Speed Revision Guide

Keith Pledger

Gareth Cole

Peter Jolly

Graham Newman

www.heinemann.co.uk
- Free online support
- Useful weblinks
- 24 hour online ordering

01865 888058

Heinemann

Heinemann is an imprint of Pearson Education Limited, a company incorporated in England and Wales, having its registered office at Edinburgh Gate, Harlow, Essex, CM20 2JE. Registered company number: 872828

www.heinemann.co.uk

Heinemann is the registered trademark of Pearson Education Limited

First published 2007

12 11 10 09 08 07
10 9 8 7 6 5 4 3 2 1

British Library Cataloguing in Publication Data is available from the British Library on request.

ISBN 978 0435 807177

Typeset by Tech-Set Ltd, Gateshead, Tyne and Wear
Cover design by Tony Richardson
Cover photo/illustration © Digital Vision
Printed in the UK at Scotprint

Acknowledgements
Every effort has been made to contact copyright holders of material reproduced in this book. Any omissions will be rectified in subsequent printings if notice is given to the publishers.

Welcome to 4-speed revision!

4-speed revision lets you plan your revision at the speed you need.

- 1st speed pages are **green** – use these to revise topics thoroughly.
- 2nd speed pages are **orange** – use the topic tests on these pages to identify your weaknesses in detail.
- 3rd speed pages are **blue** – use the subject tests on these pages to help you identify the topics you need to work on the most.
- 4th speed pages are **red** – use these to check you know the key facts.

Pages iv to vi show you how the different speed pages work – so you can choose your speed.

Contents

Revising for your GCSE maths exam

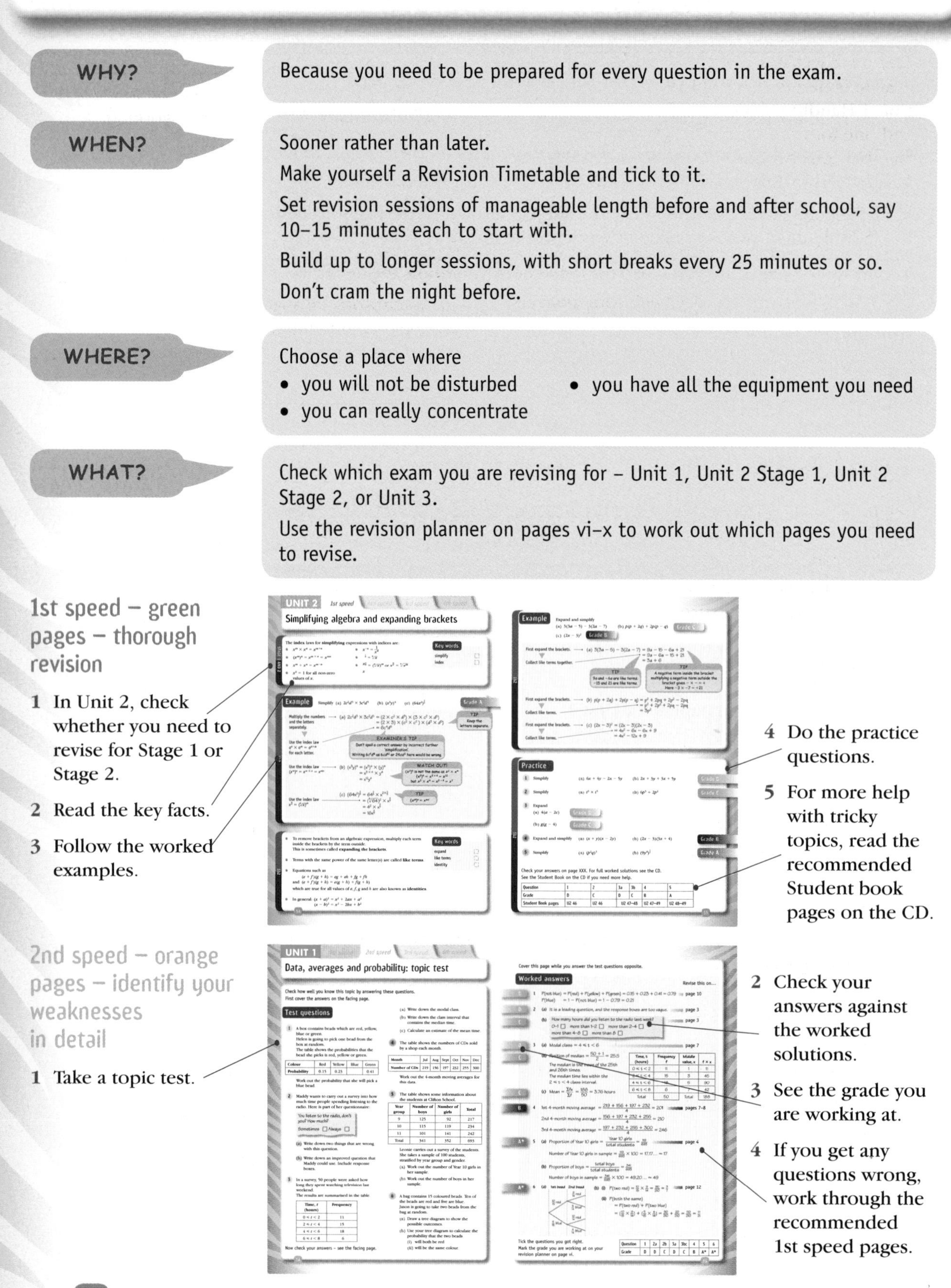

WHY?

Because you need to be prepared for every question in the exam.

WHEN?

Sooner rather than later.

Make yourself a Revision Timetable and tick to it.

Set revision sessions of manageable length before and after school, say 10–15 minutes each to start with.

Build up to longer sessions, with short breaks every 25 minutes or so.

Don't cram the night before.

WHERE?

Choose a place where

- you will not be disturbed
- you can really concentrate
- you have all the equipment you need

WHAT?

Check which exam you are revising for – Unit 1, Unit 2 Stage 1, Unit 2 Stage 2, or Unit 3.

Use the revision planner on pages vi–x to work out which pages you need to revise.

1st speed – green pages – thorough revision

1 In Unit 2, check whether you need to revise for Stage 1 or Stage 2.

2 Read the key facts.

3 Follow the worked examples.

4 Do the practice questions.

5 For more help with tricky topics, read the recommended Student book pages on the CD.

2nd speed – orange pages – identify your weaknesses in detail

1 Take a topic test.

2 Check your answers against the worked solutions.

3 See the grade you are working at.

4 If you get any questions wrong, work through the recommended 1st speed pages.

3rd speed – blue pages – focus on your top priority topics

1 Take a subject test – there is one for Unit 1 Handling data and one for each of Number, Algebra and Shape, space and measure in both Unit 2 and Unit 3.

UNIT 1 3rd speed

Handling data: subject test

Exam practice questions

1 20 people do a lap round a race track.
Here are their times to the nearest second:

62	46	39	53	28	44	65	41	48	37
36	49	51	46	39	27	60	50	45	33

(a) Draw an ordered stem and leaf diagram to show this information. Include a key.
(b) Use your stem and leaf diagram to write down the median.

2 100 students were asked how they came to school that day.
Some of the results are shown in the two-way table.

EXAMINER'S TIP
You wouldn't get a whole question on this on the Higher paper, but it might come up as part of a question.

	Car	Walk	Cycle	Total
Year 7		13	9	41
Year 8	5			22
Year 9		18		
Total	36		21	100

(a) Complete the two-way table.
(b) One of these students is picked at random.
Write down the probability that this student
(i) came to school by car (ii) is in Year 7 and cycled to school.

3 A freezer contains four flavours of ice-cream – vanilla, chocolate, strawberry and mint.
The table shows the probabilities that Marcus chooses vanilla, chocolate or mint.

Flavour	Vanilla	Chocolate	Strawberry	Mint
Probability	0.2	0.4		0.15

Work out the probability that he picks a strawberry ice-cream.

4 The table shows the heights and weights of ten students.

Weight (kg)	75	65	82	76	71	65	77	70	72	68
Height (cm)	185	182	191	188	184	166	175	178	181	180

(a) Draw a scatter graph to show this information.
(b) What type of correlation do you find?
(c) Draw a line of best fit.
(d) Use your scatter graph to estimate
(i) the weight of a student whose height is 188 cm
(ii) the height of a student whose weight is 74 kg.

26

5 The table shows information about the times it took a group of students to do their homework one day.
(a) Write down the modal class.
(b) Calculate an estimate of the mean.

Time, t (hours)	Number of students
0	3
$0 < t \leqslant 1$	14
$1 < t \leqslant 2$	17
$2 < t \leqslant 3$	5
$3 < t \leqslant 4$	1

6 Use data from the table in question 5 for this question.
(a) Construct a cumulative frequency table.
(b) Use your table to draw a cumulative frequency graph.
(c) Use your graph to find an estimate for
(i) the median
(ii) the interquartile range.

7 A bag contains ten coloured beads. Six of the beads are red and four are blue.
Bob is going to take one bead from the bag at random, put it back, and then take another bead at random.
(a) Draw a tree diagram to show the possible outcomes.
(b) Use your tree diagram to calculate the probability that the two beads
(i) will both be red
(ii) will be the same colour.

8 There are 1200 students at Shefford High School. 640 are girls and 560 are boys.
Louise wants to take a sample of 50 students, stratified by gender.
How many girls and how many boys should there be in the sample?

9 The table gives information about some people's waiting times at a bus stop.
Draw a histogram to show this data.

Waiting time, t (minutes)	Frequency
$0 < t \leqslant 10$	32
$10 < t \leqslant 15$	20
$15 < t \leqslant 30$	24
$30 < t \leqslant 35$	4
$t > 35$	0

10 A drawer contains five yellow scarves and four white scarves.
June selects a scarf from the drawer at random, and then Heather selects a scarf from the drawer at random.
Work out the probability that
(a) they both select a yellow scarf
(b) they select scarves of different colours.

Check your answers on pages 170–171. For full worked solutions see the CD.
Tick the questions you got right.

Question	1	2a	2b	3	4ab	4cd	5	6	7	8	9	10
Grade	D	E	D	D	D	C	C	B	B	A	A	A*
Revise this on page	16	2	10	10	20		7	22–23	11	4	17	12

Mark the grade you are working at on your revision planner on page vi.
Go to the pages shown to revise for the ones you got wrong.

27

2 Check your answers.

3 See the grade you are working at.

4 If you get any questions wrong check the worked solutions on the CD.

5 Then work through the recommended 1st speed pages.

4th speed – red pages – instant overview of the key facts

1 Check that you understand the key facts for each subject.

2 Read the 1st speed pages of any topics you are not sure about.

3 Learn any formulae that are not on the formulae sheet.

UNIT 1 4th speed

Handling data

Data, averages and probability

- **Two-way tables** are used to record or display information that is grouped in two categories.
- When you are writing questions for a **questionnaire**:
  - be clear what you want to find out, and what data you need
  - ask short, simple questions
  - provide **response boxes** with possible answers
  - avoid questions which are too vague, too personal, or which may influence the answer (**leading questions**).
- A **stratified sample** is one in which the population is divided into groups called strata and each of the strata is randomly sampled. The proportions of the different strata in the sample are the same as in the whole population.
- A **systematic sample** is one where every nth item is chosen.
- The **mode** of a set of data is the value which occurs the most often.
- The **median** is the middle value when the data is arranged in order of size. If there are n data values the median is the $\frac{n+1}{2}$th value.
- The **mean** of a set of data is the sum of the values divided by the number of values:
  $\text{Mean} = \frac{\text{sum of values}}{\text{number of values}}$
- The **range** of a set of data is the difference between the highest and lowest values:
  range = highest value − lowest value.
- For **grouped data**:
  - the **modal class** is the class interval with the highest frequency
  - you can state the **class interval** that contains the median
  - you can calculate an **estimate of the mean** using the middle value of each class interval.
- A **trend line** is the line of best fit drawn through the moving averages plotted on a graph.
- If the **probability** of an event happening is p, the probability of it *not* happening is $1 - p$.
- **Estimated probability** $= \frac{\text{number of successful trials}}{\text{total number of trails}}$
- If events A and B are **independent**, P(A and B) = P(A) × P(B).

28

Frequency charts and scatter graphs

- A **stem and leaf diagram** shows the shape of a distribution and keeps all the data values. It needs a **key** to show how the stem and leaf are combined.

0 | 9
1 | 2, 3
2 | 0, 7
3 | 4, 8
4 | 1
Key: 1 | 2 means 12

- You can use a stem and leaf diagram to find the **median**, the **upper and lower quartiles** and the **interquartile range**.
- A **histogram** is used to display grouped continuous data.
The areas of the rectangles are proportional to the frequencies they represent.

Waiting time, t (minutes)	Frequency
$0 < t \leqslant 10$	16
$10 < t \leqslant 15$	20
$15 < t \leqslant 20$	19
$20 < t \leqslant 30$	28
$30 < t \leqslant 60$	12
$t > 60$	0

Frequency density
Time (minutes)
0 10 20 30 40 50 60
Key

- A **scatter graph** shows the relationship between two sets of data.
- The **line of best fit** is a straight line that passes through or is close to the plotted points on a scatter graph.
- A linear relationship between two sets of data is called **correlation**.

Positive correlation Negative correlation No correlation

- A **cumulative frequency graph** can be used to estimate the median, the upper and lower quartiles and the interquartile range.

Cumulative frequency
LQ median UQ
IQR

29

Practice examination papers

1 Do the practice examination papers.

2 Use the 'Maths language in exams' page to decode what the question is asking you for.

3 For each question, how many marks is it worth? Make sure that you show all your working out to get all the marks.

4 Check your answers.

5 Check the worked solutions on the CD for any you got wrong and work through the relevant 1st speed pages.

Revision planner (pages vi–x)

The revision planner shows the contents of each page. You can use it to help you plan your revision:

1 Tick the topics you understand.

2 After each test, mark on the grade you are working at.

Maths language in exams (page xi)

Use this page to decode what a question really means.

CD

- Full text book for each unit. Use them to revise topics in more depth.
- Worked solutions to practice questions, practice exam papers and 3rd speed subject test questions.

Revision planner

Use this to help you plan your revision.

- Tick the topics you understand ✓
- After each test, mark the grade you are working at Ⓒ

UNIT 1

Speed	Topic		I need to revise	I understand ✓
1st	**Collecting and organising data (I)**	**page 2**		
	Two-way tables			
	Questionnaires			
1st	**Collecting and organising data (II)**	**page 4**		
	Surveys and samples			
1st	**Averages (I)**	**page 6**		
	Mode			
	Median			
	Mean			
	Modal class			
1st	**Averages (II)**	**page 8**		
	Four-point moving averages			
1st	**Probability (I)**	**page 10**		
	Probability			
	Estimated probability			
	Independent events			
	Tree diagrams			
1st	**Probability (II)**	**page 12**		
	Conditional probability			
2nd	**DATA, AVERAGES AND PROBABILITY topic test**		**I am working at grade** ◯	
1st	**Frequency charts (I)**	**page 16**		
	Stem and leaf diagrams			
	Quartiles and interquartile range			
	Histograms			
	Frequency polygons			
1st	**Frequency charts (II)**	**page 18**		
	Frequency density			
1st	**Scatter graphs, correlation and the RPI**	**page 20**		
	Scatter graphs			
	Correlation			
	Retail Prices Index (RPI)			
1st	**Cumulative frequency and box plots**	**page 22**		
	Cumulative frequency graphs			
	Box plots			
2nd	**FREQUENCY CHARTS AND SCATTER GRAPHS topic test**		**I am working at grade** ◯	
3rd	**HANDLING DATA subject test**		**I am working at grade** ◯	
4th	**UNIT 1 KEY FACTS**			
	UNIT 1 EXAMINATION PRACTICE PAPER			

UNIT 2

Speed	Topic	I need to revise	I understand ✓
1st	**Operations** **page 34**		
	Adding, subtracting, multiplying and dividing negative numbers		
	Factors and multiples		
	Prime numbers and prime factors		
	Highest common factors		
	Lowest common multiples		
1st	**Indices, powers and roots** **page 36**		
	Indices and powers		
	Square and cube roots		
	Multiplying and dividing powers		
	Fractional indices		
	Order of operations		
	Reciprocals		
2nd	**INTEGERS topic test**	**I am working at grade** ◯	
1st	**Decimals and rounding** **page 40**		
	Rounding to decimal places		
	Rounding to significant figures		
	Adding, subtracting, multiplying and dividing decimals		
1st	**Standard form and bounds** **page 42**		
	Standard form		
	Upper and lower bounds		
2nd	**DECIMALS AND ROUNDING topic test**	**I am working at grade** ◯	
1st	**Fractions** **page 46**		
	Adding and subtracting fractions and mixed numbers		
	Multiplying and dividing fractions and mixed numbers		
	Terminating decimals		
	Recurring decimals		
1st	**Percentages, fractions and decimals** **page 48**		
	Comparing fractions, decimals and percentages		
	Comparing proportions		
	Finding a percentage of an amount		
2nd	**FRACTIONS, DECIMALS AND PERCENTAGES topic test**	**I am working at grade** ◯	
3rd	**NUMBER subject test**	**I am working at grade** ◯	
1st	**Simplifying algebra and expanding brackets** **page 54**		
	Index laws for simplifying expressions with indices		
	Expanding brackets		
	Identities		
1st	**Factorising and algebraic fractions** **page 56**		
	Factorising with highest common factor (HCF)		
	Factorising quadratic expressions		
	Difference of two squares		
	Algebraic fractions		
2nd	**MANIPULATIVE ALGEBRA topic test**	**I am working at grade** ◯	
1st	**Patterns and sequences** **page 60**		
	nth term of a sequence		
	Arithmetic sequences		
1st	**Coordinates and algebraic line graphs** **page 62**		
	Coordinates and plotting straight line graphs		
	Equations of straight lines		

UNIT 2 (cont.)

Speed	Topic		I need to revise	I understand ✓
	Mid-points of line segments			
	3-D coordinates			
2nd	**PATTERNS, SEQUENCES, COORDINATES AND GRAPHS topic test**		**I am working at grade** ◯	
3rd	**ALGEBRA subject test**		**I am working at grade** ◯	
1st	**Working with angles**	**page 68**		
	Angle sums			
	Opposite angles			
	Alternate angles			
	Corresponding angles			
	Bearings			
2nd	**ANGLES topic test**		**I am working at grade** ◯	
	Perimeter, area, volume and measures	**page 72**		
	Conversion between units			
	Speed			
	Perimeter			
	Area			
	Volume			
	Surface area			
	Density			
2nd	**PERIMETER, AREA, VOLUME AND MEASURES topic test**		**I am working at grade** ◯	
3rd	**SHAPE, SPACE AND MEASURE subject test**		**I am working at grade** ◯	
4th	**UNIT 2 KEY FACTS**			
	UNIT 2 EXAMINATION PRACTICE PAPER			

UNIT 3

Speed	Topic		I need to revise	I understand ✓
1st	**Powers, surds and bounds**	**page 86**		
	Standard form			
	Surds and square roots			
	Surds in area and volume calculations			
	Upper and lower bounds in calculations			
1st	**Percentages**	**page 88**		
	Percentage increase and decrease			
	Simple and compound interest			
	Index numbers and price index			
2nd	**POWERS, SURDS, BOUNDS AND PERCENTAGES topic test**		**I am working at grade** ◯	
1st	**Ratio and proportion**	**page 92**		
	Ratio			
	Direct proportion			
	Unitary method			
	Scales			
2nd	**RATIO AND PROPORTION topic test**		**I am working at grade** ◯	
3rd	**NUMBER subject test**		**I am working at grade** ◯	
1st	**Graphs**	**page 98**		
	Equations of straight lines			
	Intercepts			
	Gradients			
	Parallel and perpendicular lines			

Speed	Topic		I need to revise	I understand ✓
1st	**More graphs**	**page 100**		
	Solving simultaneous equations graphically			
	Distance–time graphs			
	Speed–time graphs			
1st	**Curved graphs**	**page 102**		
	Quadratic functions			
	Parabolas			
	Solving quadratic equations			
	Cubic functions			
1st	**Transformations of graphs**	**page 104**		
	Translating graphs			
	Reflecting graphs			
	Stretching graphs			
	Scale factors			
	Vertices of graphs			
2nd	**GRAPHS topic test**		**I am working at grade** ◯	
1st	**Formulae**	**page 108**		
	Algebraic expressions and formulae			
	Substituting into formulae			
1st	**Formulae and proof**	**page 110**		
	Rearranging formulae			
	Fractional indices			
2nd	**FORMULAE topic test**		**I am working at grade** ◯	
1st	**Solving linear equations**	**page 114**		
	Expanding brackets			
	Rearranging equations			
1st	**Solving quadratic and cubic equations**	**page 116**		
	Solving quadratic equations			
	Solving quadratic equations by formulae			
	Trial and improvement method for cubic equations			
1st	**Solving simultaneous equations**	**page 118**		
	Solving simultaneous equations algebraically			
	Solving linear and quadratic equations simultaneously			
1st	**Solving inequalities**	**page 120**		
	Inequalities on the number line			
	Solving inequalities			
	Graphing inequalities			
2nd	**SOLVING EQUATIONS AND INEQUALITIES topic test**		**I am working at grade** ◯	
1st	**Proportion**	**page 124**		
	Direct proportion			
	Constant of proportionality			
	Inverse proportion			
2nd	**PROPORTION topic test**		**I am working at grade** ◯	
3rd	**ALGEBRA subject test**		**I am working at grade** ◯	
1st	**Angles, similarity and congruence**	**page 130**		
	Regular polygons			
	Sums of angles in polygons			
	Similar shapes and enlargements			
	Congruence			

Speed	Topic		I need to revise	I understand ✓
1st	**3-D shapes**	**page 132**		
	Nets			
	Prisms			
	Planes of symmetry			
	Plans			
	Elevations			
1st	**Scale drawing, locus and bearings**	**page 134**		
	Bearings			
	Loci			
	Angle bisectors			
	Perpendicular bisectors			
2nd	**2-D AND 3-D SHAPES topic test**		**I am working at grade** ◯	
1st	**Circle theorems and proof**	**page 138**		
	Tangents			
	Perpendiculars			
	Chords			
	Segments			
1st	**Perimeter, area and volume**	**page 140**		
	Perimeter and circumference			
	Volume of cylinder, cone and frustrum			
	Surface area			
2nd	**PERIMETER, AREA, VOLUME AND CIRCLES topic test**		**I am working at grade** ◯	
1st	**Reflections, rotations and enlargements**	**page 144**		
	Reflections and lines of symmetry			
	Rotations			
	Enlargements and scale factors			
1st	**Vectors and translations**	**page 146**		
	Translations			
	Vectors			
	Column vectors			
	Magnitude of vectors			
	Position vectors			
2nd	**TRANSFORMATIONS topic test**		**I am working at grade** ◯	
1st	**Basic trigonometry**	**page 150**		
	Sine			
	Cosine			
	Tangent			
	Graphs of trigonometric functions			
1st	**Pythagoras' theorem**	**page 152**		
	Using Pythagoras' theorem			
1st	**Advanced trigonometry**	**page 154**		
	Area of a triangle			
	Sine rule			
	Cosine rule			
2nd	**TRIGONOMETRY AND PYTHAGORAS' THEOREM topic test**		**I am working at grade** ◯	
3rd	**SHAPE, SPACE AND MEASURE subject test**		**I am working at grade** ◯	
4th	**UNIT 3 KEY FACTS**			
	UNIT 3 EXAMINATION PRACTICE PAPER			

Maths language in exams

When a question says...	What it means
You must show your working...	You will lose marks if you do not show how you worked out the answer.
Estimate...	Usually means round numbers to 1 significant figure and then carry out the calculation.
Calculate...	Some working out is needed – so show it!
Work out OR Find...	A written or mental calculation is needed.
Write down...	Written working out is not usually required.
Give an exact value of...	No rounding or approximations.
Give your answer to an appropriate degree of accuracy...	If the numbers in the question are given to 2 decimal places, give your answer to 2 decimal places.
Give your answer in its simplest form...	Usually means you will need to cancel a fraction or a ratio.
Simplify...	In algebra, means collect like terms together.
Solve...	Usually means find the value of x in an equation.
Expand...	Multiply out the brackets.
Factorise...	Put in brackets with common factors outside the bracket.
Measure...	Use a ruler or a protractor to measure lengths or angles accurately.
Draw an accurate diagram...	Use a ruler and protractor to draw the diagram. Lengths must be exact and angles must be accurate.
Construct, using ruler and compasses...	Draw, using a ruler as a straight edge and compasses to draw arcs. Leave your construction lines and arcs in – don't rub them out.
Sketch...	An accurate drawing is not required; a freehand drawing will be fine.
Diagram NOT accurately drawn...	Don't measure angles or sides. If you are asked to find them, you need to work them out.
Give reasons for your answer OR Explain why...	You need to write an explanation. Show any working out, or quote any laws or theorems you used, for example Pythagoras' theorem.
Use your (the) graph...	Read values from your graph and use them.
Describe fully...	Usually means transformations: • Reflection – give equation of the line of reflection (2 marks) • Rotation – give the angle, direction of turn and the centre of rotation (3 marks) • Enlargement – give the scale factor and the centre of enlargement (3 marks)
Give a reason for your answer...	In angle questions, means write a reason. For example: • angles in a triangle add up to 180° • alternate angles

Collecting and organising data (I)

- **Two-way tables** are used to record or display information that is grouped in two categories.
- **Discrete data** can be counted and can take only particular values. For example, shoe sizes 6, $6\frac{1}{2}$, 7, …
- **Continuous data** is measured and can take any value. For example, height in centimetres or weight in kilograms.

Key words

- two-way table
- discrete data
- continuous data

Example

Grade E

Lucy interviewed 100 people who buy coffee.
She asked them which type they buy most, and in which size packets.
Some of her results are given in this table:

	Instant	Beans	Ground	Total
50 g	3	1	0	
100 g	16			36
250 g	32	9		
Total		18		100

EXAMINER'S TIP
You wouldn't get a whole question on this on the Higher paper, but it might come up as part of a question.

Complete the table.

Look for a row or column with *one* missing value – cell **1**.

Fill in this value, which is the total of the three cells above 3 + 16 + 32 = 51

Now look for other rows or columns with one missing value, and complete the table, filling in the empty cells in turn.

	Instant	Beans	Ground	Total
50 g	3	1	0	3 4
100 g	16	2 8	4 12	36
250 g	32	9	6 19	5 60
Total	1 51	18	7 31	100

TIP
Look for the next row or column with one mnissing cell.

TIP
Start with a row or column with only *one* missing value.

TIP
Check that additions work both ways.

TIP
Continue to look for a row or column with one missing cell. This is the next.

- When you are writing questions for a **questionnaire**:
  – be clear what you want to find out, and what data you need
  – ask short, simple questions
  – provide **response boxes** with possible answers
  – avoid questions which are vague, too personal, or which may influence the answer (**leading questions**).

Key words

questionnaire ☐
response box ☐
leading question ☐

Example

A shopkeeper wants to find out how many chocolate bars students eat.

He uses this question in his questionnaire:

'You enjoy eating chocolate bars, don't you?'

(a) Explain why this is not a good question to ask. *Grade D*

His next question is

'How many chocolate bars have you eaten?

A few ☐ A lot ☐'

(b) Write down two things that are wrong with this question. *Grade D*

(c) Write down an improved question that the shopkeeper could use. Include response boxes. *Grade C*

The shopkeeper asks his first 10 customers on one day to complete the questionnaire.

(d) Give reasons why this way of collecting data is not appropriae. *Grade C*

Read the question. Is it a leading question?

(a) This is not a good question because it encourages the student to answer Yes. It is a leading question.

EXAMINER'S TIP
You need to explain *why* it is not a good question.

Identify two things wrong with the question. Think about how you could answer it.

(b) You cannot answer 'none' – there need to be response boxes to cover all possible answers. 'A few' and 'a lot' are too vague. The question does not specify a time period.

TIP
There are three possible answers here. You only need to give two.

Design a new question with a time period and response boxes to cover all possible answers.

(c) How many chocolate bars do you eat each week?

0 ☐ 1 ☐ 2–4 ☐ 5–10 ☐ more than 10 ☐

EXAMINER'S TIP
Give between 4 and 6 response boxes.

EXAMINER'S TIP
Make sure the response boxes don't overlap.

(d) The sample is too small. The sample is biased as only 10 customers are asked. The sample should be a random selection of customers of different types.

EXAMINER'S TIP
A sample should have at least twenty responses.

For more on collecting and organising data, including practice questions, see pages 4–5.

Collecting and organising data (II)

- The **population** is the complete set of items under consideration.
- A **sample** is part of the population selected for surveying.
- A **biased sample** is one which would be likely to have an undue influence on the outcome.
- In a **random sample** every member of the population has an equal chance of being chosen.
- A **stratified sample** is one in which the population is divided into groups called strata and each of the strata is randomly sampled. The proportions of the different strata in the sample are the same as in the whole population.
- A **systematic sample** is one where every nth item is chosen.

Key words

- population
- sample
- biased sample
- random sample
- stratified sample
- systematic sample

Example

Grade A*

The table shows information about the 542 members of a gym.

Age	Female	Male	Total
under 18	25	14	39
18–25	86	63	149
26–50	151	112	263
over 50	57	34	91

The gym wants to carry out a survey of the members.
They decide to use a sample of 80 members, stratified by age and gender.

(a) Calculate how many 18–25 year-old women they should include in the sample.

(b) Work out the total number of female members they would need to sample.

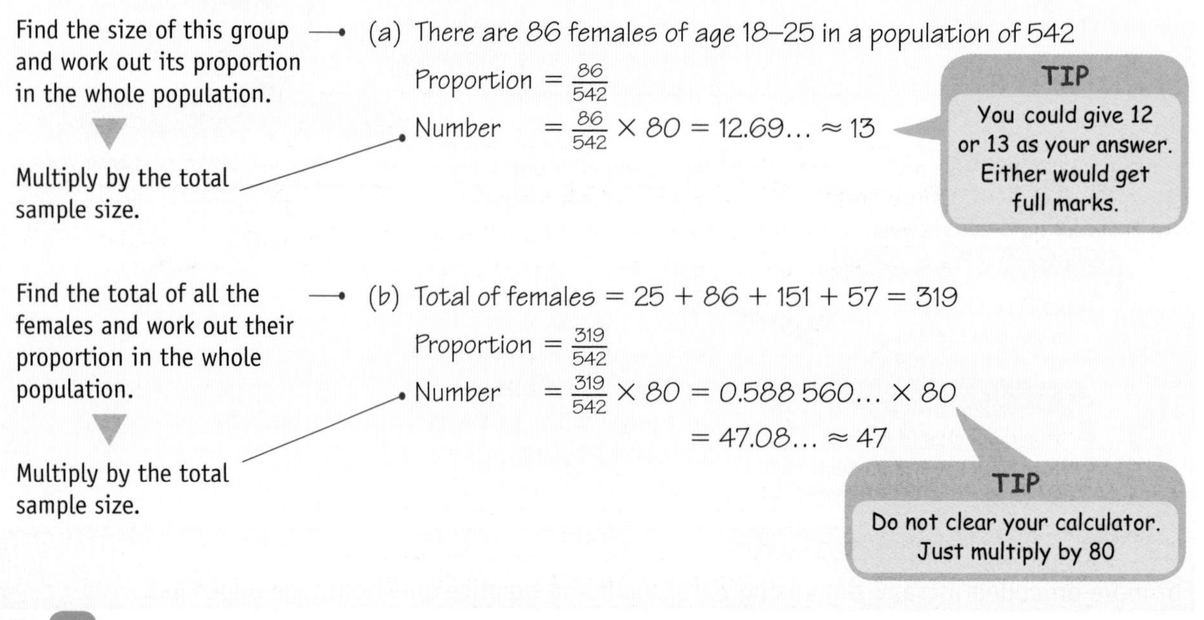

Practice

EXAMINER'S TIP

You wouldn't get a whole question on this on the Higher paper, but it might come up as part of a question.

1 Here is a mileage chart:

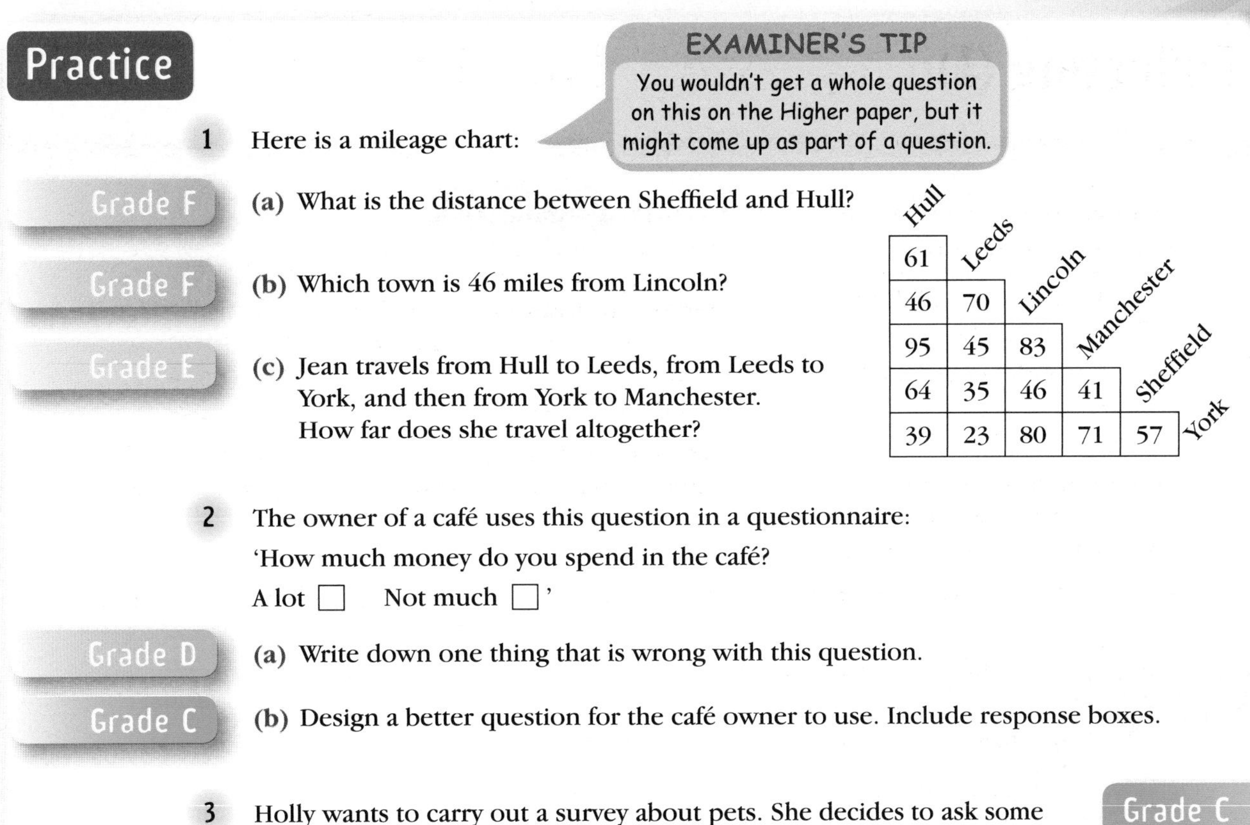

Hull					
61	Leeds				
46	70	Lincoln			
95	45	83	Manchester		
64	35	46	41	Sheffield	
39	23	80	71	57	York

Grade F **(a)** What is the distance between Sheffield and Hull?

Grade F **(b)** Which town is 46 miles from Lincoln?

Grade E **(c)** Jean travels from Hull to Leeds, from Leeds to York, and then from York to Manchester. How far does she travel altogether?

2 The owner of a café uses this question in a questionnaire:

'How much money do you spend in the café?

A lot ☐ Not much ☐ '

Grade D **(a)** Write down one thing that is wrong with this question.

Grade C **(b)** Design a better question for the café owner to use. Include response boxes.

3 Holly wants to carry out a survey about pets. She decides to ask some people whether they prefer dogs, cats, hamsters, rabbits or goldfish. Design a data collection sheet that she can use to carry out the survey. Grade C

4 There are 1500 students at Chapel High School. Grade A*
The table shows how they are distributed by year group and by gender.

Year group	Number of boys	Number of girls
7	146	153
8	135	151
9	160	147
10	149	151
11	151	157

Chris wants to take a stratified sample of 200 students.

(a) How many Year 8 boys should there be in the sample?

(b) How many Year 11 girls should there be in the sample?

Check your answers on page 168. For full worked solutions see the CD.
See the Student Book on the CD if you need more help.

Question	1ab	1c	2a	2b	3	4
Grade	F	E	D	C	C	A*
Student Book pages	U1 1–5		U1 1–5		U1 1–5	U1 7–10

Averages (I)

- The **mode** of a set of data is the value which occurs most often.
- The **median** is the middle value when the data is arranged in order of size. If there are n data values, the median is the $\frac{n+1}{2}$th value.
- The **mean** of a set of data is the sum of the values divided by the number of values:

$$\text{mean} = \frac{\text{sum of values}}{\text{number of values}}$$

- The **range** of a set of data is the difference between the highest and lowest values:

range = highest value – lowest value

Key words

- mode ☐
- median ☐
- mean ☐
- range ☐
- frequency distribution ☐

- For a **frequency distribution**:

$$\text{mean} = \frac{\sum fx}{\sum f}$$

$\sum fx$ — the sum of all the ($f \times x$) values in the distribution

$\sum f$ — the sum of the frequencies

Example

Daniel has some boxes of candles. The table gives information about the numbers of candles in each box.

Number of candles	Frequency
3	4
4	10
5	5
6	1

EXAMINER'S TIP

You wouldn't get a whole question on this on the Higher paper, but it might come up as part of a question.

Grade F **(a)** How many boxes does Daniel have?

Grade F **(b)** Write down the modal number of candles in a box.

Grade E **(c)** Work out the median number of candles in a box.

Grade D **(d)** Work out the mean number of candles in a box.

Add the numbers in the frequency column. → (a) Number of boxes = 4 + 10 + 5 + 1 = 20

The mode is the one with the highest frequency. → (b) Mode = 4

WATCH OUT!

Remember to write down the *value*. Students often write down the frequency.

The median is the middle value of the data. → (c) Position of middle value $= \frac{20+1}{2} = 10.5$

Boxes 1–4 have 3 candles.

Boxes 5–14 have 4 candles.

So boxes 10 and 11 each have 4 candles.

Median $= \frac{4+4}{2} = 4$

Add a third column to the table – work out frequency × number for each row. → (d)

Number of candles x	Frequency f	Frequency × number of candles $f \times x$
3	4	12
4	10	40
5	5	25
6	1	6
Total	20	83

TIP

This is the total number of candles.

TIP

The number of candles in 10 boxes of 4 is 10 × 4

Work out the mean. → Mean = total number of candles ÷ total number of boxes

= 83 ÷ 20 = 4.15 candles

- For **grouped data**:
  - the **modal class** is the class interval with the highest frequency
  - you can state the **class interval** that contains the median
  - you can calculate an **estimate of the mean** using the middle value of each class interval.

Key words

grouped data ☐
modal class ☐
estimate of the mean ☐

Example

The table gives information about the weights of 40 small children.

Weight, w (kg)	Frequency, f
$0 < w \leq 4$	5
$4 < w \leq 8$	13
$8 < w \leq 12$	14
$12 < w \leq 16$	8

Grade D **(a)** Write down the modal class.

Grade C **(b)** Write down the class interval in which the median lies.

Grade C **(c)** Work out an estimate for the mean weight.

Find the class interval with the highest frequency → (a) The modal class is $8 < w \leq 12$

Work out the position of the median → (b) Position of median

$$= \frac{40 + 1}{2} = 20.5$$

Find the class interval that contains the 20th and 21st values → Class interval of median is $8 < w \leq 12$

TIP
$8 < w \leq 12$ has the 19th to 32nd values.

Add two more columns to the table – work out the middle value x for each row and $f \times x$ for each row → (c)

Weight, w (kg)	Frequency f	Middle value x	$f \times x$
$0 < w \leq 4$	5	2	10
$4 < w \leq 8$	13	6	78
$8 < w \leq 12$	14	10	140
$12 < w \leq 16$	8	14	112
Total	40	Total	340

TIP
Add a row for the totals.

Work out the estimate of the mean → Estimate of mean

$$= \frac{\text{sum of (middle values} \times \text{frequencies)}}{\text{sum of frequencies}}$$

$$= \frac{340}{40} = 8.5 \text{ kg}$$

WATCH OUT!
Remember to use the *middle* values. Students often use the beginning or end of the class intervals.

- A graph showing how a given value changes over time is called a **time series** graph.
- A **moving average** smooths out the data to show the trend on a time series graph. To find the n-point moving average, calculate the mean of every n data values.
- A **trend line** is the line of best fit drawn through the moving averages plotted on a graph.

Key words

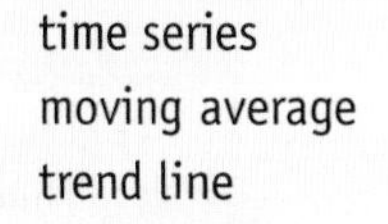

time series ☐
moving average ☐
trend line ☐

For more on averages, including practice questions, see pages 8–9.

Averages (II)

Example

Grade B

The table shows information about the quarterly sales at Mike's Bikes from January 2006 to June 2007.

(a) Calculate the 4-point moving averages for the data.

(b) What do your moving averages tell you about the trend in the sales of bicycles?

Year	Months	Number of cycles sold
2006	Jan–Mar	62
	Apr–Jun	68
	Jul–Sept	90
	Oct–Dec	108
2007	Jan–Mar	69
	Apr–Jun	74

Find the mean of the first four data values. → (a) The 4-point averages are

$$\frac{62 + 68 + 90 + 108}{4} = 82$$

Then move forwards one data value each time. →

$$\frac{68 + 90 + 108 + 69}{4} = 83.75$$

$$\frac{90 + 108 + 69 + 74}{4} = 85.25$$

TIP
Give exact values.

Relate your explanation to the context of the question. → (b) The trend in the moving averages shows that the number of bicycles sold gradually increased between early 2006 and mid 2007.

Example

Grade B

The table shows the four-point moving averages for Hassan's quarterly gas bills.

The last four bills (oldest first) were for £118, £56, £170 and £232.
Hassan's next bill is for £126.

(a) Work out the 8th four-point moving average.

(b) Plot the moving averages on a graph and add a trend line.

(c) Comment on the changes in Hassan's gas bills.

Moving average	Amount
1st	£113
2nd	£119.50
3rd	£121.50
4th	£138
5th	£139.50
6th	£141
7th	£144

Find the average of the four most recent bills. → (a) $\frac{56 + 170 + 232 + 126}{4} = £146$

Plot each pair of values on the graph with a cross. → (b)

Draw a straight line as close to as many of the points as possible.

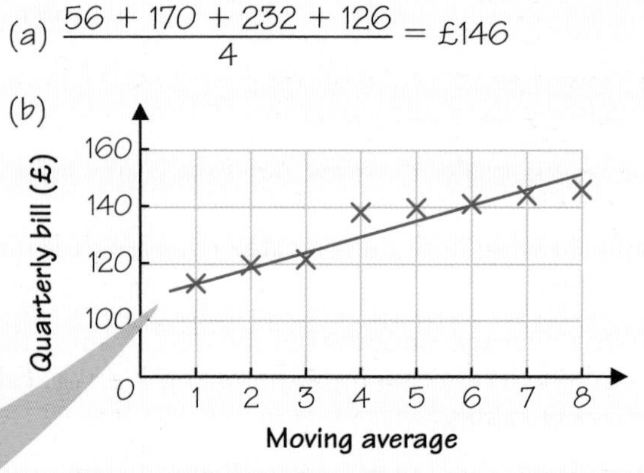

TIP
It is best to use a clear plastic ruler. There should be roughly equal numbers of points above and below the line.

The trend line is the solid red line on the graph.

Does the trend line slope up or down? → (c) Hassan's gas bills rose gradually over this period.

Practice

1 The table gives information about the number of tries scored by a rugby team in each match of the season.

Number of tries	0	1	2	3	4
Number of matches	4	6	11	8	5

Grade F **(a)** How many matches were there?

Grade F **(b)** Write down the modal number of tries in a match.

Grade E **(c)** Work out the median number of tries in a match.

EXAMINER'S TIP

You wouldn't get a whole question on this on the Higher paper, but it might come up as part of a question.

Grade D **(d)** Work out the mean number of tries in a match.

Grade D **(e)** Peter said 'The team scored an average of 5 tries per match.' Explain why this is wrong.

2 Bronwen recorded the times, in minutes, it took her to complete 50 crosswords.

Time, t (minutes)	Frequency
$10 \leqslant t < 15$	3
$15 \leqslant t < 20$	9
$20 \leqslant t < 25$	18
$25 \leqslant t < 30$	15
$30 \leqslant t < 35$	5

Grade D **(a)** Write down the modal class.

Grade C **(b)** Write down the class interval in which the median lies.

Grade C **(c)** Calculate an estimate of the mean time it took Bronwen to complete a crossword.

Grade B **3** The table shows the numbers of parcels sent out by a small shop.

Month	Apr	May	Jun	Jul	Aug	Sept	Oct	Nov
Number of parcels	33	41	25	21	29	26	30	23

Work out the first three 4-point moving averages for this data.

Grade B **4** The table gives information about the number of footballs sold by a sports shop each month in 2007.

(a) Work out the next two 3-point moving averages.

(b) Draw a graph of the moving averages and add a trend line.

(c) Comment on the changes in the number of footballs sold.

Month	Number of footballs sold	3-point moving average
January	21	
February	33	22
March	12	20
April	15	12
May	9	...
June	12	...
July	6	

Check your answers on page 168. For full worked solutions see the CD.
See the Student Book on the CD if you need more help.

Question	1ab	1c	1de	2a	2bc	3	4
Grade	F	E	D	D	C	B	B
Student Book pages	U1 33–37			U1 38–40		U1 58–60	U1 58–60

Probability (I)

- The **probability** of an event is expressed as a number from 0 to 1.
  - If an event is **impossible** its probability is 0.
  - If an event is **certain** its probability is 1.
- When there are n equally likely possible outcomes, the probability of each outcome is $\frac{1}{n}$
- If the probability of an event happening is p, the probability of it *not* happening is $1 - p$
- **Estimated probability** $= \frac{\text{number of successful trials}}{\text{total number of trials}}$

Key words

- probability ☐
- impossible ☐
- certain ☐
- estimated probability ☐

Example

Grade D

Mrs Coley chooses a drink from a machine. She can choose tea, coffee, chocolate or soup.

The table shows the probabilities that she chooses tea, coffee or soup.

Drink	Tea	Coffee	Chocolate	Soup
Probability	0.3	0.4		0.1

Work out the probability that she chooses chocolate.

Add up the probabilities that you know. → $P(\text{tea}) + P(\text{coffee}) + P(\text{soup}) = 0.3 + 0.4 + 0.1$
$= 0.8$

Subtract from 1. → $P(\text{chocolate}) = 1 - 0.8 = 0.2$

TIP
Write this value in the table and check that the probabilities add up to 1.

Example

Grade C

The probability that a biased coin lands 'heads' is 0.7

Henry is going to spin the coin 200 times.

Work out an estimate for the number of times it will land 'heads'.

Write down the rule → $P(\text{head}) = 0.7$

estimated probability $= \frac{\text{number of successful trials}}{\text{total number of trials}}$ → Estimated number of heads in 200 trials $= 0.7 \times 200$
$= 140$

WATCH OUT!
Remember to check that your answer is sensible. Students often put the decimal point in the wrong place.

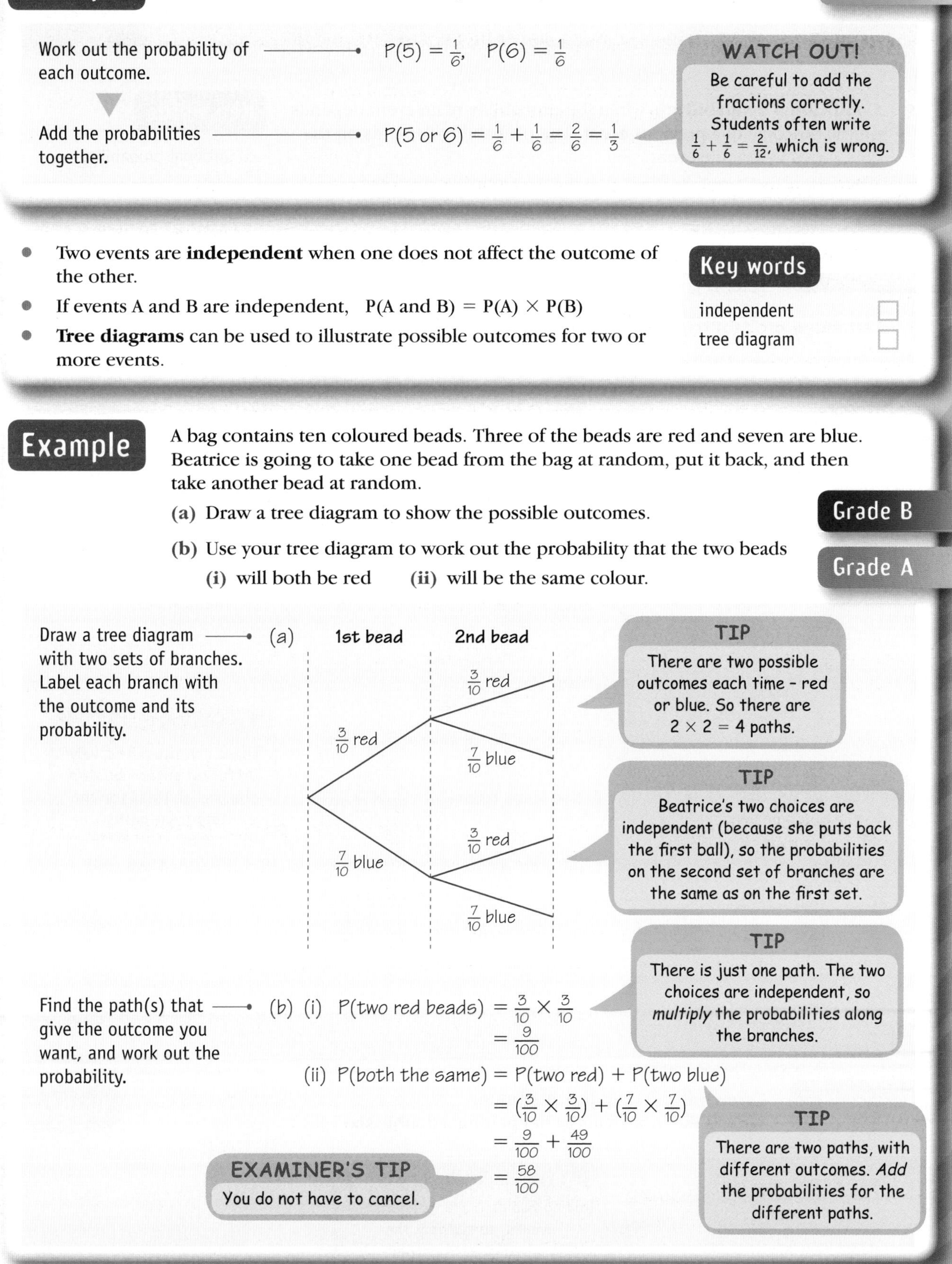

Example A fair dice is rolled. Work out the probability that it will land on 5 or 6.

Grade C

Work out the probability of each outcome. → $P(5) = \frac{1}{6}$, $P(6) = \frac{1}{6}$

Add the probabilities together. → $P(5 \text{ or } 6) = \frac{1}{6} + \frac{1}{6} = \frac{2}{6} = \frac{1}{3}$

WATCH OUT!
Be careful to add the fractions correctly. Students often write $\frac{1}{6} + \frac{1}{6} = \frac{2}{12}$, which is wrong.

- Two events are **independent** when one does not affect the outcome of the other.
- If events A and B are independent, P(A and B) = P(A) × P(B)
- **Tree diagrams** can be used to illustrate possible outcomes for two or more events.

Key words
independent ☐
tree diagram ☐

Example A bag contains ten coloured beads. Three of the beads are red and seven are blue. Beatrice is going to take one bead from the bag at random, put it back, and then take another bead at random.

Grade B

(a) Draw a tree diagram to show the possible outcomes.

Grade A

(b) Use your tree diagram to work out the probability that the two beads
(i) will both be red **(ii)** will be the same colour.

Draw a tree diagram with two sets of branches. Label each branch with the outcome and its probability. → (a)

TIP
There are two possible outcomes each time – red or blue. So there are 2 × 2 = 4 paths.

TIP
Beatrice's two choices are independent (because she puts back the first ball), so the probabilities on the second set of branches are the same as on the first set.

Find the path(s) that give the outcome you want, and work out the probability. → (b) (i) $P(\text{two red beads}) = \frac{3}{10} \times \frac{3}{10}$
$= \frac{9}{100}$

TIP
There is just one path. The two choices are independent, so *multiply* the probabilities along the branches.

(ii) $P(\text{both the same}) = P(\text{two red}) + P(\text{two blue})$
$= (\frac{3}{10} \times \frac{3}{10}) + (\frac{7}{10} \times \frac{7}{10})$
$= \frac{9}{100} + \frac{49}{100}$
$= \frac{58}{100}$

EXAMINER'S TIP
You do not have to cancel.

TIP
There are two paths, with different outcomes. *Add* the probabilities for the different paths.

For more on probability, including practice questions, see pages 12–13.

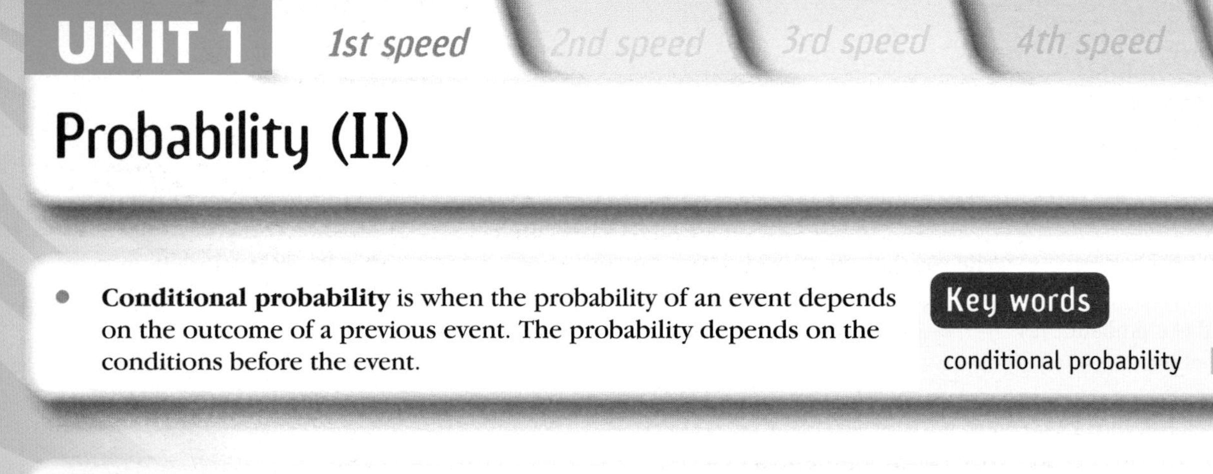

Probability (II)

- **Conditional probability** is when the probability of an event depends on the outcome of a previous event. The probability depends on the conditions before the event.

Key words

conditional probability ☐

Example

A bag contains 12 coloured beads. Seven of the beads are red and five are black. Peter is going to take two beads from the bag at random.

(a) Draw a tree diagram to show the possible outcomes. Grade A

(b) Use your tree diagram to calculate the probability that the two beads Grade A*

(i) will both be black **(ii)** will be different colours.

Draw a tree diagram. Label each branch with the outcome and its probability. → (a)

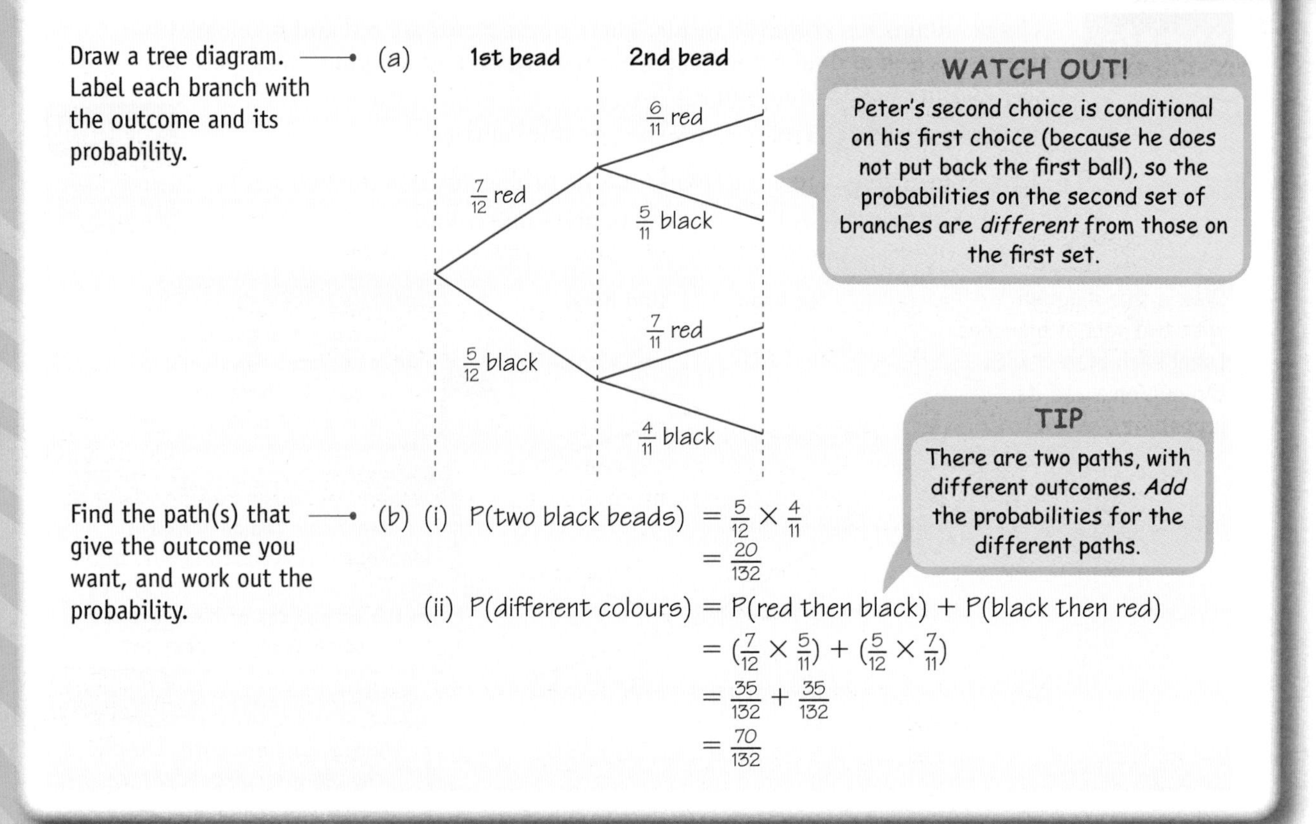

WATCH OUT!

Peter's second choice is conditional on his first choice (because he does not put back the first ball), so the probabilities on the second set of branches are *different* from those on the first set.

Find the path(s) that give the outcome you want, and work out the probability. → (b) (i) P(two black beads) $= \frac{5}{12} \times \frac{4}{11}$

$= \frac{20}{132}$

(ii) P(different colours) = P(red then black) + P(black then red)

$= (\frac{7}{12} \times \frac{5}{11}) + (\frac{5}{12} \times \frac{7}{11})$

$= \frac{35}{132} + \frac{35}{132}$

$= \frac{70}{132}$

TIP

There are two paths, with different outcomes. *Add* the probabilities for the different paths.

Example

Caren rolls two fair dice at random. Grade A*

What is the probability of her getting a double six?

P(two sixes) $= \frac{1}{6} \times \frac{1}{6}$

$= \frac{1}{36}$

TIP

Rolling two dice are independent events, so multiply the probabilities together.

Practice

Grade D

1 Monty plays a game of draughts with his friend.
In draughts, games are won, lost or drawn.
The probability that Monty loses the game is 0.25
The probability that Monty draws is 0.4
Work out the probability that Monty wins the game.

2 Emma rolls a blue dice and a red dice.

> **EXAMINER'S TIP**
> You wouldn't get a whole question on this on the Higher paper, but it might come up as part of a question.

Grade E

(a) List all the possible outcomes.

Grade C

(b) Use your list to find the probability that she gets a total score of 7.

Grade C

3 A fair spinner is made in the shape of a regular hexagon.
It can land on red, blue or yellow.
Write down the probability that the spinner will land on red.

4 A bag contains 15 coloured beads.
Six of the beads are red, four are blue and the rest are white.
Jade is going to take one bead from the bag at random, put it back, and then take another bead at random.

Grade B

(a) Draw a tree diagram to show the possible outcomes.

Grade A

(b) Use your tree diagram to calculate the probability that she picks
(i) two blue beads **(ii)** two beads of the same colour.

Grade A*

5 Helen is about to take her driving test.
The probability that she will pass at her first attempt is 0.7
If she fails on her first attempt, the probability that she will pass on her second or any subsequent attempt is 0.8

(a) Draw a tree diagram to show the possible outcomes.

(b) Work out the probability that she passes in three attempts at most.

Grade A*

6 Lindsey picks two cards at random from a normal pack of playing cards.

(a) What is the probability of a red card and a black card?

(b) What is the probability of two kings?

Check your answers on pages 168–169. For full worked solutions see the CD.
See the Student Book on the CD if you need more help.

Question	1	2a	2b	3	4a	4b	5	6
Grade	D	E	C	C	B	A	A*	A*
Student Book pages	U1 72–76	U1 71–72		U1 67–70	U1 72–76		U1 72–76	U1 72–76

Data, averages and probability: topic test

Check how well you know this topic by answering these questions.
First cover the answers on the facing page.

Test questions

1 A box contains beads which are red, yellow, blue or green.
Helen is going to pick one bead from the box at random.
The table shows the probabilities that the bead she picks is red, yellow or green.

Colour	Red	Yellow	Blue	Green
Probability	0.15	0.23		0.41

Work out the probability that she will pick a blue bead.

2 Maddy wants to carry out a survey into how much time people spending listening to the radio. Here is part of her questionnaire:

(a) Write down two things that are wrong with this question.

(b) Write down an improved question that Maddy could use. Include response boxes.

3 In a survey, 50 people were asked how long they spent watching television last weekend.
The results are summarised in the table.

Time, t (hours)	Frequency
$0 \leq t < 2$	11
$2 \leq t < 4$	15
$4 \leq t < 6$	18
$6 \leq t < 8$	6

(a) Write down the modal class.

(b) Write down the class interval that contains the median time.

(c) Calculate an estimate of the mean time.

4 The table shows the numbers of CDs sold by a shop each month.

Month	Jul	Aug	Sept	Oct	Nov	Dec
Number of CDs	219	156	197	232	255	300

Work out the 4-month moving averages for this data.

5 The table shows some information about the students at Clifton School.

Year group	Number of boys	Number of girls	Total
9	125	92	217
10	115	119	234
11	101	141	242
Total	341	352	693

Leonie carries out a survey of the students. She takes a sample of 100 students, stratified by year group and gender.

(a) Work out the number of Year 10 girls in her sample.

(b) Work out the number of boys in her sample.

6 A bag contains 15 coloured beads. Ten of the beads are red and five are blue.
Jason is going to take two beads from the bag at random.

(a) Draw a tree diagram to show the possible outcomes.

(b) Use your tree diagram to calculate the probability that the two beads

(i) will both be red

(ii) will be the same colour.

Now check your answers – see the facing page.

Cover this page while you answer the test questions opposite.

Worked answers

Revise this on...

D

1 P(not blue) = P(red) + P(yellow) + P(green) = 0.15 + 0.23 + 0.41 = 0.79 page 10

P(blue) = 1 − P(not blue) = 1 − 0.79 = 0.21

D

2 (a) It is a leading question, and the response boxes are too vague. page 3

C

(b) How many hours did you listen to the radio last week? page 3

0–1 ☐ more than 1–2 ☐ more than 2–4 ☐

more than 4–8 ☐ more than 8 ☐

D

3 (a) Modal class = $4 \leqslant t < 6$ page 7

C

(b) Position of median = $\frac{50+1}{2} = 25.5$

The median is the mean of the 25th and 26th times.

The median time lies within the $2 \leqslant t < 4$ class interval.

C

(c) Mean = $\frac{\Sigma fx}{\Sigma f} = \frac{188}{50} = 3.76$ hours

Time, t (hours)	Frequency f	Middle value, x	f × x
$0 \leqslant t < 2$	11	1	11
$2 \leqslant t < 4$	15	3	45
$4 \leqslant t < 6$	18	5	90
$6 \leqslant t < 8$	6	7	42
Total	50	Total	188

B

4 1st 4-month moving average = $\frac{219 + 156 + 197 + 232}{4} = 201$ pages 7–8

2nd 4-month moving average = $\frac{156 + 197 + 232 + 255}{4} = 210$

3rd 4-month moving average = $\frac{197 + 232 + 255 + 300}{4} = 246$

A*

5 (a) Proportion of Year 10 girls = $\frac{\text{Year 10 girls}}{\text{total students}} = \frac{119}{693}$ page 4

Number of Year 10 girls in sample = $\frac{119}{693} \times 100 = 17.17\ldots \approx 17$

(b) Proportion of boys = $\frac{\text{total boys}}{\text{total students}} = \frac{341}{693}$

Number of boys in sample = $\frac{341}{693} \times 100 = 49.20\ldots \approx 49$

A*

6 (a) 1st bead 2nd bead

$\frac{10}{15}$ red → $\frac{9}{14}$ red, $\frac{5}{14}$ blue

$\frac{5}{15}$ blue → $\frac{10}{14}$ red, $\frac{4}{14}$ blue

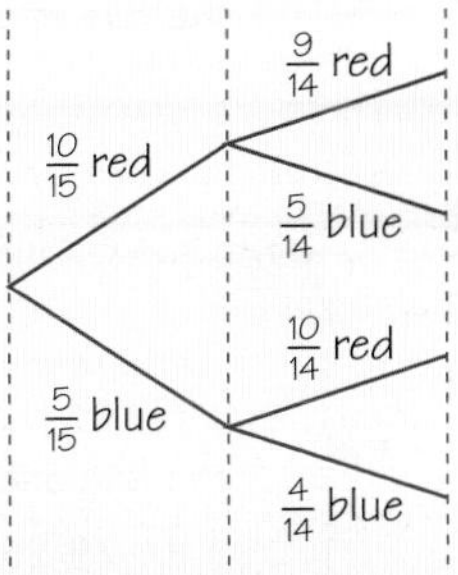

(b) (i) P(two red) = $\frac{10}{15} \times \frac{9}{14} = \frac{90}{210} = \frac{3}{7}$ page 12

(ii) P(both the same)

= P(two red) + P(two blue)

= $(\frac{10}{15} \times \frac{9}{14}) + (\frac{5}{15} \times \frac{4}{14}) = \frac{90}{210} + \frac{20}{210} = \frac{110}{210} = \frac{11}{21}$

Tick the questions you got right.
Mark the grade you are working at on your revision planner on page vi.

Question	1	2a	2b	3a	3bc	4	5	6
Grade	D	D	C	D	C	B	A*	A*

Frequency charts (I)

- A **stem and leaf diagram** shows the shape of a distribution and keeps all the data values. It needs a **key** to show how the stem and leaf are combined.
- You can use a stem and leaf diagram to find the median, the upper and lower quartiles, and the interquartile range:
  - the **median** is the $\frac{n+1}{2}$th value
  - the **lower quartile** is the $\frac{n+1}{4}$th value
  - the **upper quartile** is the $\frac{3(n+1)}{4}$th value
  - the **interquartile range** is the difference between the upper and lower quartiles:

    interquartile range = upper quartile − lower quartile

Key words

- stem and leaf diagram
- key
- median
- lower quartile
- upper quartile
- interquartile range

Example

Grade D

Jane throws a dart 20 times. Here are her scores:

15	30	23	9	37	42	49	36	49	55
66	57	69	62	38	31	20	46	17	37

(a) Draw an ordered stem and leaf diagram to show these scores. Include a key.
(b) Find the median. (c) Find the interquartile range.

Write the data as a stem and leaves.
Use the tens as a stem.
The units are the leaves.

(a)

0	9
1	5, 7
2	3, 0
3	0, 7, 6, 8, 1, 7
4	2, 9, 9, 6
5	5, 7
6	6, 9, 2

This question asks for an *ordered* stem and leaf diagram, so now write the leaves in order.

0	9
1	5, 7
2	0, 3
3	0, 1, 6, 7, 7, 8
4	2, 6, 9, 9
5	5, 7
6	2, 6, 9

Key: 1 | 5 means 15

WATCH OUT!
Remember the key! Students often forget, and lose 1 mark.

Use the ordered stem and leaf diagram to find the 'middle value' of the data.

(b) Position of middle value $= \frac{20+1}{2} = 10.5$

10th value = 37, 11th value = 38

Median $= \frac{37+38}{2} = 37.5$

TIP
If the number of data values is even, the median is the mean of the two middle values.

The lower quartile is the $\frac{n+1}{4}$th value.

(c) Position of lower quartile $= \frac{20+1}{4} = 5\frac{1}{4}$

5th value = 23, 6th value = 30

Lower quartile $= 23 + \frac{1}{4}$ of $(30 - 23)$
$= 23 + 1.75 = 24.75$

TIP
The lower quartile is a quarter of the way between the 5th and 6th values:

23 24 25 26 27 28 29 30
LQ

The upper quartile is the $\frac{3(n+1)}{4}$th value.

Position of upper quartile $= \frac{3(20+1)}{4} = 15\frac{3}{4}$

15th value = 49, 16th value = 55

Upper quartile $= 49 + \frac{3}{4}$ of $(55 - 49)$

$= 49 + 4.5 = 53.5$

The interquartile range is the difference between the quartiles.

Interquartile range $= 53.5 - 24.75$

$= 28.75$

TIP

The upper quartile is three quarters of the way between the 15th and 16th values:

49 50 51 52 53 54 55

UQ

- A **histogram** is used to display grouped continuous data. The areas of the rectangles are proportional to the frequencies they represent.
- A **frequency polygon** can be used to show the general pattern of data. Plot the frequency for each class interval against the mid-point of that interval.

Key words

histogram

frequency polygon

Example

Grade C

The table shows the times, in minutes, it takes some people to finish a crossword.

Draw a frequency polygon for this data.

Time, t (minutes)	Frequency
$0 \leq t < 5$	4
$5 \leq t < 10$	8
$10 \leq t < 15$	13
$15 \leq t < 20$	16
$20 \leq t < 25$	6
$25 \leq t < 30$	3

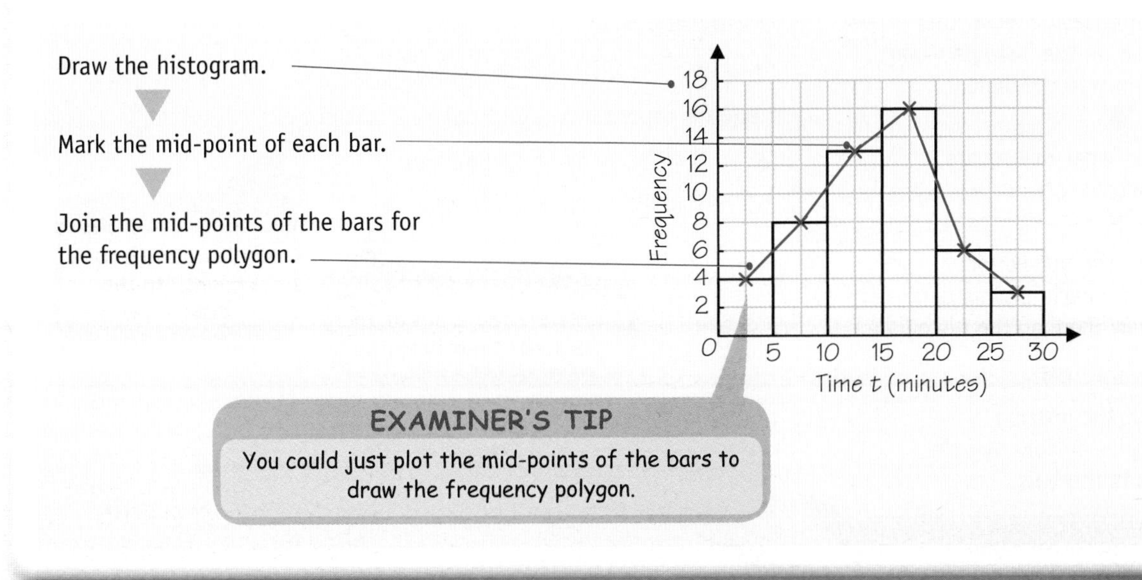

EXAMINER'S TIP

You could just plot the mid-points of the bars to draw the frequency polygon.

For more on frequency charts, including practice questions, see pages 18–19.

Frequency charts (II)

- In a histogram with unequal class intervals, the vertical axis shows the **frequency density**:

$$\text{frequency density} = \frac{\text{frequency}}{\text{class width}}$$

Key words

frequency density

Example

Grade A*

Kanti recorded the waiting times for patients at a doctor's surgery.
Some of the information is shown in the unfinished table and histogram.

Waiting time, t (minutes)	Frequency
$0 < t \leqslant 10$	
$10 < t \leqslant 15$	20
$15 < t \leqslant 20$	
$20 < t \leqslant 30$	28
$30 < t \leqslant 60$	12
$t > 60$	0

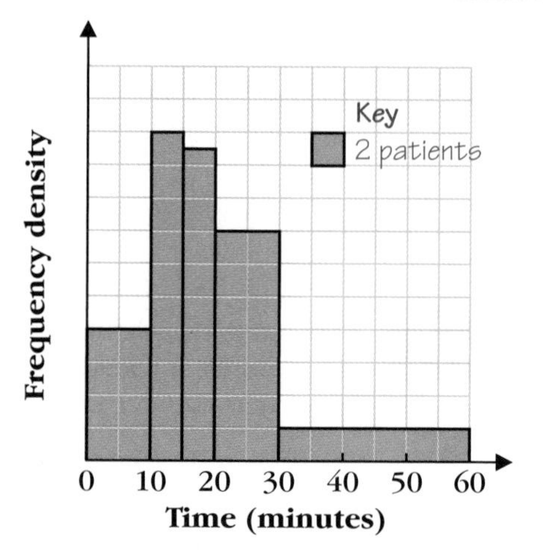

(a) Use the information in the table to complete the histogram.
(b) Use the information in the histogram to complete the frequency table.

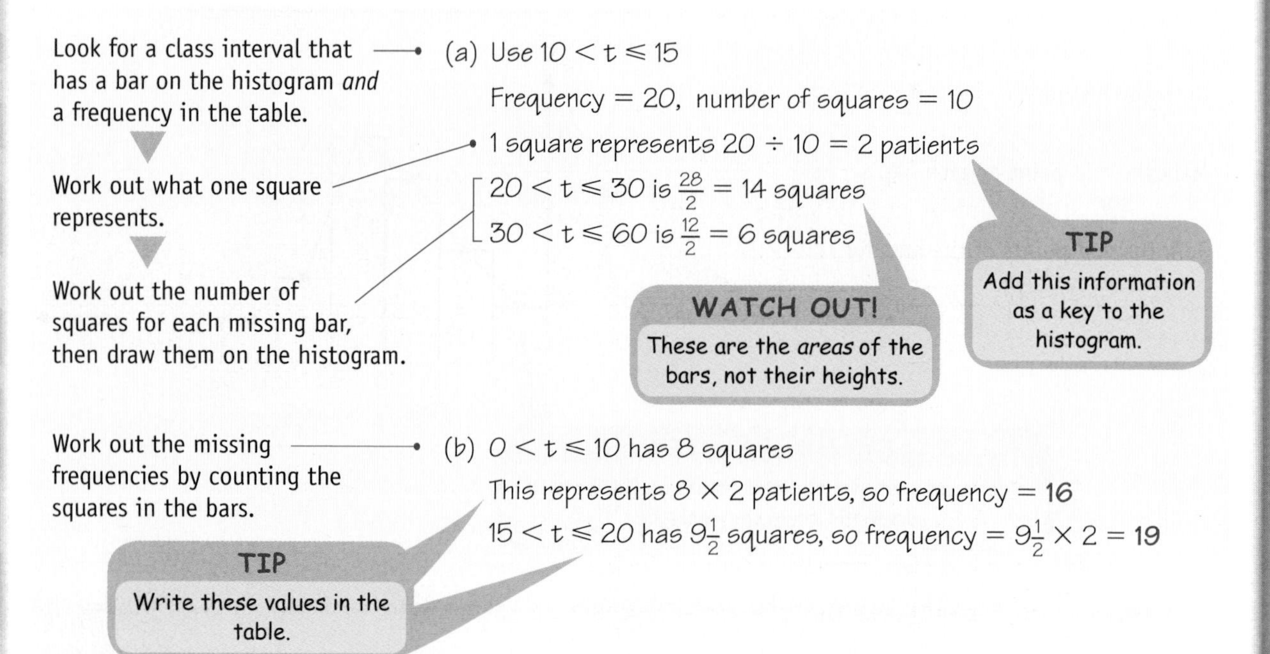

Practice

1 20 people were asked

'What are the last two digits of your telephone number?'

Here are the results:

08	12	38	24	00	47	07	19	03	02
31	09	31	22	15	11	03	29	13	06

Grade D **(a)** Draw an ordered stem and leaf diagram to represent this data. Include a key.

Grade D **(b)** Find the median.

Grade B **(c)** Find the interquartile range

Grade C **2** The table shows some people's weights in kilograms. Draw a frequency polygon for this data.

Weight, w (kilograms)	Frequency
$50 \leqslant w < 55$	2
$55 \leqslant w < 60$	5
$60 \leqslant w < 65$	11
$65 \leqslant w < 70$	13
$70 \leqslant w < 75$	7
$75 \leqslant w < 80$	2

Grade A* **3** The incomplete table and histogram give some information about the ages of people who live in a street.

Age, x (years)	Frequency
$0 < x \leqslant 10$	80
$10 < x \leqslant 20$	120
$20 < x \leqslant 45$	
$45 < x \leqslant 60$	60
$60 < x \leqslant 80$	
$x > 80$	0

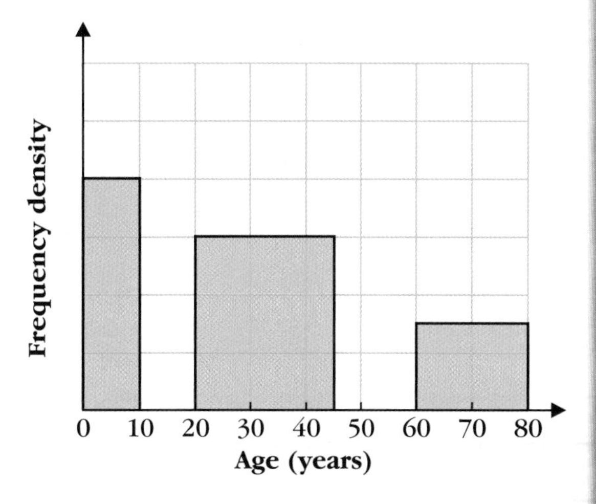

(a) Complete the frequency table.

(b) Complete the histogram.

Check your answers on page 169. For full worked solutions see the CD.
See the Student Book on the CD if you need more help.

Question	1ab	1c	2	3
Grade	D	B	C	A*
Student Book pages	U1 13–15, 33–37	U1 33–37	U1 15–18	U1 19–25

Scatter graphs, correlation and the RPI

- A **scatter graph** shows the relationship between two sets of data.
- The **line of best fit** is a straight line that passes through or is close to the plotted points on a scatter graph.
- A linear relationship between two sets of data is called **correlation**.

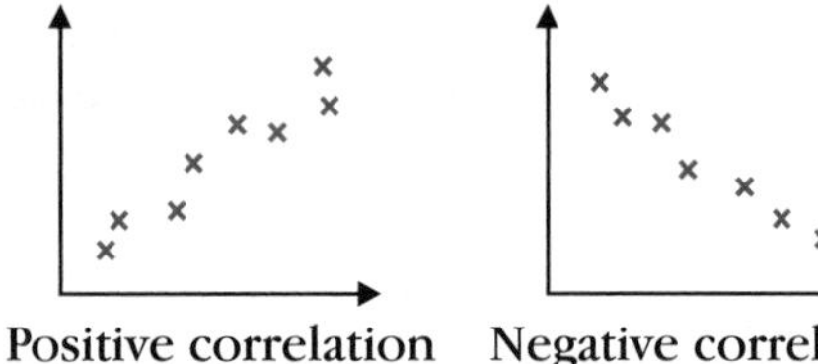

Positive correlation

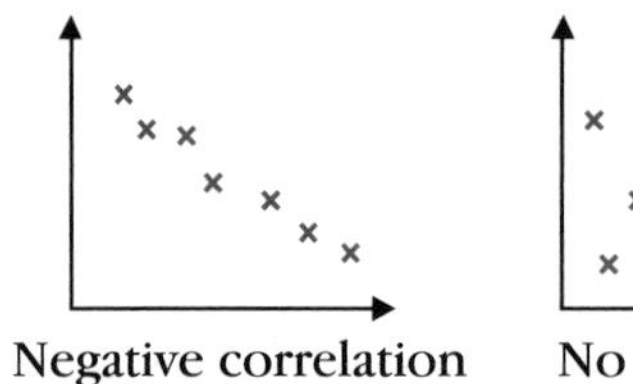

Negative correlation

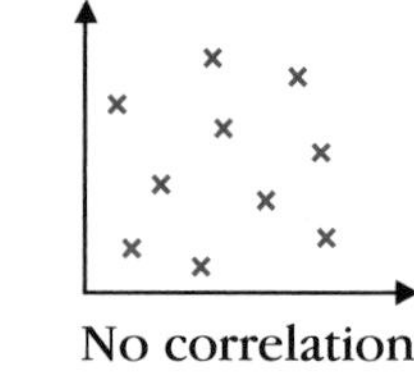

No correlation

- A line of best fit can be used to **estimate** other data values.

Key words

- scatter graph
- correlation
- positive correlation
- negative correlation
- no correlation
- line of best fit
- estimate

Example

The table shows the numbers of pages in nine books and their weight in grams.

Number of pages	65	115	85	125	100	75	145	125	90
Weight (g)	150	260	170	280	220	170	310	260	200

Grade D **(a)** Draw a scatter graph to represent this data.

Grade D **(b)** Describe the relationship between the number of pages and the weight.

Grade C **(c)** Draw a line of best fit on your scatter graph.

Grade C **(d)** Use your line of best fit to estimate
(i) the number of pages in a book of weight 265 g
(ii) the weight of a book with 110 pages.

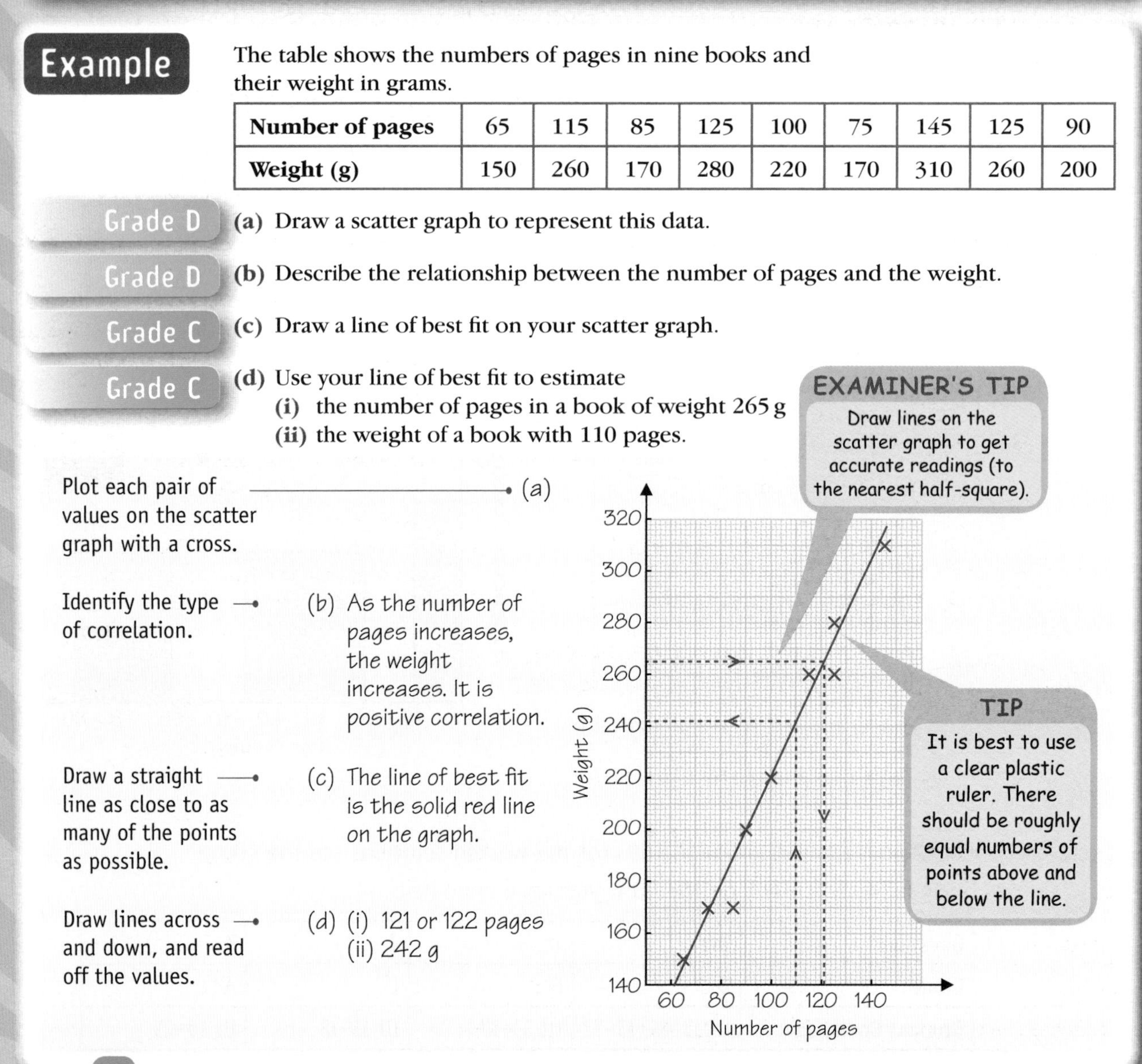

- The **Retail Prices Index (RPI)** is calculated each month by the government. It measures changes in the prices of goods and services.

Key words
Retail Prices Index (RPI) ☐

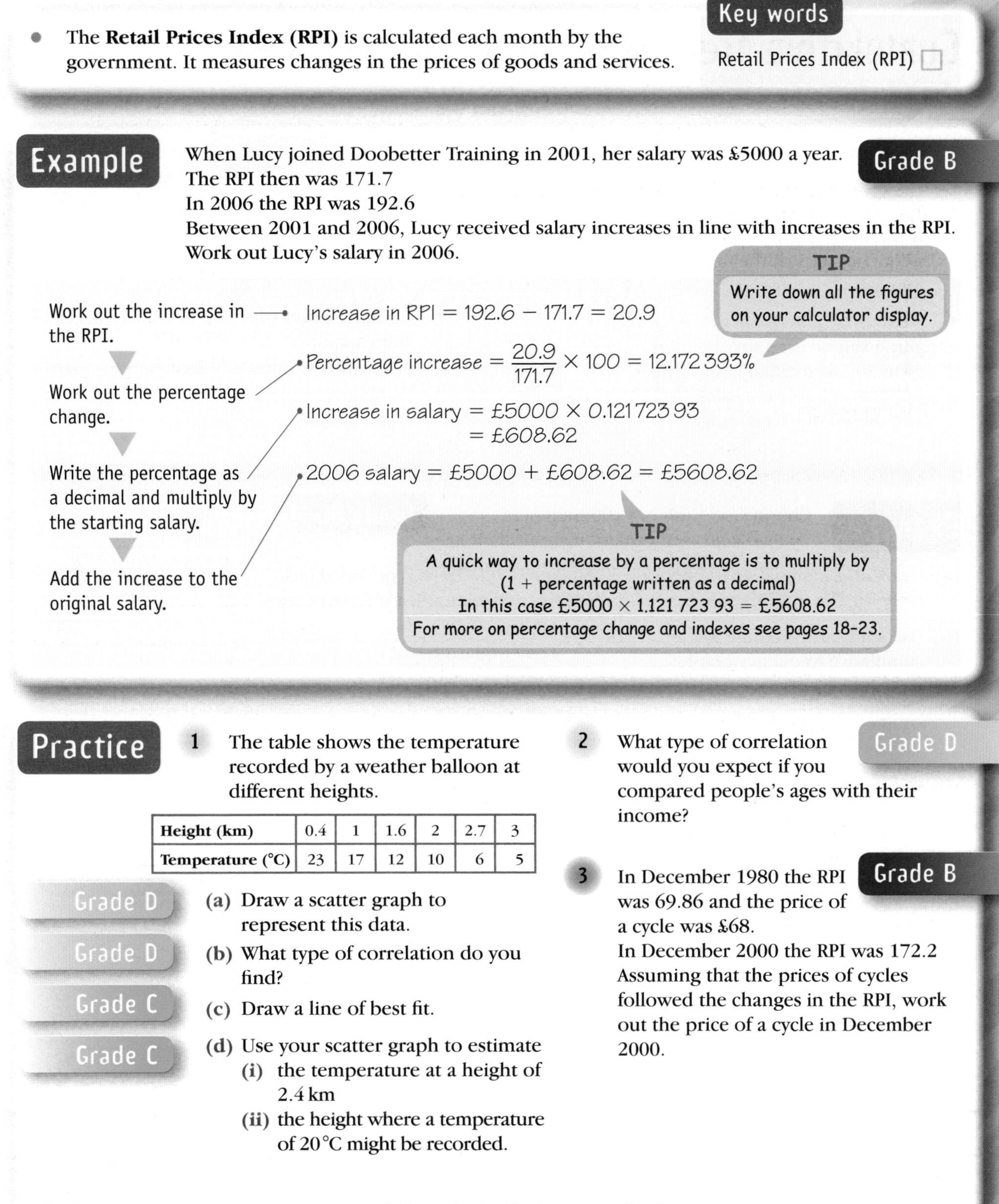

Example

Grade B

When Lucy joined Doobetter Training in 2001, her salary was £5000 a year.
The RPI then was 171.7
In 2006 the RPI was 192.6
Between 2001 and 2006, Lucy received salary increases in line with increases in the RPI.
Work out Lucy's salary in 2006.

Work out the increase in the RPI. → Increase in RPI = 192.6 − 171.7 = 20.9

Work out the percentage change. → Percentage increase = $\frac{20.9}{171.7} \times 100 = 12.172\,393\%$

TIP Write down all the figures on your calculator display.

Write the percentage as a decimal and multiply by the starting salary. → Increase in salary = £5000 × 0.121 723 93
= £608.62

Add the increase to the original salary. → 2006 salary = £5000 + £608.62 = £5608.62

TIP
A quick way to increase by a percentage is to multiply by (1 + percentage written as a decimal)
In this case £5000 × 1.121 723 93 = £5608.62
For more on percentage change and indexes see pages 18–23.

Practice

1 The table shows the temperature recorded by a weather balloon at different heights.

Height (km)	0.4	1	1.6	2	2.7	3
Temperature (°C)	23	17	12	10	6	5

Grade D **(a)** Draw a scatter graph to represent this data.

Grade D **(b)** What type of correlation do you find?

Grade C **(c)** Draw a line of best fit.

Grade C **(d)** Use your scatter graph to estimate
(i) the temperature at a height of 2.4 km
(ii) the height where a temperature of 20 °C might be recorded.

2 What type of correlation would you expect if you compared people's ages with their income? Grade D

3 In December 1980 the RPI was 69.86 and the price of a cycle was £68. Grade B
In December 2000 the RPI was 172.2
Assuming that the prices of cycles followed the changes in the RPI, work out the price of a cycle in December 2000.

Check your answers on page 169. For full worked solutions see the CD.
See the Student Book on the CD if you need more help.

Question	1ab	1cd	2	3
Grade	D	C	D	B
Student Book pages	U1 55–58		U1 55–58	U1 61–63

Cumulative frequency and box plots

- The **cumulative frequency** is the total frequency up to a particular class boundary.
- You can display data in a **cumulative frequency graph** by plotting the cumulative frequency against the upper class boundary for each class interval.
- A cumulative frequency graph can be used to estimate the median, the upper and lower quartiles, and the interquartile range.

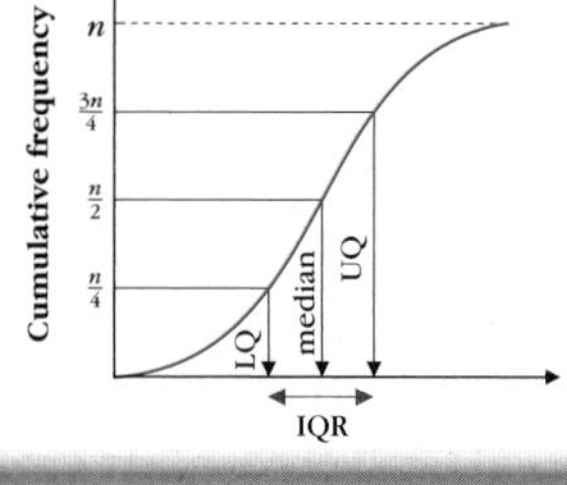

Key words

- cumulative frequency ☐
- cumulative frequency graph ☐
- box plot ☐

- To draw a **box plot** you need five pieces of information:
  - the lowest value
  - the lower quartile
  - the median
  - the upper quartile
  - the highest value.

Example

Grade B

Clifton Golf Club has 200 members.
The table gives information about their ages.

Age, a (years)	Frequency
$0 < a \leq 10$	8
$10 < a \leq 20$	26
$20 < a \leq 30$	32
$30 < a \leq 40$	45
$40 < a \leq 50$	37
$50 < a \leq 60$	29
$60 < a \leq 70$	16
$70 < a \leq 80$	7

(a) Draw up a cumulative frequency table for this information.

(b) Use your table to draw a cumulative frequency graph.

(c) Use your graph to estimate
  (i) the median
  (ii) the interquartile range.

(d) Estimate the percentage of members aged between 35 and 55.

Shefford Golf Club has provided this information about their members' ages:

Median = 42
Lower quartile = 29
Upper quartile = 62
Lowest value = 0
Highest value = 75

(e) Draw box plots for the two sets of data.

(f) Compare the distribution of members' ages in the two clubs.

TIP
Cumulative means a running total.

Work out the total frequency up to each upper class boundary. → (a)

Age, a (years)	Cumulative frequency
$0 < a \leq 10$	8
$0 < a \leq 20$	34
$0 < a \leq 30$	66
$0 < a \leq 40$	111
$0 < a \leq 50$	148
$0 < a \leq 60$	177
$0 < a \leq 70$	193
$0 < a \leq 80$	200

Plot each cumulative frequency against the *upper* class boundary. → (b)

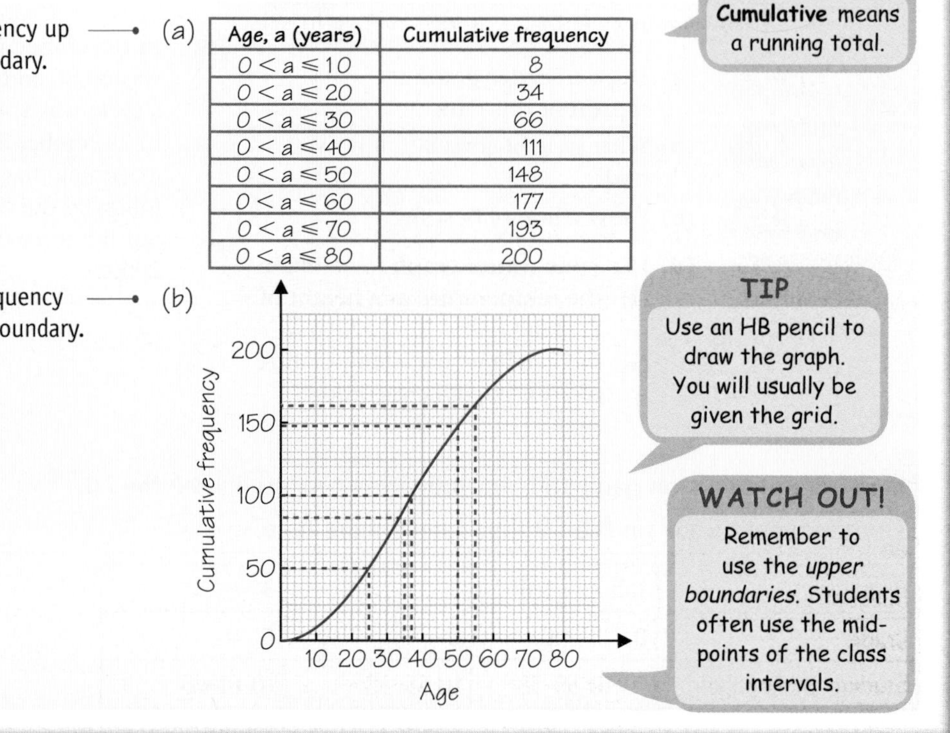

TIP
Use an HB pencil to draw the graph. You will usually be given the grid.

WATCH OUT!
Remember to use the *upper boundaries*. Students often use the mid-points of the class intervals.

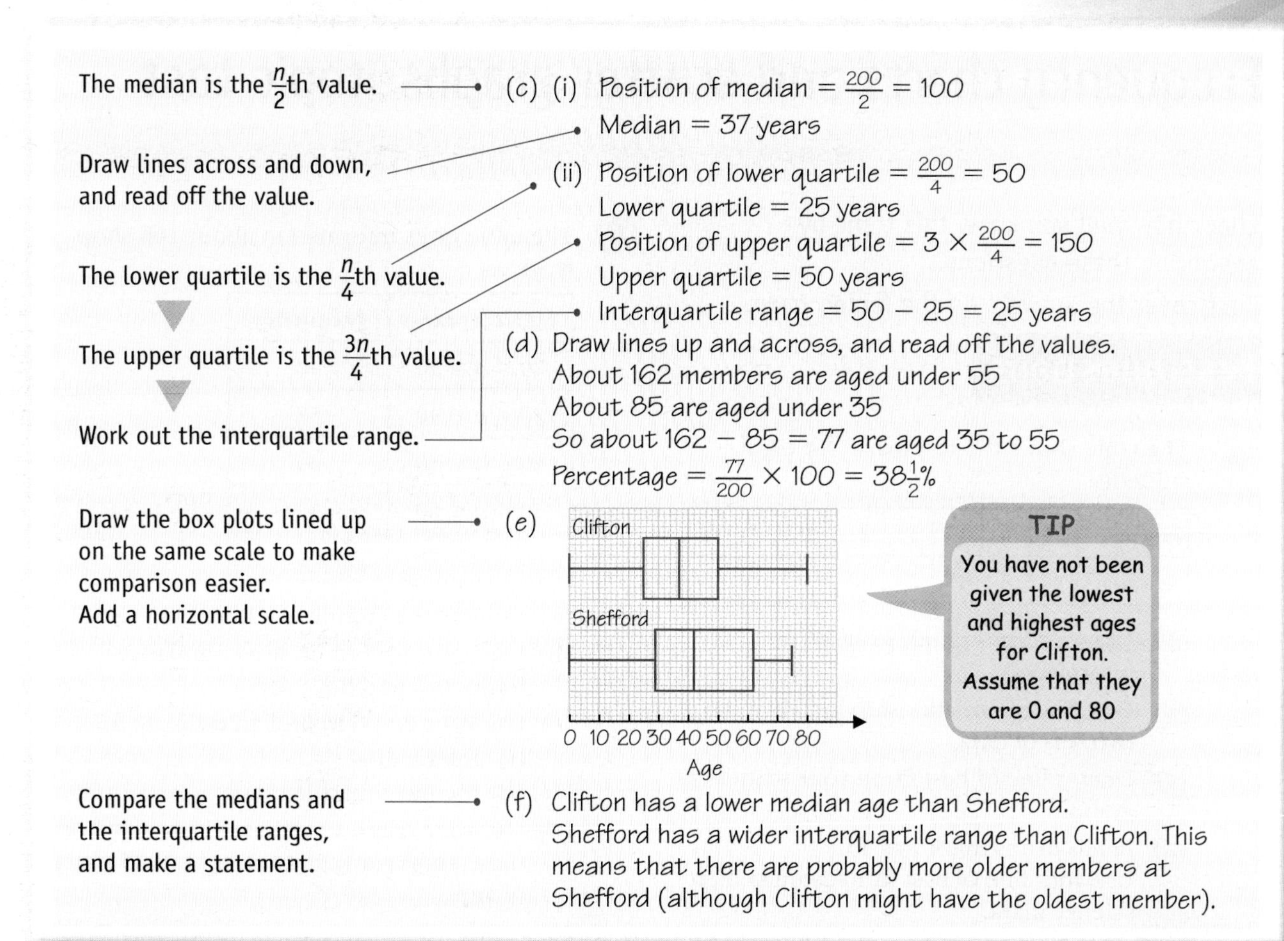

The median is the $\frac{n}{2}$th value. → (c) (i) Position of median $= \frac{200}{2} = 100$

Draw lines across and down, and read off the value. → Median = 37 years

(ii) Position of lower quartile $= \frac{200}{4} = 50$

Lower quartile = 25 years

The lower quartile is the $\frac{n}{4}$th value. → Position of upper quartile $= 3 \times \frac{200}{4} = 150$

Upper quartile = 50 years

The upper quartile is the $\frac{3n}{4}$th value.

Work out the interquartile range. → Interquartile range = 50 − 25 = 25 years

(d) Draw lines up and across, and read off the values.
About 162 members are aged under 55
About 85 are aged under 35
So about 162 − 85 = 77 are aged 35 to 55
Percentage $= \frac{77}{200} \times 100 = 38\frac{1}{2}\%$

Draw the box plots lined up on the same scale to make comparison easier. Add a horizontal scale. → (e)

Compare the medians and the interquartile ranges, and make a statement. → (f) Clifton has a lower median age than Shefford. Shefford has a wider interquartile range than Clifton. This means that there are probably more older members at Shefford (although Clifton might have the oldest member).

Practice

1 Grade B

The grouped frequency table gives information about the weekly rainfall at Heathrow Airport in 2005.

Weekly rainfall, d (mm)	Number of weeks
$0 \leqslant d < 10$	20
$10 \leqslant d < 20$	18
$20 \leqslant d < 30$	6
$30 \leqslant d < 40$	4
$40 \leqslant d < 50$	2
$50 \leqslant d < 60$	2

(a) Construct a cumulative frequency table.
(b) Draw the cumulative frequency curve.
(c) Estimate the median weekly rainfall.
(d) Estimate the interquartile range for the rainfall.
(e) Draw a box plot for the distribution.
(f) Estimate the number of weeks in which the rainfall was less than 15 mm.

2 Grade B

The grouped frequency table gives information about the amounts of daily sunshine in Downtown in August 2006.

Hours of sunshine s	Number of days
$0 \leqslant s < 2$	2
$2 \leqslant s < 4$	1
$4 \leqslant s < 6$	3
$6 \leqslant s < 8$	8
$8 \leqslant s < 10$	11
$10 \leqslant s < 12$	4
$12 \leqslant s < 14$	2

(a) Construct a cumulative frequency table.
(b) Draw the cumulative frequency graph.
(c) Use your graph to estimate the median amount of daily sunshine in Downtown in August 2006. Make your method clear.
(d) Estimate the interquartile range for the number of hours of sunshine.

In Ashwell during 2006, the median amount of daily sunshine was 10.2 hours and the interquartile range was 3 hours.

(e) Compare the distributions of hours of daily sunshine in Downtown and Ashwell for August 2006.

Check your answers on pages 169–170.
For full worked solutions see the CD.
See the Student Book on the CD if you need more help.

Question	1	2
Grade	B	B
Student Book pages	U1 41–44	U1 44–48

Frequency charts and scatter graphs: topic test

Check how well you know this topic by answering these questions.

First cover the answers on the facing page.

Test questions

1 The table shows the test marks for eight students.

Maths	25	6	17	33	21	10	17	28
Science	20	8	15	29	22	9	19	30

(a) Draw a scatter graph to show this information.

(b) Describe the relationship between the two sets of data.

(c) Draw a line of best fit on your scatter graph.

(d) Jane's maths mark was 15.
Use your line of best fit to estimate her science mark.

2 An internet company recorded the number of orders it received on each of 30 days.
Here are the results:

18 48 35 12 43 26 40 14 26
16 26 13 58 39 36 13 38 57
16 38 29 44 29 26 44 26 51
52 24 15

Represent this data using an ordered stem and leaf diagram. Include a key.

3 The table gives information about the examination marks of 100 students.

Mark	Number of students
1–10	2
11–20	7
21–30	10
31–40	15
41–50	27
51–60	22
61–70	13
71–80	4

Draw a frequency polygon for this data.

4 The table gives information about 160 shop workers.

Age, a (years)	Frequency
$20 < a \leq 30$	42
$30 < a \leq 40$	58
$40 < a \leq 50$	34
$50 < a \leq 60$	17
$60 < a \leq 70$	9

(a) Construct a cumulative frequency table.

(b) Use your table to draw a cumulative frequency graph.

(c) Use your graph to find an estimate for
(i) the median
(ii) the interquartile range.

(d) Use your graph to estimate how many of these shop workers are less than 45 years of age.

5 A school has 80 teachers.
The youngest is aged 22 and the oldest is 59.
The median age is 43 years, the lower quartile is 38 and the upper quartile is 52.
Draw a box plot to show this information.

6 The incomplete table and histogram give some information about the numbers of DVDs sold in a shop.

DVD price, p (£)	Frequency
$0 < p \leq 5$	5
$5 < p \leq 10$	
$10 < p \leq 20$	40
$20 < p \leq 40$	

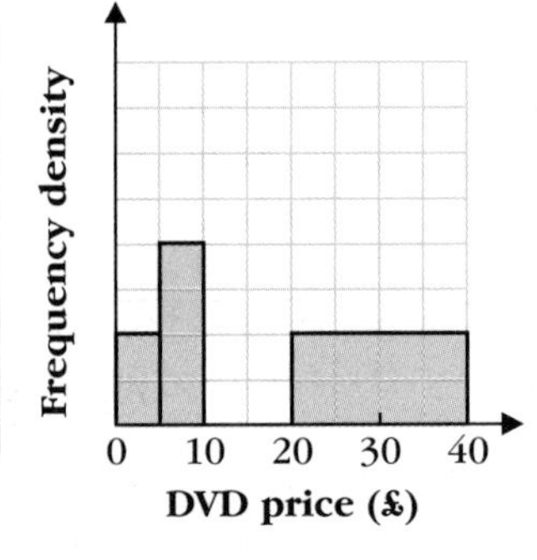

(a) Use the information from the histogram to complete the frequency table.

(b) Use the information from the table to complete the histogram.

Now check your answers – see the facing page.

Cover this page while you answer the test questions opposite.

Worked answers

Revise this on...

D

1 (a), (c)

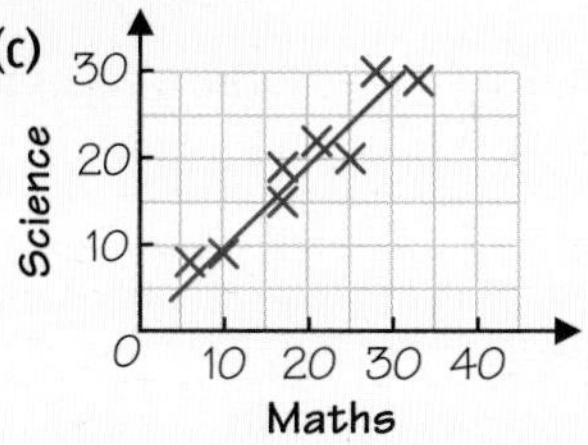

page 20

(b) The higher the maths mark, the higher the science mark. It is positive correlation.

(c) See graph opposite. C

(d) 15 C

D

2

1	2, 3, 3, 4, 5, 6, 6, 8
2	4, 6, 6, 6, 6, 6, 9, 9
3	5, 6, 8, 8, 9
4	0, 3, 4, 4, 8
5	1, 2, 7, 8

Key: 2 | 4 means 24

page 16

C **3**

page 17

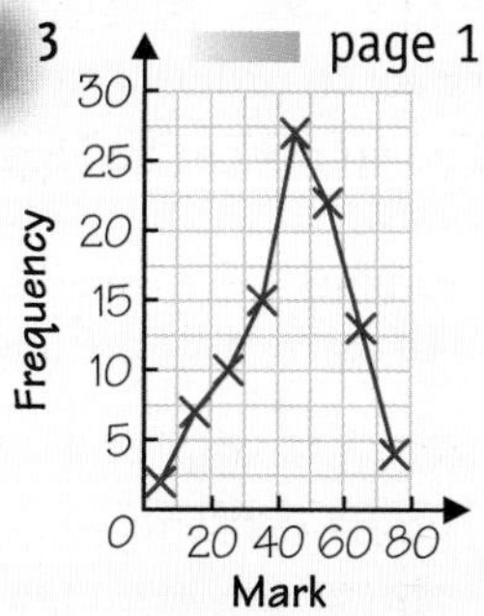

B

4 (a)

Age, a (years)	Cumulative frequency
$20 < a \leq 30$	42
$20 < a \leq 40$	100
$20 < a \leq 50$	134
$20 < a \leq 60$	151
$20 < a \leq 70$	160

(b)

pages 22–23

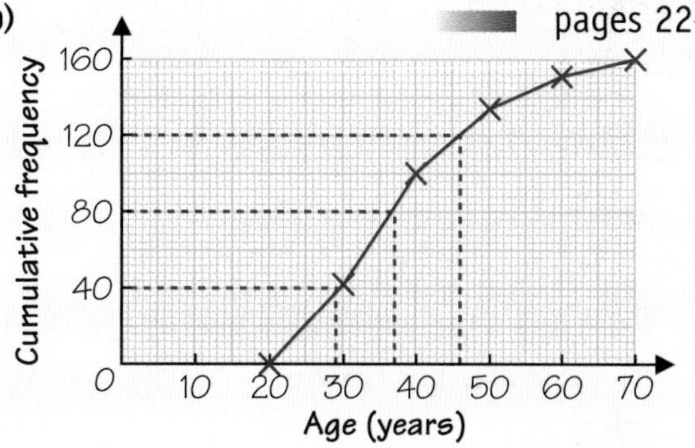

(c) (i) $n = 160 \rightarrow$ position of median $= \frac{160}{2} = 80 \rightarrow$ median $= 37$

(ii) Position of lower quartile $= \frac{160}{4} = 40$ $\rightarrow$ lower quartile $= 29$

Position of upper quartile $= 3 \times \frac{160}{4}$ 120 $\rightarrow$ upper quartile $= 46$

Interquartile range $= 46 - 29 = 17$ years

(d) About 116 workers (from graph)

B

5

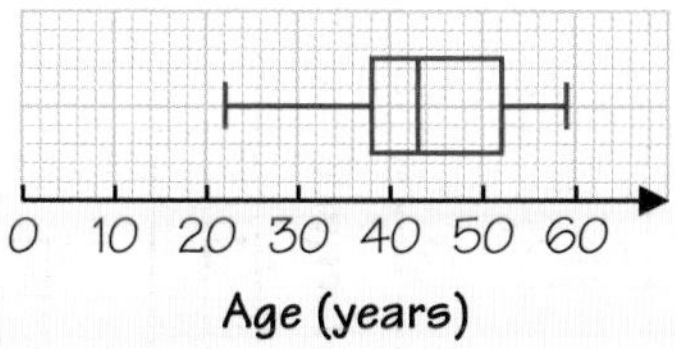

pages 22–23

A*

6 (a)

page 18

DVD price, p (£)	Frequency
$0 < p \leq 5$	5
$5 < p \leq 10$	10
$10 < p \leq 20$	40
$20 < p \leq 40$	20

(b)

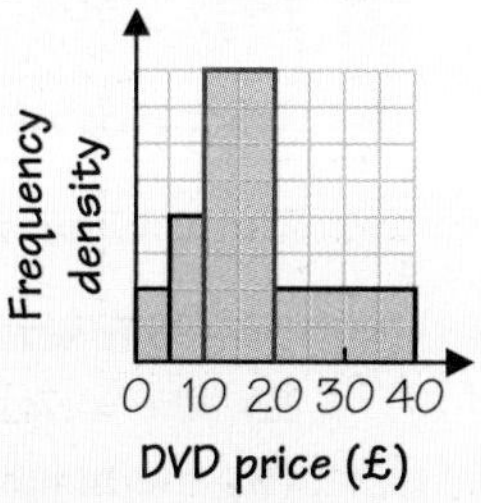

Tick the questions you got right.

Question	1ab	1cd	2	3	4	5	6
Grade	D	C	D	C	B	B	A*

Mark the grade you are working at on your revision planner on page vi.

Handling data: subject test

Exam practice questions

1 20 people do a lap round a race track.
Here are their times to the nearest second:

62	46	39	53	28	44	65	41	48	37
36	49	51	46	39	27	60	50	45	33

(a) Draw an ordered stem and leaf diagram to show this information. Include a key.

(b) Use your stem and leaf diagram to write down the median.

2 100 students were asked how they came to school that day.
Some of the results are shown in the two-way table.

EXAMINER'S TIP
You wouldn't get a whole question on this on the Higher paper, but it might come up as part of a question.

	Car	Walk	Cycle	Total
Year 7		13	9	41
Year 8	5			22
Year 9		18		
Total	36		21	100

(a) Complete the two-way table.

(b) One of these students is picked at random.
Write down the probability that this student

(i) came to school by car **(ii)** is in Year 7 and cycled to school.

3 A freezer contains four flavours of ice-cream – vanilla, chocolate, strawberry and mint.
The table shows the probabilities that Marcus chooses vanilla, chocolate or mint.

Flavour	Vanilla	Chocolate	Strawberry	Mint
Probability	0.2	0.4		0.15

Work out the probability that he picks a strawberry ice-cream.

4 The table shows the heights and weights of ten students.

Weight (kg)	75	65	82	76	71	65	77	70	72	68
Height (cm)	185	182	191	188	184	166	175	178	181	180

(a) Draw a scatter graph to show this information.

(b) What type of correlation do you find?

(c) Draw a line of best fit.

(d) Use your scatter graph to estimate

(i) the weight of a student whose height is 188 cm

(ii) the height of a student whose weight is 74 kg.

5 The table shows information about the times it took a group of students to do their homework one day.

(a) Write down the modal class.

(b) Calculate an estimate of the mean.

Time, t (hours)	Number of students
0	3
$0 < t \leqslant 1$	14
$1 < t \leqslant 2$	17
$2 < t \leqslant 3$	5
$3 < t \leqslant 4$	1

6 Use data from the table in question **5** for this question.

(a) Construct a cumulative frequency table.

(b) Use your table to draw a cumulative frequency graph.

(c) Use your graph to find an estimate for

(i) the median

(ii) the interquartile range.

7 A bag contains ten coloured beads. Six of the beads are red and four are blue.
Bob is going to take one bead from the bag at random, put it back, and then take another bead at random.

(a) Draw a tree diagram to show the possible outcomes.

(b) Use your tree diagram to calculate the probability that the two beads

(i) will both be red

(ii) will be the same colour.

8 There are 1200 students at Shefford High School. 640 are girls and 560 are boys.
Louise wants to take a sample of 50 students, stratified by gender.
How many girls and how many boys should there be in the sample?

9 The table gives information about some people's waiting times at a bus stop.

Draw a histogram to show this data.

Waiting time, t (minutes)	Frequency
$0 < t \leqslant 10$	32
$10 < t \leqslant 15$	20
$15 < t \leqslant 30$	24
$30 < t \leqslant 35$	4
$t \geqslant 35$	0

10 A drawer contains five yellow scarves and four white scarves.
June selects a scarf from the drawer at random, and then Heather selects a scarf from the drawer at random.
Work out the probability that

(a) they both select a yellow scarf

(b) they select scarves of different colours.

Check your answers on pages 170–171. For full worked solutions see the CD.
Tick the questions you got right.

Question	1	2a	2b	3	4ab	4cd	5	6	7	8	9	10
Grade	D	E	D	D	D	C	C	B	B	A	A	A*
Revise this on page	16	2	10	10	20		7	22–23	11	4	17	12

Mark the grade you are working at on your revision planner on page vi.
Go to the pages shown to revise for the ones you got wrong.

Handling data

Data, averages and probability

- **Two-way tables** are used to record or display information that is grouped in two categories.
- When you are writing questions for a **questionnaire**:
  - be clear what you want to find out, and what data you need.
  - ask short, simple questions.
  - provide **response boxes** with possible answers.
  - avoid questions which are too vague, too personal, or which may influence the answer (**leading questions**).
- A **stratified sample** is one in which the population is divided into groups called strata and each of the strata is randomly sampled. The proportions of the different strata in the sample are the same as in the whole population.
- A **systematic sample** is one where every *n*th item is chosen.
- The **mode** of a set of data is the value which occurs the most often.
- The **median** is the middle value when the data is arranged in order of size. If there are n data values the median is the $\frac{n+1}{2}$th value.
- The **mean** of a set of data is the sum of the values divided by the number of values:

  $$\text{Mean} = \frac{\text{sum of values}}{\text{number of values}}$$

- The **range** of a set of data is the difference between the highest and lowest values:

  range = highest value − lowest value.
- For **grouped data**:
  - the **modal class** is the class interval with the highest frequency.
  - you can state the **class interval** that contains the median.
  - you can calculate an **estimate of the mean** using the middle value of each class interval.
- A **trend line** is the line of best fit drawn through the moving averages plotted on a graph.
- If the **probability** of an event happening is p, the probability of it *not* happening is $1 - p$.
- **Estimated probability** $= \frac{\text{number of successful trials}}{\text{total number of trails}}$
- If events A and B are **independent**, P(A and B) = P(A) × P(B).

Frequency charts and scatter graphs

- A **stem and leaf diagram** shows the shape of a distribution and keeps all the data values. It needs a **key** to show how the stem and leaf are combined.

0	9
1	2, 3
2	0, 7
3	4, 8
4	1

Key: 1 | 2 means 12

- You can use a stem and leaf diagram to find the **median**, the **upper and lower quartiles** and the **interquartile range**.
- A **histogram** is used to display grouped continuous data.
  The areas of the rectangles are proportional to the frequencies they represent.

Waiting time, t (minutes)	Frequency
$0 < t \leq 10$	16
$10 < t \leq 15$	20
$15 < t \leq 20$	19
$20 < t \leq 30$	28
$30 < t \leq 60$	12
$t > 60$	0

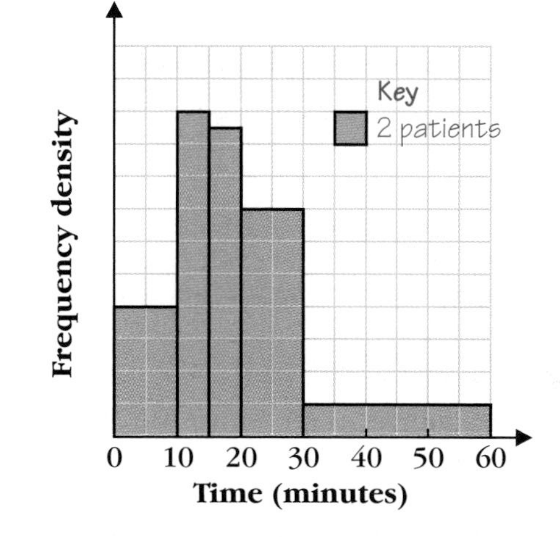

- A **scatter graph** shows the relationship between two sets of data.
- The **line of best fit** is a straight line that passes through or is close to the plotted points on a scatter graph.
- A linear relationship between two sets of data is called **correlation**.

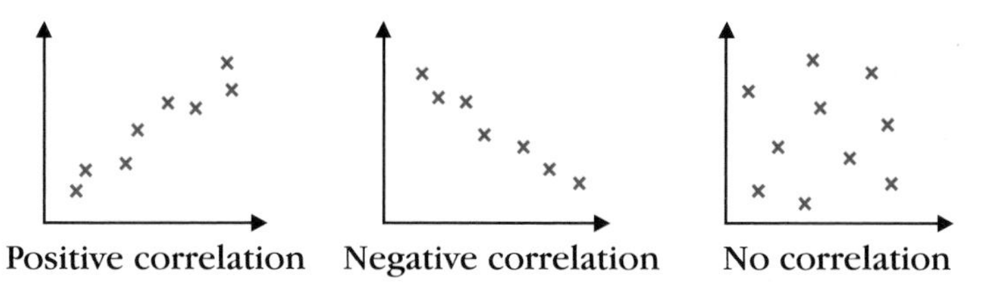

- A **cumulative frequency graph** can be used to estimate the median, the upper and lower quartiles and the interquartile range.

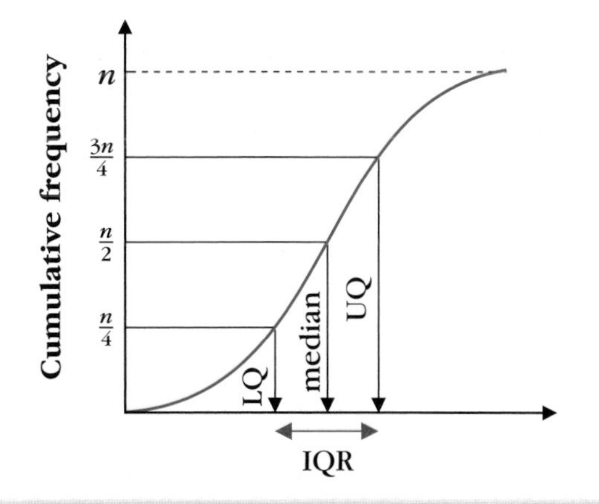

Unit 1 Examination practice paper

A formulae sheet can be found on page 167.

Section A (calculator)

1 A packet contains seeds that can produce red, yellow, orange or blue flowers.

The table shows each of the probabilities that a seed taken at random from the packet will produce a flower that is red or yellow or orange.

Colour of flower	red	yellow	orange	blue
Probability	0.2	0.3	0.3	

A seed is taken at random from the packet.

(a) Work out the probability that the seed, will produce a blue flower.

The packet contains 150 seeds.

(b) Work out the number of seeds that will produce a red flower.

(4 marks)

2 These are the marks of some students in a Maths test.

32	45	15	36	33
13	17	44	27	26
24	48	29	18	31

Draw an ordered stem and leaf diagram for these marks.

You must include a key.

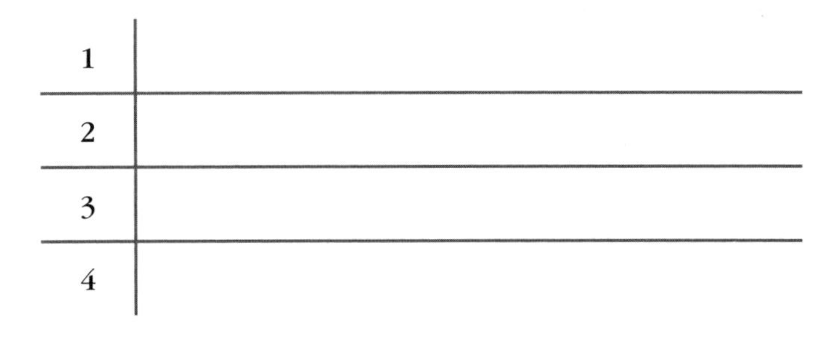

Key:

(3 marks)

3 Terri recorded the time, in minutes, taken to complete his last 30 homeworks.

This table gives the information about the times.

Time, t (minutes)	Frequency, f		
$10 < t \leq 15$	2		
$15 < t \leq 20$	5		
$20 < t \leq 25$	10		
$25 < t \leq 30$	8		
$30 < t \leq 35$	5		

(a) Calculate the class interval in which the median lies.

(b) Calculate an estimate of the mean time it took Terri to complete each homework.

(6 marks)

4 The table shows the number of boys and the number of girls in each year group at Springfield Secondary School.

There are 500 boys and 500 girls in the school.

Year group	Number of boys	Number of girls
7	100	100
8	150	50
9	100	100
10	50	150
11	100	100
Total	500	500

Joyce took a stratified sample of 50 girls, by year group.

Work out the number of Year 8 girls in her sample.

(2 marks)

Check your answers on page 171. For full worked solutions see the CD.

Section B (non-calculator)

1 Mrs Green makes cakes.

She wants to find out what people think of the cakes she makes.

She uses this question on her questionnaire.

> You do like my cakes don't you?

(a) Write down what is wrong with this question.

(b) Mrs Green wants to find out how many cakes people eat.

Design a suitable question for her questionnaire.

You must include some response boxes.

(3 marks)

2 The scatter graph shows the Science mark and the Maths mark for 15 students.

(a) What type of correlation does this scatter graph show?

(b) Draw a line of best fit on the scatter graph.

(c) Sophie's Science mark was 36. Use your line of best fit to estimate Sophie's Maths mark.

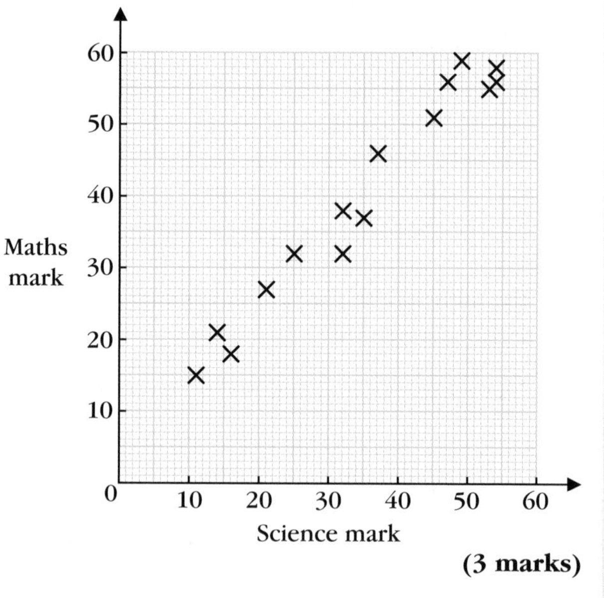

(3 marks)

3 Here is the cumulative frequency curve of the weights of 120 girls at Mayfield Secondary School.

Use the cumulative frequency curve to find an estimate for the

(a) median weight of the Year 10 girls,

(b) interquartile range of the Year 10 girls.

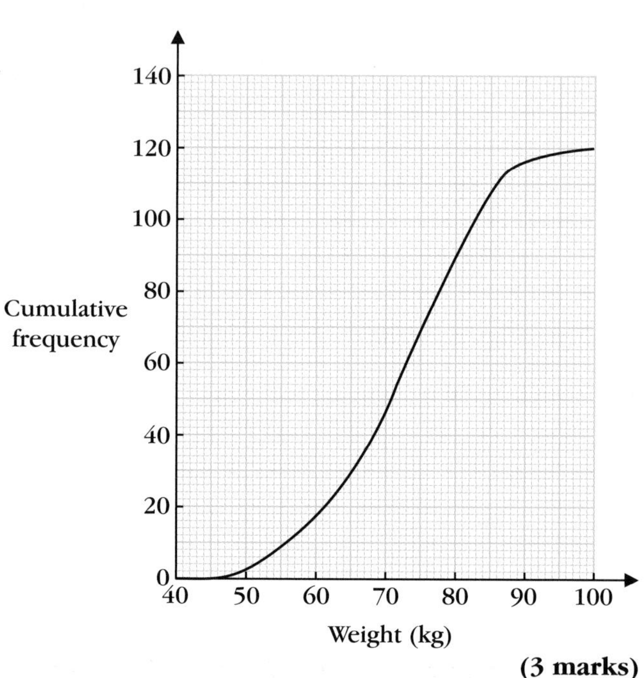

(3 marks)

4 The unfinished table and histogram show information about the weight, w grams, of fish that Alan caught one day.

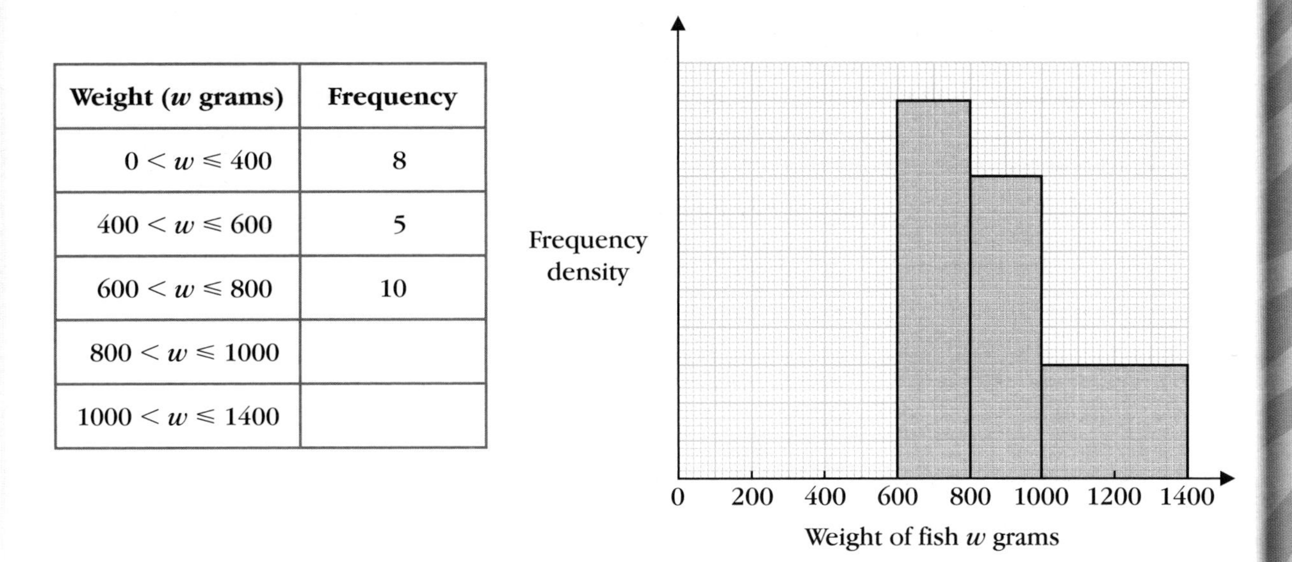

Weight (w grams)	Frequency
$0 < w \leqslant 400$	8
$400 < w \leqslant 600$	5
$600 < w \leqslant 800$	10
$800 < w \leqslant 1000$	
$1000 < w \leqslant 1400$	

(a) Use the information in the histogram to complete the table.

(b) Use the information in the table to complete the histogram. **(4 marks)**

5 The probability that Saba will pass her driving test at the first attempt is $\frac{7}{10}$

If she fails her driving test the first time, the probability that she passes her test on the second occasion is $\frac{9}{10}$

What is the probability that Saba will pass her driving test on one of the first two occasions?

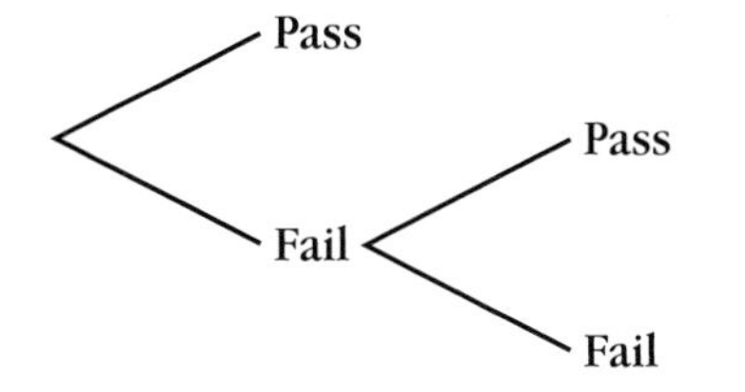

(2 marks)

Check your answers on page 171. For full worked solutions see the CD.

Operations

STAGE 1

- Adding a **negative** number has the same effect as subtracting the **positive** number, for example $4 + -1 = 4 - 1 = 3$
- Subtracting a negative number has the same effect as adding the positive number, for example $2 - -3 = 2 + 3 = 5$
- These tables show the signs you get when you multiply or divide one number by another:

+	×	+	=	+
+	×	−	=	−
−	×	+	=	−
−	×	−	=	+

+	÷	+	=	+
+	÷	−	=	−
−	÷	+	=	−
−	÷	−	=	+

Two **like signs** give a +, two **unlike signs** give a −

Key words

- negative
- positive
- like signs
- unlike signs

1

- **Product** and **times** are words that mean **multiplication** (×).
- **Sharing** and **goes into** are words that mean **division** (÷).
- A **factor** of a number is a whole number that divides exactly into the number. The factors include 1 and the number itself.
- A **multiple** of a number is the result of multiplying the number by a positive whole number.
- A **prime number** is a number greater than 1 which has only two factors: itself and 1.
- Any factor of a number that is a prime number is a **prime factor**.
- Writing a number as the **product of its prime factors** is called writing it in prime factor form.

Key words

- factor
- multiple
- prime number
- prime factor
- product of prime factors

1

Example

Grade C

Write each of these numbers as a product of its prime factors.
(a) 12 **(b)** 18

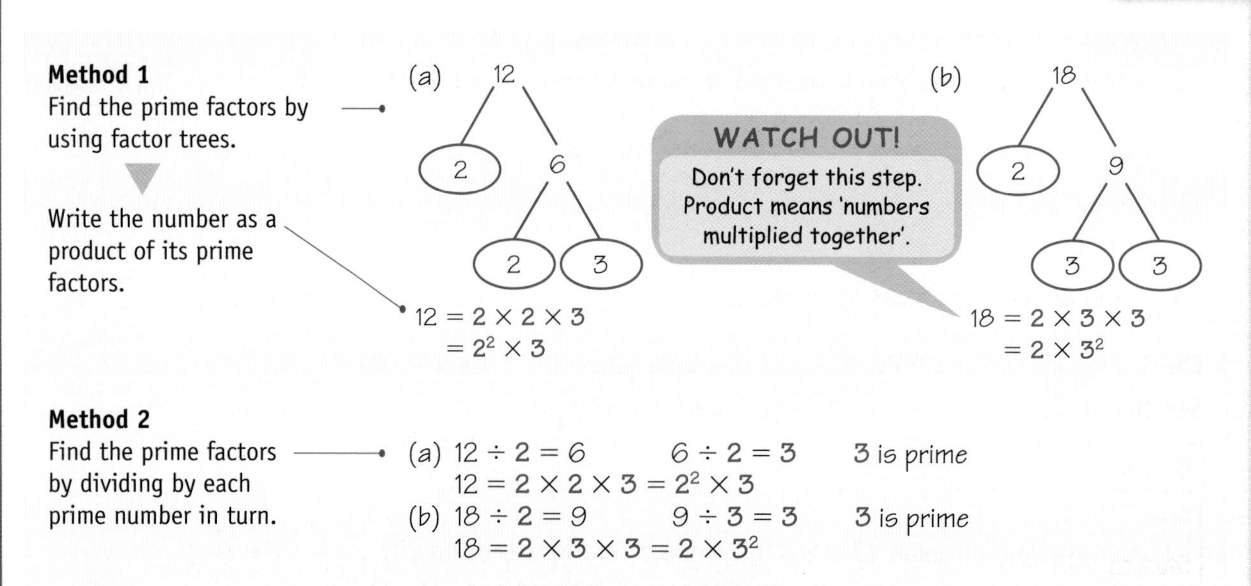

- The **highest common factor** (HCF) of two whole numbers is the highest factor that is common to both of them.
- The **lowest common multiple** (LCM) of two whole numbers is the lowest number that is a multiple of them both.

Key words

highest common factor (HCF) ☐
lowest common multiple (LCM) ☐

Example

Grade C

(a) Find the highest common factor (HCF) of 12 and 18
(b) Find the lowest common multiple (LCM) of 12 and 18

Write each number as a product of its prime factors.
Circle the common factors and multiply them together.

(a) $12 = (2) \times 2 \times (3)$
$18 = (2) \times 3 \times (3)$
The HCF is $2 \times 3 = 6$

List the first few multiples of each number.
Select the lowest number that appears in both lists.

(b) 12 24 (36) 48 60 (72)
18 (36) 54 (72)
Common multiples are 36 and 72. The LCM is 36

Practice

EXAMINER'S TIP
You wouldn't get a whole question on this on the Higher paper, but it might come up as part of a question.
Make sure you can do long multiplication and long division.

1 Work out these additions and subtractions. Do not use a calculator. (Grade F)
(a) $3 - 7$ **(b)** $-8 + 3$
(c) $-2 - 6$ **(d)** $5 - -3$

2 Work out these multiplications and divisions. Do not use a calculator. (Grade E)
(a) 7×-6 **(b)** $18 \div -3$
(c) $-33 \div 11$ **(d)** $-12 \div -4$

3 Work out these multiplications and divisions. Do not use a calculator. (Grade E)
(a) 205×34 **(b)** 546×53 **(c)** 475×28
(d) $375 \div 15$ **(e)** $392 \div 16$ **(f)** $744 \div 24$

4 Write each of these numbers as a product of its prime factors. (Grade C)
(a) 16 **(b)** 24

5 Find the HCF and LCM of 16 and 24 (Grade C)

Check your answers on page 171. For full worked solutions see the CD.
See the Student Book on the CD if you need more help.

Question	1	2	3	4	5
Grade	F	E	E	C	C
Student Book pages	U2 2–4	U2 2–4	–	U2 11–12	U2 11–12

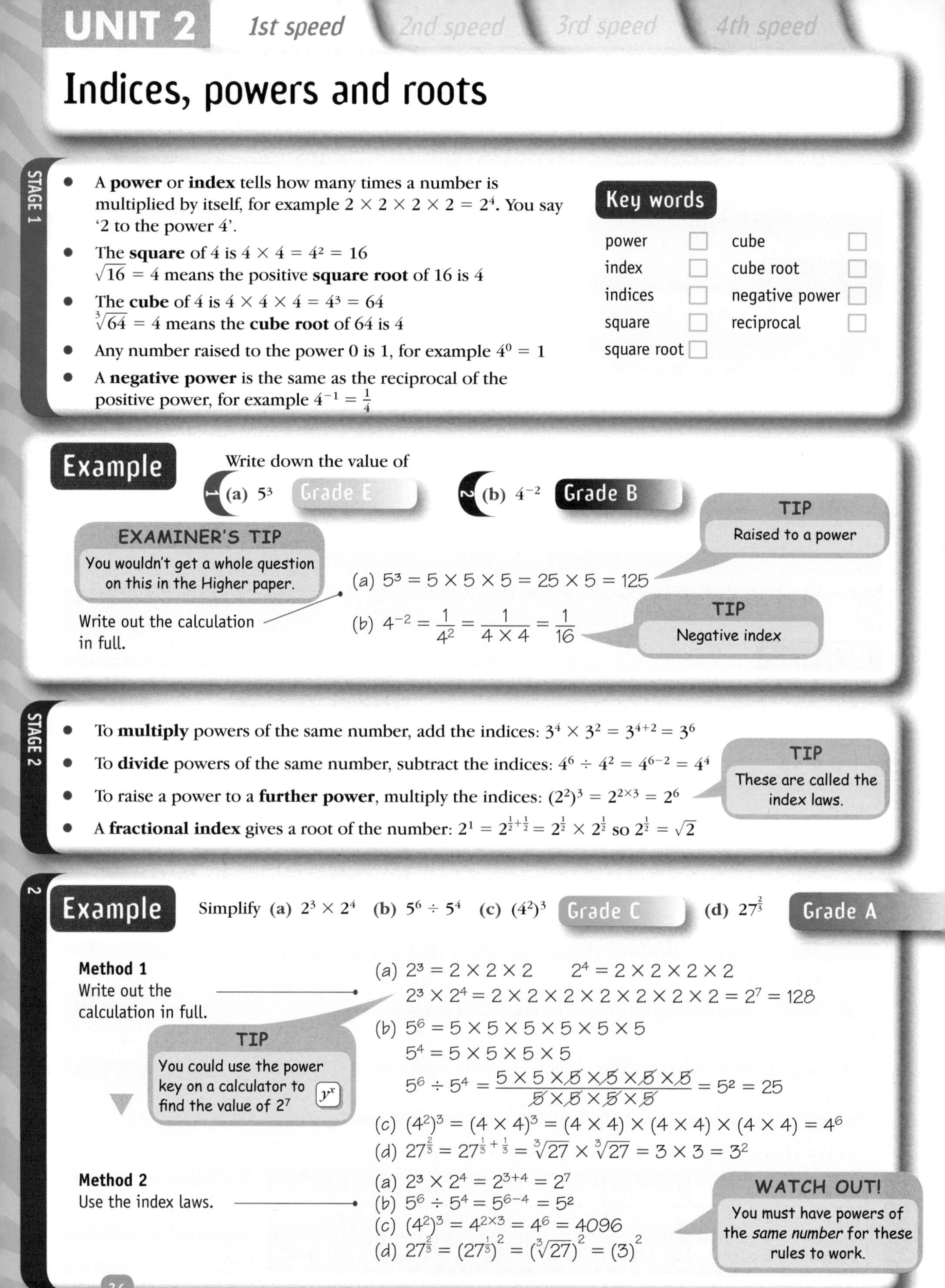

Indices, powers and roots

STAGE 1

- A **power** or **index** tells how many times a number is multiplied by itself, for example $2 \times 2 \times 2 \times 2 = 2^4$. You say '2 to the power 4'.
- The **square** of 4 is $4 \times 4 = 4^2 = 16$
  $\sqrt{16} = 4$ means the positive **square root** of 16 is 4
- The **cube** of 4 is $4 \times 4 \times 4 = 4^3 = 64$
  $\sqrt[3]{64} = 4$ means the **cube root** of 64 is 4
- Any number raised to the power 0 is 1, for example $4^0 = 1$
- A **negative power** is the same as the reciprocal of the positive power, for example $4^{-1} = \frac{1}{4}$

Key words

power	☐	cube	☐
index	☐	cube root	☐
indices	☐	negative power	☐
square	☐	reciprocal	☐
square root	☐		

Example

Write down the value of

1 (a) 5^3 Grade E   2 (b) 4^{-2} Grade B

EXAMINER'S TIP
You wouldn't get a whole question on this in the Higher paper.

Write out the calculation in full.

(a) $5^3 = 5 \times 5 \times 5 = 25 \times 5 = 125$

TIP Raised to a power

(b) $4^{-2} = \frac{1}{4^2} = \frac{1}{4 \times 4} = \frac{1}{16}$

TIP Negative index

STAGE 2

- To **multiply** powers of the same number, add the indices: $3^4 \times 3^2 = 3^{4+2} = 3^6$
- To **divide** powers of the same number, subtract the indices: $4^6 \div 4^2 = 4^{6-2} = 4^4$
- To raise a power to a **further power**, multiply the indices: $(2^2)^3 = 2^{2 \times 3} = 2^6$
- A **fractional index** gives a root of the number: $2^1 = 2^{\frac{1}{2}+\frac{1}{2}} = 2^{\frac{1}{2}} \times 2^{\frac{1}{2}}$ so $2^{\frac{1}{2}} = \sqrt{2}$

TIP These are called the index laws.

Example

2 Simplify (a) $2^3 \times 2^4$ (b) $5^6 \div 5^4$ (c) $(4^2)^3$ Grade C (d) $27^{\frac{2}{3}}$ Grade A

Method 1
Write out the calculation in full.

(a) $2^3 = 2 \times 2 \times 2$ $2^4 = 2 \times 2 \times 2 \times 2$
$2^3 \times 2^4 = 2 \times 2 \times 2 \times 2 \times 2 \times 2 \times 2 = 2^7 = 128$

TIP You could use the power key on a calculator to find the value of 2^7 y^x

(b) $5^6 = 5 \times 5 \times 5 \times 5 \times 5 \times 5$
$5^4 = 5 \times 5 \times 5 \times 5$
$5^6 \div 5^4 = \frac{5 \times 5 \times \cancel{5} \times \cancel{5} \times \cancel{5} \times \cancel{5}}{\cancel{5} \times \cancel{5} \times \cancel{5} \times \cancel{5}} = 5^2 = 25$

(c) $(4^2)^3 = (4 \times 4)^3 = (4 \times 4) \times (4 \times 4) \times (4 \times 4) = 4^6$

(d) $27^{\frac{2}{3}} = 27^{\frac{1}{3}+\frac{1}{3}} = \sqrt[3]{27} \times \sqrt[3]{27} = 3 \times 3 = 3^2$

Method 2
Use the index laws.

(a) $2^3 \times 2^4 = 2^{3+4} = 2^7$
(b) $5^6 \div 5^4 = 5^{6-4} = 5^2$
(c) $(4^2)^3 = 4^{2 \times 3} = 4^6 = 4096$
(d) $27^{\frac{2}{3}} = (27^{\frac{1}{3}})^2 = (\sqrt[3]{27})^2 = (3)^2$

WATCH OUT! You must have powers of the *same number* for these rules to work.

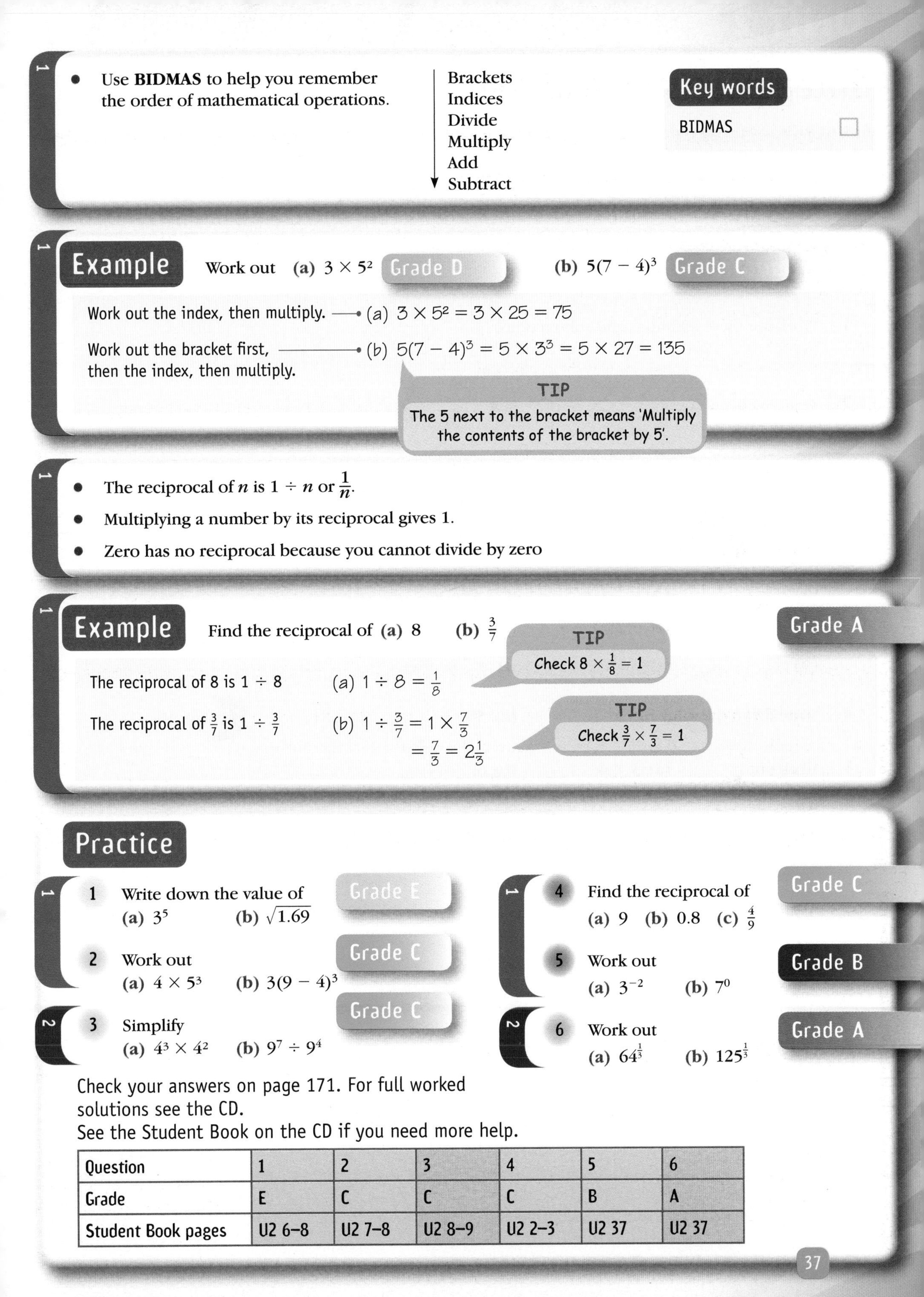

- Use **BIDMAS** to help you remember the order of mathematical operations.

Brackets
Indices
Divide
Multiply
Add
Subtract

Key words

BIDMAS ☐

Example

Work out **(a)** 3×5^2 Grade D **(b)** $5(7 - 4)^3$ Grade C

Work out the index, then multiply. → (a) $3 \times 5^2 = 3 \times 25 = 75$

Work out the bracket first, then the index, then multiply. → (b) $5(7 - 4)^3 = 5 \times 3^3 = 5 \times 27 = 135$

TIP
The 5 next to the bracket means 'Multiply the contents of the bracket by 5'.

- The reciprocal of n is $1 \div n$ or $\frac{1}{n}$.
- Multiplying a number by its reciprocal gives 1.
- Zero has no reciprocal because you cannot divide by zero

Example

Find the reciprocal of **(a)** 8 **(b)** $\frac{3}{7}$ Grade A

The reciprocal of 8 is $1 \div 8$ (a) $1 \div 8 = \frac{1}{8}$

TIP
Check $8 \times \frac{1}{8} = 1$

The reciprocal of $\frac{3}{7}$ is $1 \div \frac{3}{7}$ (b) $1 \div \frac{3}{7} = 1 \times \frac{7}{3}$
$= \frac{7}{3} = 2\frac{1}{3}$

TIP
Check $\frac{3}{7} \times \frac{7}{3} = 1$

Practice

1 Write down the value of Grade E
(a) 3^5 **(b)** $\sqrt{1.69}$

2 Work out Grade C
(a) 4×5^3 **(b)** $3(9 - 4)^3$

3 Simplify Grade C
(a) $4^3 \times 4^2$ **(b)** $9^7 \div 9^4$

4 Find the reciprocal of Grade C
(a) 9 **(b)** 0.8 **(c)** $\frac{4}{9}$

5 Work out Grade B
(a) 3^{-2} **(b)** 7^0

6 Work out Grade A
(a) $64^{\frac{1}{3}}$ **(b)** $125^{\frac{1}{3}}$

Check your answers on page 171. For full worked solutions see the CD.
See the Student Book on the CD if you need more help.

Question	1	2	3	4	5	6
Grade	E	C	C	C	B	A
Student Book pages	U2 6–8	U2 7–8	U2 8–9	U2 2–3	U2 37	U2 37

Integers: topic test

Check how well you know this topic by answering these questions.
First cover the answers on the facing page.

Test questions

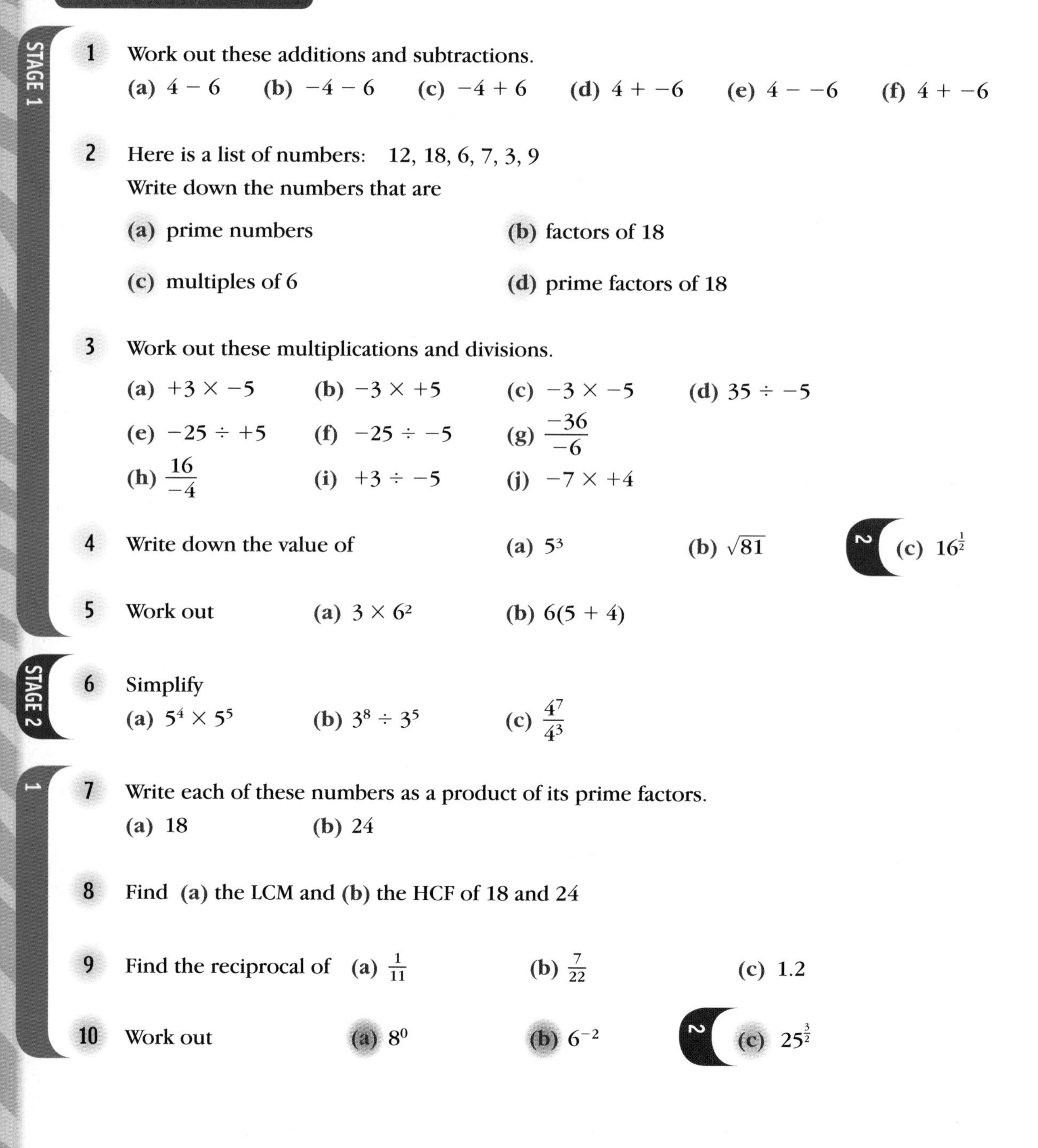

STAGE 1

1 Work out these additions and subtractions.
(a) $4 - 6$ (b) $-4 - 6$ (c) $-4 + 6$ (d) $4 + -6$ (e) $4 - -6$ (f) $4 + -6$

2 Here is a list of numbers: 12, 18, 6, 7, 3, 9
Write down the numbers that are
(a) prime numbers (b) factors of 18
(c) multiples of 6 (d) prime factors of 18

3 Work out these multiplications and divisions.
(a) $+3 \times -5$ (b) $-3 \times +5$ (c) -3×-5 (d) $35 \div -5$
(e) $-25 \div +5$ (f) $-25 \div -5$ (g) $\frac{-36}{-6}$
(h) $\frac{16}{-4}$ (i) $+3 \div -5$ (j) $-7 \times +4$

4 Write down the value of (a) 5^3 (b) $\sqrt{81}$ 2 (c) $16^{\frac{1}{2}}$

5 Work out (a) 3×6^2 (b) $6(5 + 4)$

STAGE 2

6 Simplify
(a) $5^4 \times 5^5$ (b) $3^8 \div 3^5$ (c) $\frac{4^7}{4^3}$

1

7 Write each of these numbers as a product of its prime factors.
(a) 18 (b) 24

8 Find (a) the LCM and (b) the HCF of 18 and 24

9 Find the reciprocal of (a) $\frac{1}{11}$ (b) $\frac{7}{22}$ (c) 1.2

10 Work out (a) 8^0 (b) 6^{-2} 2 (c) $25^{\frac{3}{2}}$

Now check your answers – see the facing page.

Cover this page while you answer the test questions opposite.

Worked answers

Revise this on...

1 (a) -2 (b) -10 (c) 2 (d) -2 (e) 10 (f) -2 page 34 F

2 (a) 3 and 7 (b) 3, 6, 9 and 18 (c) 6, 12 and 18 (d) 3 page 34 F/E

3 (a) -15 (b) -15 (c) 15 (d) -7 (e) -5 page 34 E
(f) 5 (g) 6 (h) -4 (i) -0.6 (j) -28

4 (a) 125 (b) 9 (c) 4 page 36 E

5 (a) 108 (b) 54 page 37 E

6 (a) $5^{4+5} = 5^9$ (b) $3^{8-5} = 3^3$ (c) $4^{7-3} = 4^4$ page 36 C

7 (a) $18 = 2 \times 3 \times 3$ (b) $24 = 2 \times 2 \times 2 \times 3$ page 34 C

8 (a) Multiples of 18: 18, 36, 54, 72, 90
Multiples of 24: 24, 48, 72, 96
The LCM is 72
(b) The HCF is $2 \times 3 = 6$ page 35 C

9 (a) 11 (b) $\frac{22}{7}$ (c) $\frac{5}{6}$ page 37 C

10 (a) 1 (b) $\frac{1}{36}$ (c) 125 page 36 B/A

Tick the questions you got right.

Question	1	2ab	2cd	3	4ab	4c	5	6	7	8	9	10ab	10c
Grade	F	F	E	E	E	E	E	C	C	C	C	B	A

Mark the grade you are working at on your revision planner on page vii.

Decimals and rounding

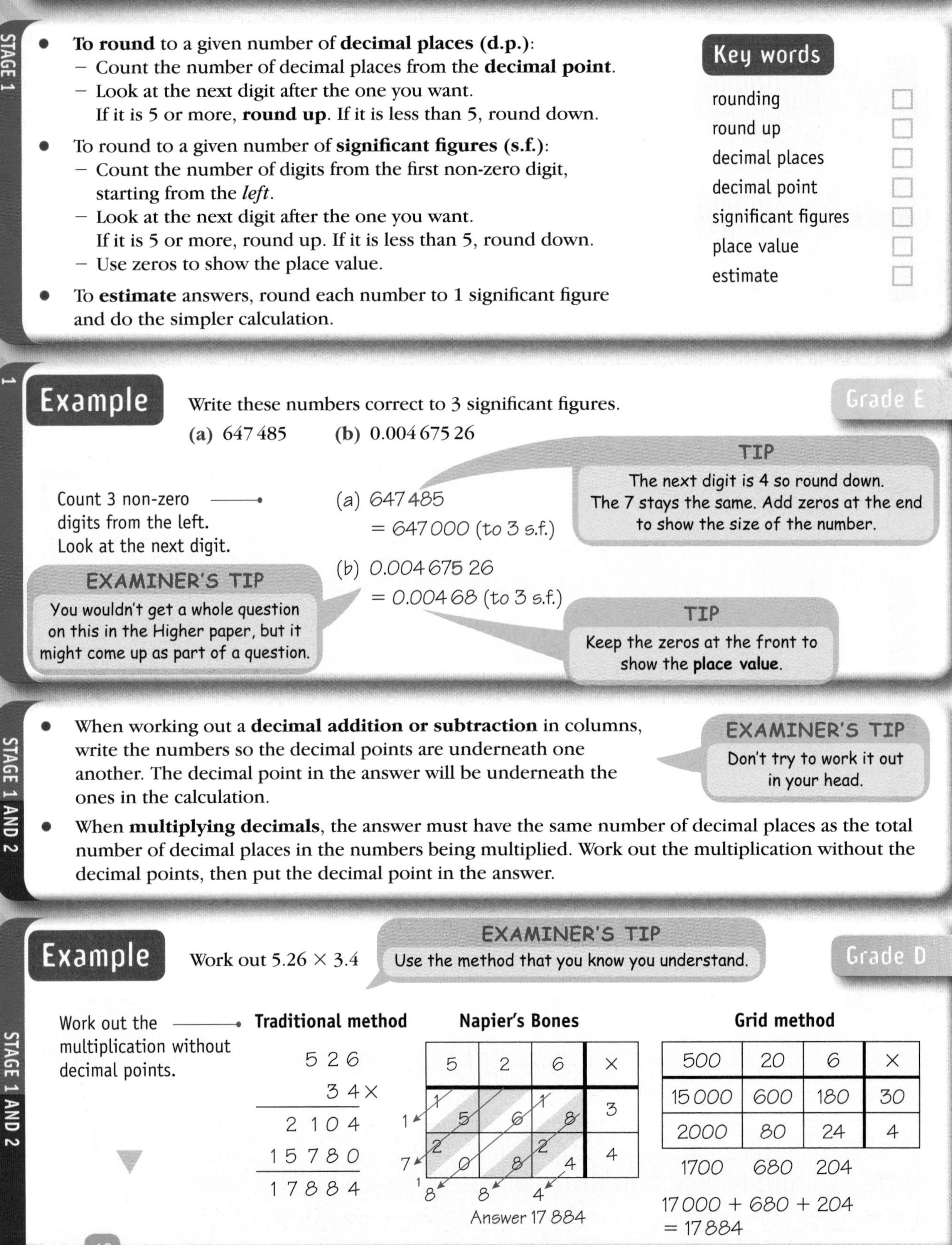

STAGE 1

- **To round** to a given number of **decimal places (d.p.)**:
  - Count the number of decimal places from the **decimal point**.
  - Look at the next digit after the one you want.
    If it is 5 or more, **round up**. If it is less than 5, round down.
- To round to a given number of **significant figures (s.f.)**:
  - Count the number of digits from the first non-zero digit, starting from the *left*.
  - Look at the next digit after the one you want.
    If it is 5 or more, round up. If it is less than 5, round down.
  - Use zeros to show the place value.
- To **estimate** answers, round each number to 1 significant figure and do the simpler calculation.

Key words

rounding	☐
round up	☐
decimal places	☐
decimal point	☐
significant figures	☐
place value	☐
estimate	☐

1

Example

Grade E

Write these numbers correct to 3 significant figures.
(a) 647 485 **(b)** 0.004 675 26

Count 3 non-zero digits from the left. Look at the next digit.

(a) 647 485
= 647 000 (to 3 s.f.)

TIP
The next digit is 4 so round down. The 7 stays the same. Add zeros at the end to show the size of the number.

(b) 0.004 675 26
= 0.004 68 (to 3 s.f.)

TIP
Keep the zeros at the front to show the **place value**.

EXAMINER'S TIP
You wouldn't get a whole question on this in the Higher paper, but it might come up as part of a question.

STAGE 1 AND 2

- When working out a **decimal addition or subtraction** in columns, write the numbers so the decimal points are underneath one another. The decimal point in the answer will be underneath the ones in the calculation.
- When **multiplying decimals**, the answer must have the same number of decimal places as the total number of decimal places in the numbers being multiplied. Work out the multiplication without the decimal points, then put the decimal point in the answer.

EXAMINER'S TIP
Don't try to work it out in your head.

STAGE 1 AND 2

Example

Grade D

Work out 5.26 × 3.4

EXAMINER'S TIP
Use the method that you know you understand.

Work out the multiplication without decimal points.

Traditional method

```
   5 2 6
     3 4×
 -------
   2 1 0 4
 1 5 7 8 0
 ---------
 1 7 8 8 4
```

Napier's Bones

5	2	6	×
1/5	0/6	1/8	3
2/0	0/8	2/4	4

Diagonal sums: 1, 7, 8, 8, 4

Answer 17 884

Grid method

500	20	6	×
15 000	600	180	30
2000	80	24	4
1700	680	204	

17 000 + 680 + 204
= 17 884

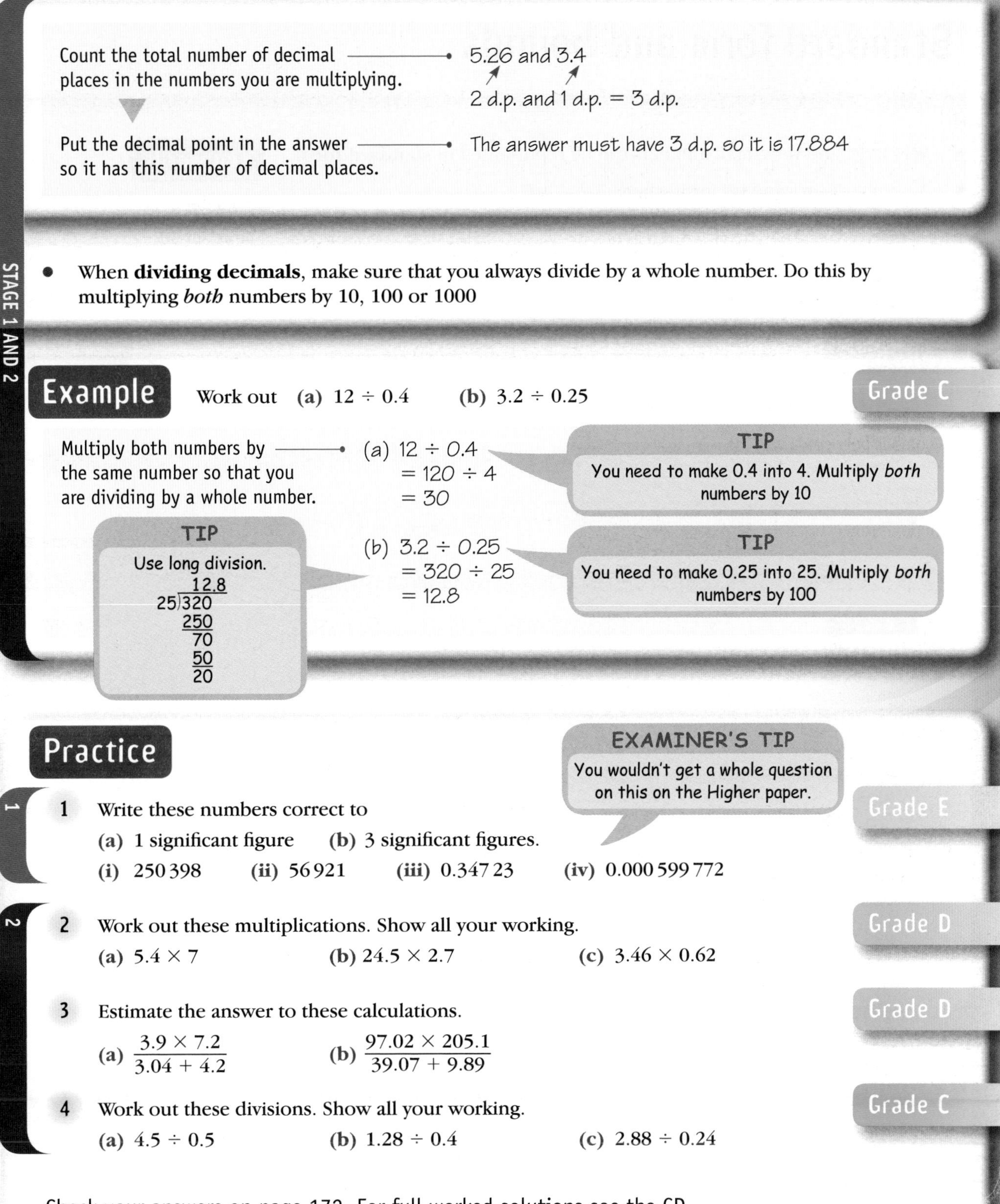

Count the total number of decimal places in the numbers you are multiplying. → 5.26 and 3.4
2 d.p. and 1 d.p. = 3 d.p.

Put the decimal point in the answer so it has this number of decimal places. → The answer must have 3 d.p. so it is 17.884

STAGE 1 AND 2

- When **dividing decimals**, make sure that you always divide by a whole number. Do this by multiplying *both* numbers by 10, 100 or 1000

Example

Work out **(a)** $12 \div 0.4$ **(b)** $3.2 \div 0.25$

Grade C

Multiply both numbers by the same number so that you are dividing by a whole number. →

(a) $12 \div 0.4$
$= 120 \div 4$
$= 30$

TIP
You need to make 0.4 into 4. Multiply *both* numbers by 10

(b) $3.2 \div 0.25$
$= 320 \div 25$
$= 12.8$

TIP
You need to make 0.25 into 25. Multiply *both* numbers by 100

TIP
Use long division.

```
    12.8
25)320
   250
    70
    50
    20
```

Practice

EXAMINER'S TIP
You wouldn't get a whole question on this on the Higher paper.

1 Write these numbers correct to (Grade E)

(a) 1 significant figure **(b)** 3 significant figures.

(i) 250 398 **(ii)** 56 921 **(iii)** 0.347 23 **(iv)** 0.000 599 772

2 Work out these multiplications. Show all your working. (Grade D)

(a) 5.4×7 **(b)** 24.5×2.7 **(c)** 3.46×0.62

3 Estimate the answer to these calculations. (Grade D)

(a) $\dfrac{3.9 \times 7.2}{3.04 + 4.2}$ **(b)** $\dfrac{97.02 \times 205.1}{39.07 + 9.89}$

4 Work out these divisions. Show all your working. (Grade C)

(a) $4.5 \div 0.5$ **(b)** $1.28 \div 0.4$ **(c)** $2.88 \div 0.24$

Check your answers on page 172. For full worked solutions see the CD.
See the Student Book on the CD if you need more help.

Question	1	2	3	4
Grade	E	D	D	C
Student Book pages	U2 4–6	U3 1–2	U2 5–6, 43–44	U3 1–2

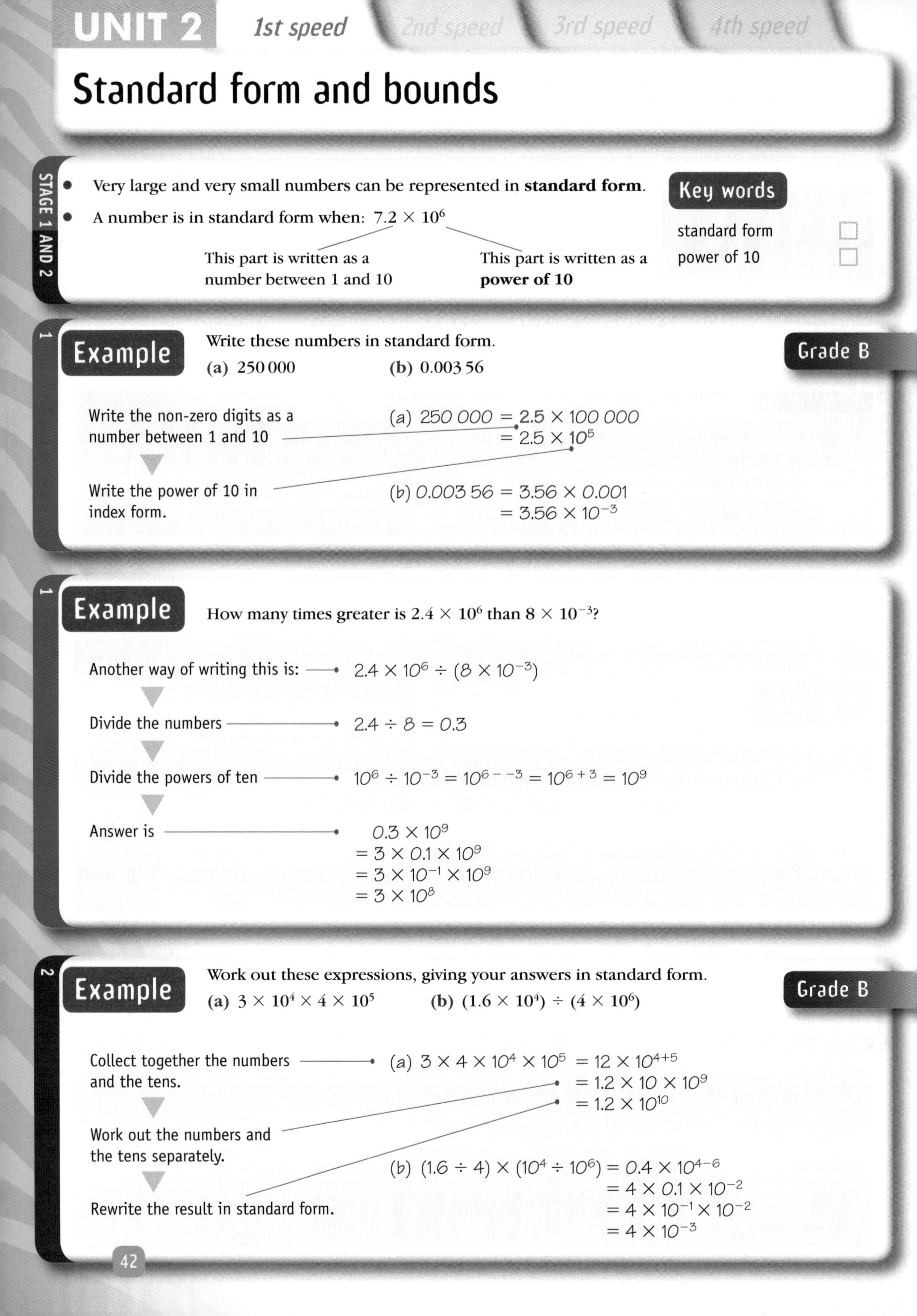

Standard form and bounds

STAGE 1 AND 2

- Very large and very small numbers can be represented in **standard form**.
- A number is in standard form when: 7.2×10^6

This part is written as a number between 1 and 10

This part is written as a **power of 10**

Key words

standard form ☐

power of 10 ☐

1

Example

Grade B

Write these numbers in standard form.

(a) 250 000 **(b)** 0.003 56

Write the non-zero digits as a number between 1 and 10

Write the power of 10 in index form.

(a) $250\,000 = 2.5 \times 100\,000$
$= 2.5 \times 10^5$

(b) $0.003\,56 = 3.56 \times 0.001$
$= 3.56 \times 10^{-3}$

1

Example

How many times greater is 2.4×10^6 than 8×10^{-3}?

Another way of writing this is: → $2.4 \times 10^6 \div (8 \times 10^{-3})$

Divide the numbers → $2.4 \div 8 = 0.3$

Divide the powers of ten → $10^6 \div 10^{-3} = 10^{6 - -3} = 10^{6+3} = 10^9$

Answer is → 0.3×10^9
$= 3 \times 0.1 \times 10^9$
$= 3 \times 10^{-1} \times 10^9$
$= 3 \times 10^8$

2

Example

Grade B

Work out these expressions, giving your answers in standard form.

(a) $3 \times 10^4 \times 4 \times 10^5$ **(b)** $(1.6 \times 10^4) \div (4 \times 10^6)$

Collect together the numbers and the tens.

Work out the numbers and the tens separately.

Rewrite the result in standard form.

(a) $3 \times 4 \times 10^4 \times 10^5 = 12 \times 10^{4+5}$
$= 1.2 \times 10 \times 10^9$
$= 1.2 \times 10^{10}$

(b) $(1.6 \div 4) \times (10^4 \div 10^6) = 0.4 \times 10^{4-6}$
$= 4 \times 0.1 \times 10^{-2}$
$= 4 \times 10^{-1} \times 10^{-2}$
$= 4 \times 10^{-3}$

- The **upper bound** is the maximum possible value of a measurement or result of a calculation.
- The **lower bound** is the minimum possible value of a measurement or result of a calculation.

Key words

upper bound ☐

lower bound ☐

Example

Grade A

Write down the upper and lower bounds for these measurements.

(a) 45 mm to the nearest mm

(b) 5.6 seconds to the nearest tenth of a second

Add half a unit to find the upper bound.

Subtract half a unit to find the lower bound.

(a) Upper bound = 45 + 0.5 = 45.5 mm
Lower bound = 45 − 0.5 = 44.5 mm

(b) Upper bound = 5.6 + 0.05 = 5.65 s
Lower bound = 5.6 − 0.05 = 5.55 s

TIP
The measurement is to the nearest 0.1 s so 'half a unit' is 0.05 s here.

Practice

1 Write these numbers in standard form. **Grade B**

(a) 356 000 **(b)** 4567

(c) 450 000 000 **(d)** 0.000 45

(e) 0.000 000 15 **(f)** 0.012

2 Change these numbers from standard form to ordinary numbers. **Grade B**

(a) 3.6×10^4

(b) 3.6×10^7

(c) 4.55×10^2

(d) 2.4×10^{-3}

(e) 2.45×10^{-6}

(f) 6.75×10^{-4}

3 Evaluate these expressions, giving your answers in standard form. **Grade B**

(a) $5 \times 10^4 \times 4 \times 10^3$

(b) $6 \times 10^2 \times 9 \times 10^6$

(c) $(2 \times 10^4) \div (4 \times 10^5)$

(d) $(8 \times 10^9) \div (2 \times 10^8)$

4 Write down the upper and lower bounds for these measurements. **Grade A**

(a) 4.65 m to the nearest cm

(b) 3.4 seconds to the nearest 0.1 s

(c) 16.32 *l* to the nearest m*l*

(d) 3.85 kg to the nearest ????????

Check your answers on page 172. For full worked solutions see the CD.
See the Student Book on the CD if you need more help.

Question	1	2	3	4
Grade	B	B	B	A
Student Book pages	U2 37–39	U2 37–39	U2 40–41	U2 48–49

Decimals and rounding: topic test

Check how well you know this topic by answering these questions.
First cover the answers on the facing page.

Test questions

STAGE 1 AND 2

1 Work out
(a) $23.02 - 14.83$ (b) $2.79 + 3.43$
(c) $36 - 28.7$ (d) $142.8 \div 7$
(e) 1.77×4 (f) 3.47×5
(g) $20.304 \div 8$

1

2 Round these numbers to the number of decimal places given in brackets.
(a) 4.925 (2) (b) 32.71 (1)
(c) 4.995 (2) (d) 3.699 (1)
(e) 100.0599 (3) (f) 5.922 (1)

3 Write these numbers to the number of significant figures given in brackets.
(a) 25 460 (2) (b) 0.005 69 (2)
(c) 312.99 (4) (d) 93 695 (2)
(e) 93 000 000 (1) (f) 5.0099 (3)

STAGE 1 AND 2

4 Work out
(a) 0.3×0.2 (b) 0.09×0.7
(c) $0.8 \div 0.2$ (d) $1.05 \div 0.05$
(e) $10.8 \div 0.4$ (f) 3.2×0.6

1

5 Estimate the answer to these calculations.
(a) $\dfrac{790 + 8.7}{197.6}$ (b) $\dfrac{4750 \times 9.7}{385 - 299.6}$

6 Write these numbers in standard form.
(a) 56 000 (b) 45 670 000
(c) 350 000 000 000 (d) 0.000 35
(e) 0.000 000 451 (f) 0.005

1

7 These numbers are written in standard form.
Write them as ordinary numbers.
(a) 3.2×10^2 (b) 5.6×10^5
(c) 4.75×10^7 (d) 2.7×10^{-2}
(e) 5.45×10^{-5} (f) 1.7×10^{-3}

2

8 Evaluate these expressions, giving your answers in standard form.
(a) $6 \times 10^4 \times 4.2 \times 10^3$
(b) $(2 \times 10^2) \div (5 \times 10^5)$

9 Work out
(a) $2.4 \times 10^6 \times 2 \times 10^{-3}$
(b) $2.4 \times 10^6 \div (4 \times 10^2)$
(c) $6 \times 10^6 \times (2.1 \times 10^3)$
(d) $9 \times 10^6 \div (1.5 \times 10^{-5})$
Give your answer in standard form.

1

10 Write the upper bound and lower bound for
(a) 450 000 (correct to 3 s.f.)
(b) 7.1×10^{-5} (correct to 2 s.f.)

11 Write down the upper and lower bounds of these measurements.
(a) 5.8 cm to one decimal place
(b) 12.4 cm to the nearest millimetre
(c) 100 m correct to the nearest millimetre
(d) 10 seconds to the nearest one hundredth second
(e) 1 km to the nearest centimetre

Now check your answers – see the facing page.

Cover this page while you answer the test questions opposite.

Worked answers

Revise this on...

1&2

1 (a) 8.19 (b) 6.22 (c) 7.3 (d) 20.4 (e) 7.08 (f) 17.35 (g) 2.538 — pages 40–41 F

1

2 (a) 4.93 (b) 32.7 (c) 5.00 (d) 3.7 (e) 100.060 (f) 5.9 — page 40 F

3 (a) 25 000 (b) 0.0057 (c) 313.0 (d) 94 000 (e) 90 000 000 (f) 5.01 — page 40 E

1&2

4 (a) 0.06 (b) 0.063 (c) $8 \div 2 = 4$ (d) $105 \div 5 = 21$ (e) $108 \div 4 = 27$ (f) 1.92 — pages 40–41 D

1

5 (a) $\frac{800}{200} = 4$ (b) $\frac{5000 \times 10}{400 - 300} = \frac{50\,000}{100} = 500$ — page 40 D

6 (a) 5.6×10^4 (b) 4.567×10^7 (c) 3.5×10^{11} (d) 3.5×10^{-4} (e) 4.51×10^{-7} (f) 5×10^{-3} — page 42 B

7 (a) 320 (b) 560 000 (c) 47 500 000 (d) 0.027 (e) 0.000 054 5 (f) 0.0017 — page 42 B

2

8 (a) $6 \times 4.2 \times 10^{4+3} = 25.2 \times 10^7 = 2.52 \times 10^8$ — page 42 B
(b) $(2 \div 5) \times 10^{2-5} = 0.4 \times 10^{-3} = 4 \times 0.1 \times 10^{-3} = 4 \times 10^{-4}$

9 (a) $2.4 \times 2 \times 10^{6+-3} = 4.8 \times 10^3$ — page 42 B
(b) $(2.4 \div 4) \times 10^{6-2} = 0.6 \times 10^4 = 6 \times 10^3$
(c) $6 \times 2.1 \times 10^{6+3} = 12.6 \times 10^9 = 1.26 \times 10^{10}$
(d) $(9 \div 1.5) \times 10^{6--5} = 6 \times 10^{11}$

1

10 (a) Upper bound 450 500, lower bound 449 500 — page 43 A
(b) Upper bound 7.15×10^{-5}, lower bound 7.05×10^{-5}

11 (a) Upper bound 5.85, lower bound 5.75 — page 43 A
(b) Upper bound 12.45, lower bound 12.55
(c) Upper bound 100.0005 m, lower bound 99.9995 m
(d) Upper bound 10.005 s, lower bound 9.995 s
(e) Upper bound 1.00005 km, lower bound 0.99995 km

Tick the questions you got right.

Question	1	2	3	4	5	6	7	8	9	10	11
Grade	F	F	E	D	D	B	B	B	B	A	A

Mark the grade you are working at on your revision planner on page vii.

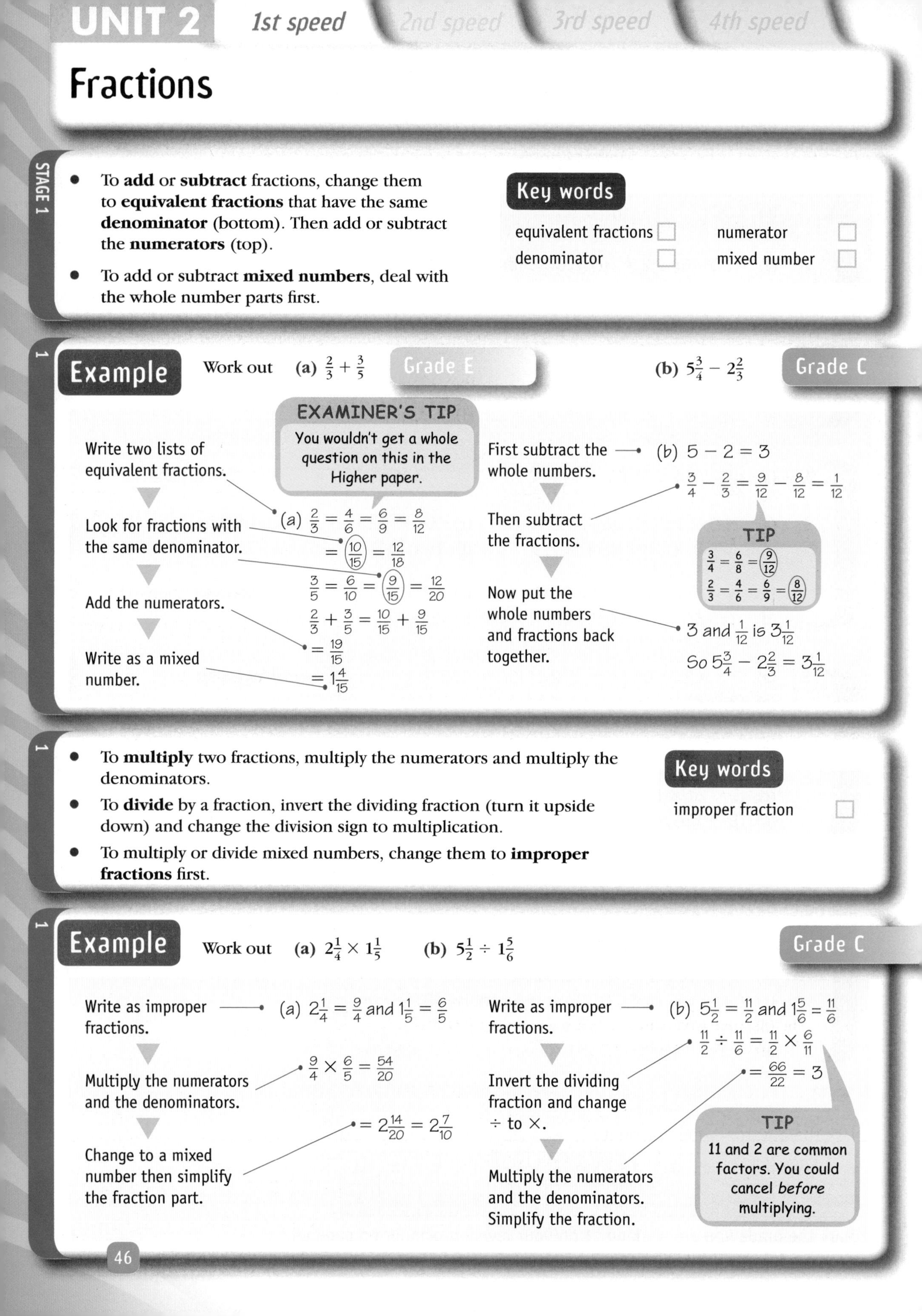

Fractions

STAGE 1

- To **add** or **subtract** fractions, change them to **equivalent fractions** that have the same **denominator** (bottom). Then add or subtract the **numerators** (top).
- To add or subtract **mixed numbers**, deal with the whole number parts first.

Key words

equivalent fractions ☐ numerator ☐
denominator ☐ mixed number ☐

1

Example

Work out (a) $\frac{2}{3} + \frac{3}{5}$ Grade E (b) $5\frac{3}{4} - 2\frac{2}{3}$ Grade C

EXAMINER'S TIP
You wouldn't get a whole question on this in the Higher paper.

Write two lists of equivalent fractions.

(a) $\frac{2}{3} = \frac{4}{6} = \frac{6}{9} = \frac{8}{12} = \frac{10}{15} = \frac{12}{18}$

$\frac{3}{5} = \frac{6}{10} = \frac{9}{15} = \frac{12}{20}$

Look for fractions with the same denominator.

Add the numerators.

$\frac{2}{3} + \frac{3}{5} = \frac{10}{15} + \frac{9}{15}$

$= \frac{19}{15}$

Write as a mixed number.

$= 1\frac{4}{15}$

First subtract the whole numbers.

(b) $5 - 2 = 3$

Then subtract the fractions.

$\frac{3}{4} - \frac{2}{3} = \frac{9}{12} - \frac{8}{12} = \frac{1}{12}$

TIP
$\frac{3}{4} = \frac{6}{8} = \frac{9}{12}$
$\frac{2}{3} = \frac{4}{6} = \frac{6}{9} = \frac{8}{12}$

Now put the whole numbers and fractions back together.

3 and $\frac{1}{12}$ is $3\frac{1}{12}$

So $5\frac{3}{4} - 2\frac{2}{3} = 3\frac{1}{12}$

1

- To **multiply** two fractions, multiply the numerators and multiply the denominators.
- To **divide** by a fraction, invert the dividing fraction (turn it upside down) and change the division sign to multiplication.
- To multiply or divide mixed numbers, change them to **improper fractions** first.

Key words

improper fraction ☐

1

Example

Work out (a) $2\frac{1}{4} \times 1\frac{1}{5}$ (b) $5\frac{1}{2} \div 1\frac{5}{6}$ Grade C

Write as improper fractions.

(a) $2\frac{1}{4} = \frac{9}{4}$ and $1\frac{1}{5} = \frac{6}{5}$

Multiply the numerators and the denominators.

$\frac{9}{4} \times \frac{6}{5} = \frac{54}{20}$

Change to a mixed number then simplify the fraction part.

$= 2\frac{14}{20} = 2\frac{7}{10}$

Write as improper fractions.

(b) $5\frac{1}{2} = \frac{11}{2}$ and $1\frac{5}{6} = \frac{11}{6}$

Invert the dividing fraction and change ÷ to ×.

$\frac{11}{2} \div \frac{11}{6} = \frac{11}{2} \times \frac{6}{11}$

Multiply the numerators and the denominators. Simplify the fraction.

$= \frac{66}{22} = 3$

TIP
11 and 2 are common factors. You could cancel *before* multiplying.

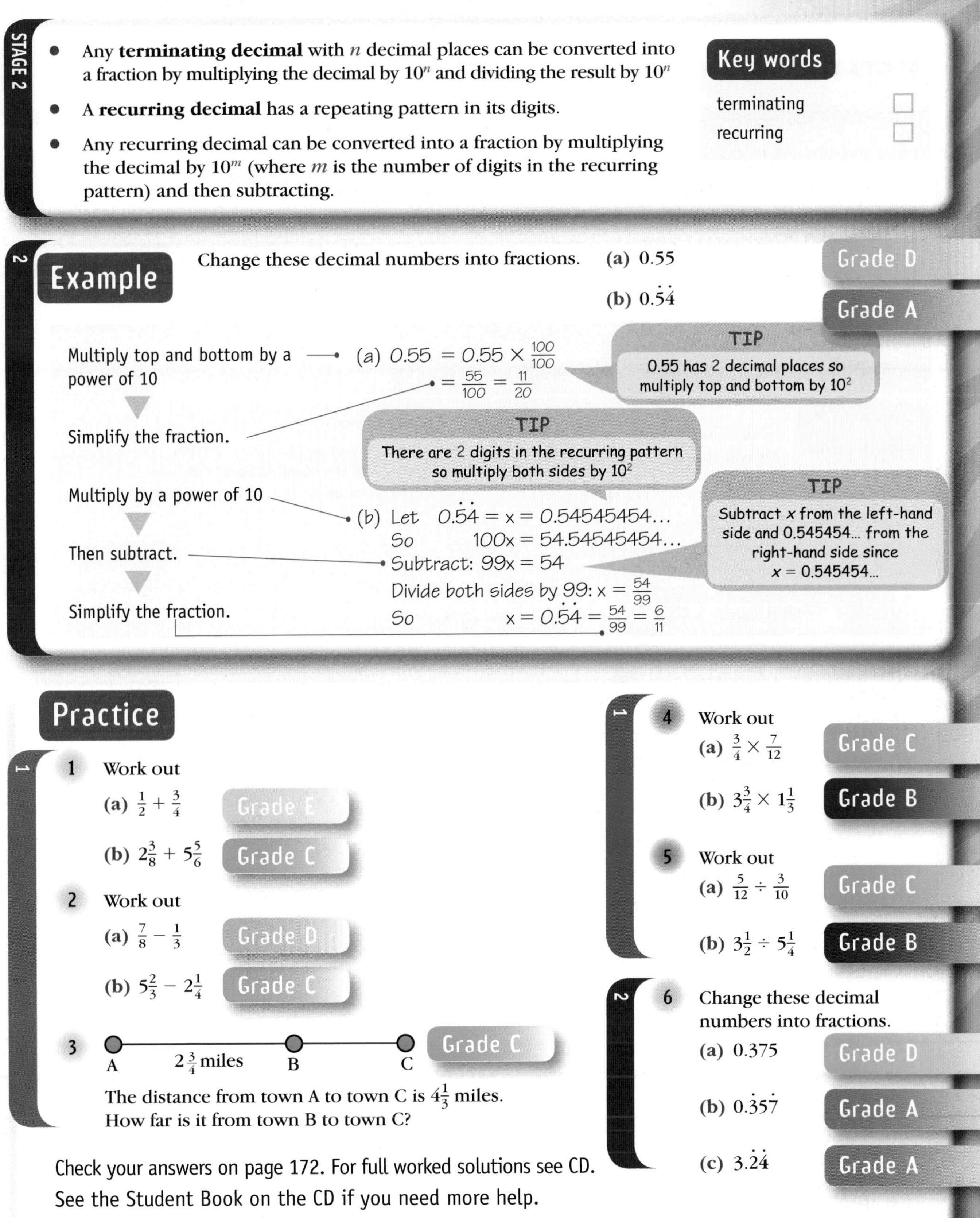

STAGE 2

- Any **terminating decimal** with n decimal places can be converted into a fraction by multiplying the decimal by 10^n and dividing the result by 10^n
- A **recurring decimal** has a repeating pattern in its digits.
- Any recurring decimal can be converted into a fraction by multiplying the decimal by 10^m (where m is the number of digits in the recurring pattern) and then subtracting.

Key words

- terminating ☐
- recurring ☐

Example

Change these decimal numbers into fractions.

(a) 0.55 — Grade D

(b) $0.\dot{5}\dot{4}$ — Grade A

Multiply top and bottom by a power of 10 → (a) $0.55 = 0.55 \times \frac{100}{100}$

Simplify the fraction. → $= \frac{55}{100} = \frac{11}{20}$

TIP: 0.55 has 2 decimal places so multiply top and bottom by 10^2

TIP: There are 2 digits in the recurring pattern so multiply both sides by 10^2

Multiply by a power of 10 → (b) Let $0.\dot{5}\dot{4} = x = 0.54545454...$

So $100x = 54.54545454...$

Then subtract. → Subtract: $99x = 54$

Divide both sides by 99: $x = \frac{54}{99}$

Simplify the fraction. → So $x = 0.\dot{5}\dot{4} = \frac{54}{99} = \frac{6}{11}$

TIP: Subtract x from the left-hand side and 0.545454... from the right-hand side since $x = 0.545454...$

Practice

1 Work out

(a) $\frac{1}{2} + \frac{3}{4}$ — Grade E

(b) $2\frac{3}{8} + 5\frac{5}{6}$ — Grade C

2 Work out

(a) $\frac{7}{8} - \frac{1}{3}$ — Grade D

(b) $5\frac{2}{3} - 2\frac{1}{4}$ — Grade C

3 — Grade C

A —— $2\frac{3}{4}$ miles —— B —— C

The distance from town A to town C is $4\frac{1}{3}$ miles. How far is it from town B to town C?

4 Work out

(a) $\frac{3}{4} \times \frac{7}{12}$ — Grade C

(b) $3\frac{3}{4} \times 1\frac{1}{3}$ — Grade B

5 Work out

(a) $\frac{5}{12} \div \frac{3}{10}$ — Grade C

(b) $3\frac{1}{2} \div 5\frac{1}{4}$ — Grade B

6 Change these decimal numbers into fractions.

(a) 0.375 — Grade D

(b) $0.\dot{3}5\dot{7}$ — Grade A

(c) $3.\dot{2}\dot{4}$ — Grade A

Check your answers on page 172. For full worked solutions see CD.
See the Student Book on the CD if you need more help.

Question	1a	1b	2a	2b	3	4a	4b	5a	5b	6a	6bc
Grade	E	C	D	C	C	C	B	C	B	D	A
Student Book pages	U2 19		U2 19		U2 19	U2 20–21		U2 20–21		U2 26–27	

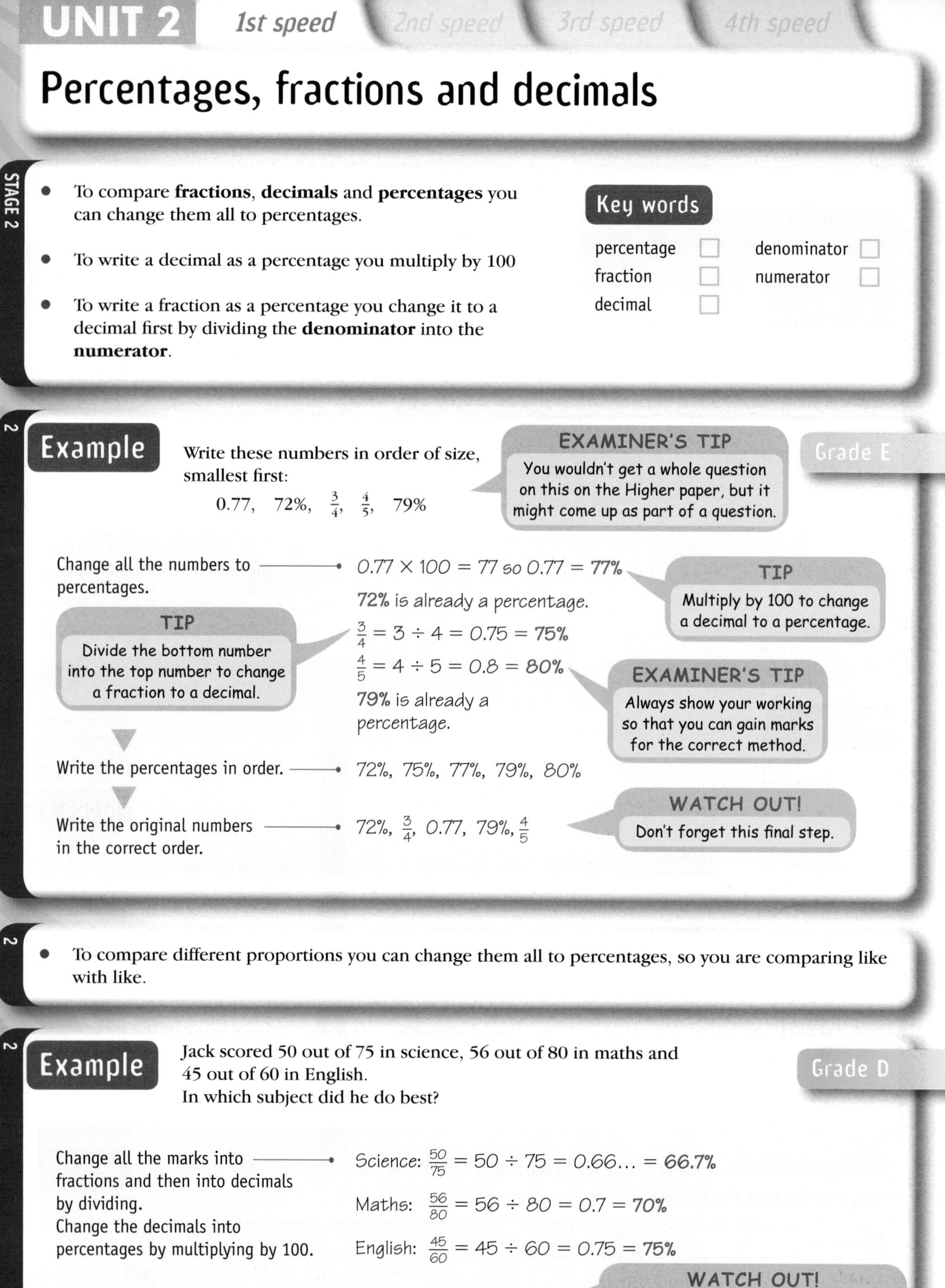

Percentages, fractions and decimals

STAGE 2

- To compare **fractions**, **decimals** and **percentages** you can change them all to percentages.
- To write a decimal as a percentage you multiply by 100
- To write a fraction as a percentage you change it to a decimal first by dividing the **denominator** into the **numerator**.

Key words

percentage ☐
fraction ☐
decimal ☐
denominator ☐
numerator ☐

2

Example

Grade E

Write these numbers in order of size, smallest first:

0.77, 72%, $\frac{3}{4}$, $\frac{4}{5}$, 79%

EXAMINER'S TIP
You wouldn't get a whole question on this on the Higher paper, but it might come up as part of a question.

Change all the numbers to percentages. → $0.77 \times 100 = 77$ so $0.77 = 77\%$

72% is already a percentage.

$\frac{3}{4} = 3 \div 4 = 0.75 = 75\%$

$\frac{4}{5} = 4 \div 5 = 0.8 = 80\%$

79% is already a percentage.

TIP
Multiply by 100 to change a decimal to a percentage.

TIP
Divide the bottom number into the top number to change a fraction to a decimal.

EXAMINER'S TIP
Always show your working so that you can gain marks for the correct method.

Write the percentages in order. → 72%, 75%, 77%, 79%, 80%

Write the original numbers in the correct order. → 72%, $\frac{3}{4}$, 0.77, 79%, $\frac{4}{5}$

WATCH OUT!
Don't forget this final step.

2

- To compare different proportions you can change them all to percentages, so you are comparing like with like.

2

Example

Grade D

Jack scored 50 out of 75 in science, 56 out of 80 in maths and 45 out of 60 in English.
In which subject did he do best?

Change all the marks into fractions and then into decimals by dividing.
Change the decimals into percentages by multiplying by 100. →

Science: $\frac{50}{75} = 50 \div 75 = 0.66... = 66.7\%$

Maths: $\frac{56}{80} = 56 \div 80 = 0.7 = 70\%$

English: $\frac{45}{60} = 45 \div 60 = 0.75 = 75\%$

He did best in English.

WATCH OUT!
Don't forget to write the final answer.

- To find a **percentage** of an amount you can:
  - change the percentage to a fraction and multiply *or*
  - change the percentage to a decimal and multiply *or*
  - work from 10%

Key words

percentage ☐

Example

Grade E

Find 15% of £60

Write percentage as a fraction out of 100 and × → $\frac{15}{100} \times 60 = \frac{15 \times 60}{100} = \frac{900}{100} = 9 \quad = £9$

Write percentage as a decimal and × → $0.15 \times 60 = 9 \quad = £9$

Find 10% first by ÷ 10
Find 5% by ÷ 10% by 2
Add 10% and 5%
→ $10\% \text{ of } 60 = 60 \div 10 = 6$

$5\% \text{ of } 60 \text{ is } \frac{1}{2} \text{ of } 10\% = 6 \div 2 = 3$

$6 + 3 = 9 \quad = £9$

Practice

Grade D

1 Write these numbers in order of size, smallest first:

67%, $\frac{2}{3}$, $\frac{3}{5}$, 0.65, 63%, 62.5%

Grade D

2 Bobbi scored 45 out of 60 in French, 60 out of 90 in German and 60 out of 80 in Spanish. In which language did she do best?

Grade D

3 Rashmi bought a TV for £240 plus VAT at 17.5%. How much was the VAT?

TIP

To find $17\frac{1}{2}\%$ of an amount
Find 10%
Find 5% ($\frac{1}{2}$ of 10%)
Find $2\frac{1}{2}\%$ ($\frac{1}{2}$ of 5%)
Total = $17\frac{1}{2}\%$

Grade D

4 Work out

(a) 35% of 64

(b) 45% of £3.20

(c) 21% of £150

(d) 32% of 2.4 m

(e) 120% of £80

(f) 2.5% of £80

Check your answers on page 172. For full worked solutions see the CD.
See the Student Book on the CD if you need more help.

Question	1	2	3	4
Grade	D	D	D	D
Student Book pages	U2 27–28, 30–32	U2 18	U2 32–33	U2 32–33

Fractions, decimals and percentages: topic test

Check how well you know this topic by answering these questions.
First cover the answers on the facing page.

Test questions

EXAMINER'S TIP
You wouldn't get a whole question on this in the Higher paper, but it might come up as part of a question.

STAGE 1

1 Work out

(a) $\frac{7}{10} + \frac{7}{10} + \frac{3}{10}$ **(b)** $2\frac{1}{2} + 3\frac{3}{4}$ **(c)** $2\frac{7}{8} + 3\frac{1}{2}$

(d) $4\frac{1}{5} + 3\frac{1}{4}$ **(e)** $3\frac{1}{3} + 4\frac{2}{5}$ **(f)** $2\frac{5}{6} + 3\frac{2}{3}$

(g) $4\frac{5}{9} + 2\frac{1}{3}$ **(h)** $1\frac{11}{12} + 2\frac{3}{4}$ **(i)** $3\frac{2}{5} + 1\frac{4}{9}$

EXAMINER'S TIP
You wouldn't get a whole question on this on the Higher paper, but it might come up as part of a question.

STAGE 2

2 Write these numbers in order of size, smallest first:
37%, $\frac{1}{3}$, $\frac{3}{10}$, 0.35, 33%, 32.5%

1

3 Work out the perimeter of this garden.

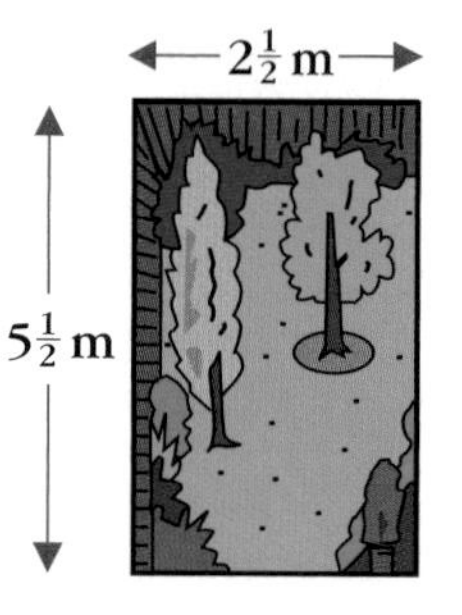

2

4 Julie scored 36 out of 60 in English, 54 out of 90 in history and 52 out of 80 in geography.
In which subject did she do best?

1

5 Work out **(a)** $\frac{3}{4} - \frac{3}{8}$ **(b)** $3\frac{1}{3} - 1\frac{5}{6}$ **(c)** $3\frac{7}{8} - 1\frac{5}{6}$

(d) $4\frac{3}{8} - 2\frac{1}{2}$ **(e)** $5\frac{1}{5} - 3\frac{2}{3}$ **(f)** $4\frac{5}{9} - 3\frac{5}{6}$

6 A glass holds a sixth of a bottle of lemonade.
How many glasses can be filled from $2\frac{3}{4}$ bottles of lemonade?

7 Work out **(a)** $2\frac{1}{4} \times 3\frac{2}{5}$ **(b)** $6\frac{3}{8} \div 2\frac{1}{2}$ **(c)** $5\frac{1}{2} \div 2\frac{1}{4}$

(d) $4\frac{1}{2} \times 2\frac{1}{3}$ **(e)** $3\frac{1}{2}(1\frac{1}{4} - \frac{3}{5})$

8 A bag which contains sweets weighs $2\frac{2}{5}$ lb. The sweets weigh $2\frac{1}{4}$ lb.
What does the bag weigh?

2

9 Write these decimal numbers as fractions.
(a) $0.\dot{3}\dot{6}$ **(b)** $0.12\dot{7}$ **(c)** $0.1\dot{3}\dot{6}$

Now check your answers – see the facing page.

Cover this page while you answer the test questions opposite.

Worked answers

Revise this on...

1 (a) $\frac{17}{10} = 1\frac{7}{10}$ (b) $5\frac{2+3}{4} = 5\frac{5}{4} = 6\frac{1}{4}$ page 46 E

(c) $5\frac{7+4}{8} = 5\frac{11}{8} = 6\frac{3}{8}$ (d) $7\frac{4+5}{20} = 7\frac{9}{20}$ D

(e) $7\frac{5+6}{15} = 7\frac{11}{15}$ (f) $5\frac{5+4}{6} = 5\frac{9}{6} = 6\frac{3}{6} = 6\frac{1}{2}$ C

(g) $6\frac{5+3}{9} = 6\frac{8}{9}$ (h) $3\frac{11+9}{12} = 3\frac{20}{12} = 4\frac{8}{12} = 4\frac{2}{3}$

(i) $4\frac{18+20}{45} = 4\frac{38}{45}$ B

2 Change into percentages: page 48 E

37%, $\frac{1}{3} = 33.3...\%$, $\frac{3}{10} = 30\%$, $0.35 = 35\%$, 33%, 32.5%

The order is $\frac{3}{10}$, 32.5%, 33%, $\frac{1}{3}$, 0.35, 37%

3 Perimeter $= 5\frac{1}{2} + 5\frac{1}{2} + 2\frac{1}{2} + 2\frac{1}{2} = 16$ m page 46 D

4 English $\frac{36}{60} \times 100 = 60\%$, history $\frac{54}{90} \times 100 = 60\%$, page 48 D

geography $\frac{52}{80} \times 100 = 65\%$

He did best in geography.

5 (a) $\frac{6-3}{8} = \frac{3}{8}$ (b) $2\frac{2-5}{6} = 1\frac{6+2-5}{6} = 1\frac{3}{6} = 1\frac{1}{2}$ page 46 D

(c) $2\frac{21-20}{24} = 2\frac{1}{24}$ (d) $2\frac{3-4}{8} = 1\frac{8+3-4}{8} = 1\frac{7}{8}$ C

(e) $2\frac{3-10}{15} = 1\frac{15+3-10}{15} = 1\frac{8}{15}$ (f) $1\frac{10-15}{18} = \frac{18+10-15}{18} = \frac{13}{18}$

6 $2\frac{3}{4} \div \frac{1}{6} = \frac{11}{4} \div \frac{1}{6} = \frac{11}{4} \times \frac{6}{1} = \frac{66}{4} = 16\frac{1}{2} \rightarrow 16$ whole glasses page 46 C

7 (a) $\frac{9}{4} \times \frac{17}{5} = \frac{153}{20} = 7\frac{13}{20}$ C (b) $\frac{51}{8} \div \frac{5}{2} = \frac{51}{8} \times \frac{2}{5} = \frac{51}{20} = 2\frac{11}{20}$ page 46 B

(c) $\frac{11}{2} \div \frac{9}{4} = \frac{11}{2} \times \frac{4}{9} = \frac{22}{9} = 2\frac{4}{9}$ (d) $\frac{9}{2} \times \frac{7}{3} = \frac{3}{2} \times \frac{7}{1} = \frac{21}{2} = 10\frac{1}{2}$

(e) $3\frac{1}{2}\left(1\frac{5-12}{20}\right) = 3\frac{1}{2}\left(\frac{20+5-12}{20}\right) = \frac{7}{2} \times \frac{13}{20} = \frac{91}{40} = 2\frac{11}{20}$

8 Weight of bag $= 2\frac{2}{5} - 2\frac{1}{4} = \frac{8-5}{20} = \frac{3}{20}$ lb page 46 B

9 (a)
$x = 0.\dot{3}\dot{6}$
$100x = 36.\dot{3}\dot{6}$
$99x = 36$
$x = \frac{36}{99} = \frac{4}{11}$

(b)
$x = 0.\dot{1}2\dot{7}$
$1000x = 127.\dot{1}2\dot{7}$
$999x = 127$
$x = \frac{127}{999}$

(c)
$x = 0.1\dot{3}\dot{6}$
$100x = 13.6\dot{3}\dot{6}$
$99x = 13.5$
$990x = 135$
$x = \frac{135}{990} = \frac{27}{198} = \frac{3}{22}$

page 47 A

Tick the questions you got right.

Question	1ab	1cd	1efgh	1i	2	3	4	5ab	5cdef	6	7a	7bcde	8	9
Grade	E	D	C	B	E	D	D	D	C	C	C	B	B	A

Mark the grade you are working at on your revision planner on page vii.

Number: subject test

Use these questions to check that you understand the key facts for Number, before you try the Examination Practice Paper on pages 148–156.

Exam practice questions

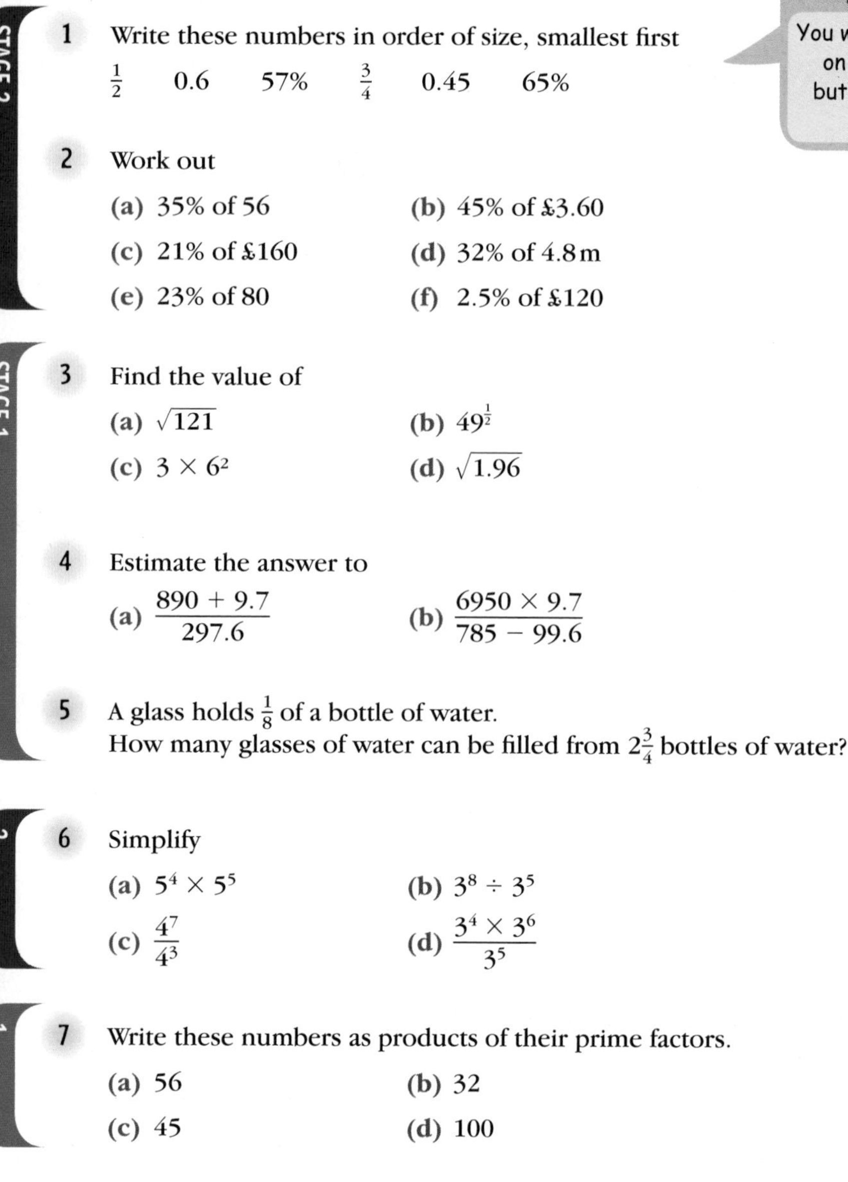

EXAMINER'S TIP

You wouldn't get a whole question on this on the Higher paper, but it might come as part of a question.

1 Write these numbers in order of size, smallest first

$\frac{1}{2}$ 0.6   57% $\frac{3}{4}$ 0.45   65%

2 Work out

(a) 35% of 56   **(b)** 45% of £3.60

(c) 21% of £160   **(d)** 32% of 4.8 m

(e) 23% of 80   **(f)** 2.5% of £120

3 Find the value of

(a) $\sqrt{121}$ **(b)** $49^{\frac{1}{2}}$

(c) 3×6^2 **(d)** $\sqrt{1.96}$

4 Estimate the answer to

(a) $\dfrac{890 + 9.7}{297.6}$ **(b)** $\dfrac{6950 \times 9.7}{785 - 99.6}$

5 A glass holds $\frac{1}{8}$ of a bottle of water.
How many glasses of water can be filled from $2\frac{3}{4}$ bottles of water?

6 Simplify

(a) $5^4 \times 5^5$ **(b)** $3^8 \div 3^5$

(c) $\dfrac{4^7}{4^3}$ **(d)** $\dfrac{3^4 \times 3^6}{3^5}$

7 Write these numbers as products of their prime factors.

(a) 56   **(b)** 32

(c) 45   **(d)** 100

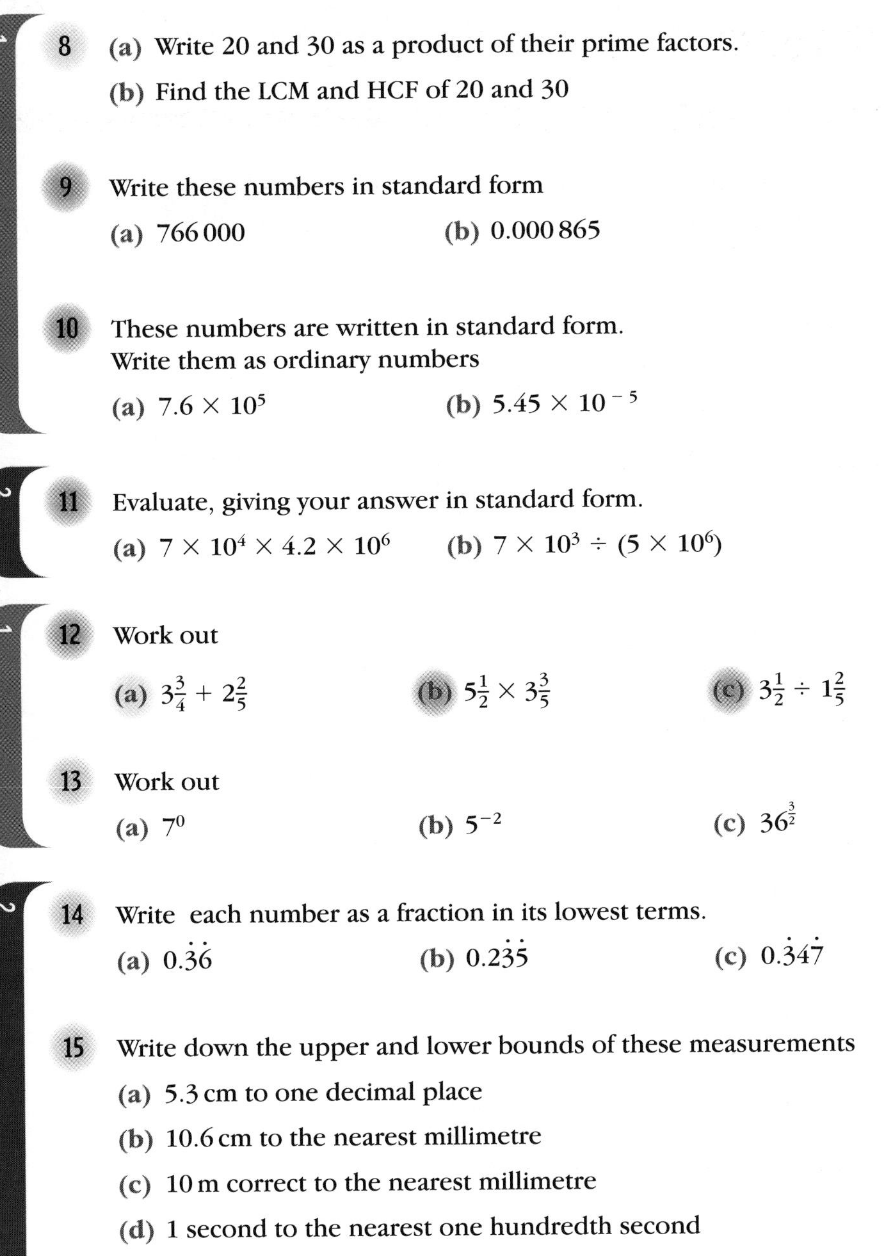

8 (a) Write 20 and 30 as a product of their prime factors.

(b) Find the LCM and HCF of 20 and 30

9 Write these numbers in standard form

(a) 766 000 (b) 0.000 865

10 These numbers are written in standard form.
Write them as ordinary numbers

(a) 7.6×10^5 (b) 5.45×10^{-5}

11 Evaluate, giving your answer in standard form.

(a) $7 \times 10^4 \times 4.2 \times 10^6$ (b) $7 \times 10^3 \div (5 \times 10^6)$

12 Work out

(a) $3\frac{3}{4} + 2\frac{2}{5}$ (b) $5\frac{1}{2} \times 3\frac{3}{5}$ (c) $3\frac{1}{2} \div 1\frac{2}{5}$

13 Work out

(a) 7^0 (b) 5^{-2} (c) $36^{\frac{3}{2}}$

14 Write each number as a fraction in its lowest terms.

(a) $0.\dot{3}\dot{6}$ (b) $0.\dot{2}3\dot{5}$ (c) $0.\dot{3}4\dot{7}$

15 Write down the upper and lower bounds of these measurements

(a) 5.3 cm to one decimal place

(b) 10.6 cm to the nearest millimetre

(c) 10 m correct to the nearest millimetre

(d) 1 second to the nearest one hundredth second

(e) 1 km to the nearest metre

Check your answers on page 172. For full worked solutions see the CD.
Tick the questions you got right.

Question	1	2	3	4	5	6	7	8	9	10	11	12a	12b	13	14	15
Grade	E	D	D	D	C	C	C	C	B	B	B	C	B	A	A	A
Revise this on pages	48	49	36–37	40	46	36	34–35	34–35	42	42	42	46		36	47	43

Mark the grade you are working at on your revision planner on page vii.
Go to the pages shown to revise for the ones you got wrong.

Simplifying algebra and expanding brackets

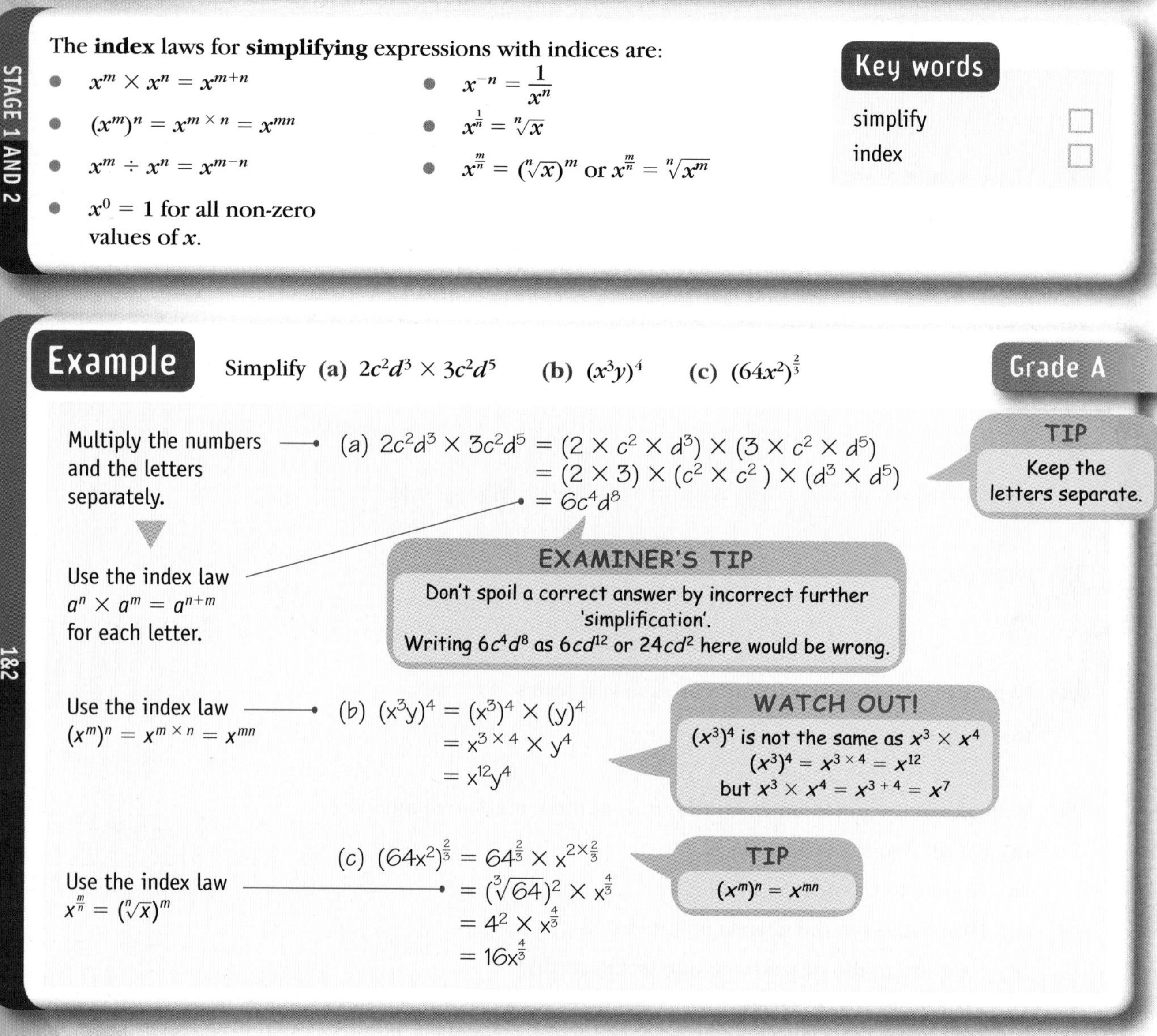

STAGE 1 AND 2

The **index** laws for **simplifying** expressions with indices are:

- $x^m \times x^n = x^{m+n}$
- $(x^m)^n = x^{m \times n} = x^{mn}$
- $x^m \div x^n = x^{m-n}$
- $x^0 = 1$ for all non-zero values of x.
- $x^{-n} = \dfrac{1}{x^n}$
- $x^{\frac{1}{n}} = \sqrt[n]{x}$
- $x^{\frac{m}{n}} = (\sqrt[n]{x})^m$ or $x^{\frac{m}{n}} = \sqrt[n]{x^m}$

Key words

simplify ☐
index ☐

Example

Simplify (a) $2c^2d^3 \times 3c^2d^5$ (b) $(x^3y)^4$ (c) $(64x^2)^{\frac{2}{3}}$

Grade A

Multiply the numbers and the letters separately.

(a) $2c^2d^3 \times 3c^2d^5 = (2 \times c^2 \times d^3) \times (3 \times c^2 \times d^5)$
$= (2 \times 3) \times (c^2 \times c^2) \times (d^3 \times d^5)$
$= 6c^4d^8$

Use the index law $a^n \times a^m = a^{n+m}$ for each letter.

TIP
Keep the letters separate.

EXAMINER'S TIP
Don't spoil a correct answer by incorrect further 'simplification'.
Writing $6c^4d^8$ as $6cd^{12}$ or $24cd^2$ here would be wrong.

Use the index law $(x^m)^n = x^{m \times n} = x^{mn}$

(b) $(x^3y)^4 = (x^3)^4 \times (y)^4$
$= x^{3 \times 4} \times y^4$
$= x^{12}y^4$

WATCH OUT!
$(x^3)^4$ is not the same as $x^3 \times x^4$
$(x^3)^4 = x^{3 \times 4} = x^{12}$
but $x^3 \times x^4 = x^{3+4} = x^7$

Use the index law $x^{\frac{m}{n}} = (\sqrt[n]{x})^m$

(c) $(64x^2)^{\frac{2}{3}} = 64^{\frac{2}{3}} \times x^{2 \times \frac{2}{3}}$
$= (\sqrt[3]{64})^2 \times x^{\frac{4}{3}}$
$= 4^2 \times x^{\frac{4}{3}}$
$= 16x^{\frac{4}{3}}$

TIP
$(x^m)^n = x^{mn}$

1&2

- To remove brackets from an algebraic expression, multiply each term inside the brackets by the term outside.
  This is sometimes called **expanding the brackets**.
- Terms with the same power of the same letter(s) are called **like terms**.
- Equations such as
  $(e + f)(g + h) = eg + eh + fg + fh$
  and $(e + f)(g + h) = e(g + h) + f(g + h)$
  which are true for all values of e, f, g and h are also known as **identities**.
- In general: $(x + a)^2 = x^2 + 2ax + a^2$
  $(x - b)^2 = x^2 - 2bx + b^2$

Key words

expand ☐
like terms ☐
identity ☐

Example

Expand and simplify

(a) $3(3a - 5) - 3(2a - 7)$ (b) $p(p + 2q) + 2p(p - q)$ Grade C

(c) $(2x - 3)^2$ Grade B

First expand the bracket. → (a) $3(3a - 5) - 3(2a - 7) = 9a - 15 - 6a + 21$

Collect like terms together. → $= 9a - 6a - 15 + 21$

$= 3a + 6$

TIP

$9a$ and $-6a$ are like terms.
-15 and 21 are like terms.

TIP

A negative term inside the bracket multiplying a negative term outside the bracket gives $- \times - = +$
Here $-3 \times -7 = +21$

First expand the brackets. → (b) $p(p + 2q) + 2p(p - q) = p^2 + 2pq + 2p^2 - 2pq$

Collect like terms. → $= p^2 + 2p^2 + 2pq - 2pq$

$= 3p^2$

First expand the brackets. → (c) $(2x - 3)^2 = (2x - 3)(2x - 3)$

Collect like terms. → $= 4x^2 - 6x - 6x + 9$

$= 4x^2 - 12x + 9$

1&2

Practice

1 Simplify (a) $6x + 4y - 2x - 5y$ (b) $2x + 3y + 3x + 5y$ Grade D

2 Simplify (a) $t^3 \times t^5$ (b) $6p^5 \div 2p^2$ Grade C

3 Expand

(a) $4(a - 2c)$ Grade D

(b) $g(g - 4)$ Grade C

4 Expand and simplify (a) $(x + y)(x - 2y)$ (b) $(2x - 3)(3x + 4)$ Grade B

5 Simplify (a) $(p^2q)^3$ (b) $(9y^4)^{\frac{1}{2}}$ Grade A

Check your answers on page 172. For full worked solutions see the CD.
See the Student Book on the CD if you need more help.

Question	1	2	3a	3b	4	5
Grade	D	C	D	C	B	A
Student Book pages	U2 46	U2 46	U2 47–48		U2 47–49	U2 48–49

1&2

Factorising and algebraic fractions

STAGE 1 AND 2

- **Factorising** is the opposite of removing brackets.
- To factorise an expression completely, the **highest common factor (HCF)** must appear outside the brackets.
- To factorise $ax^2 + bx$, take out the highest common factor (HCF).
- To factorise a **quadratic expression** such as $ax^2 + bx + c$, start by looking for two numbers whose product is ac and whose sum is b.

2

- $x^2 - y^2 = (x - y)(x + y)$ is called the **difference of two squares**.

1&2

- $x^2 + 2ax + a^2 = (x + a)^2$
- $x^2 - 2ax + a^2 = (x - a)^2$

Key words

- factorise ☐
- highest common factor (HCF) ☐
- difference of two squares ☐
- quadratic expression ☐

STAGE 1 AND 2

Example

Factorise these expressions completely.

Grade D (a) $x^2 + 6x$

Grade C (b) $3c^3 - 6c^2$

Grade B (c) $x^2 + 2x - 15$

Grade A (d) $4t^2 - 9w^4$

Find any common factors. → (a) $x^2 + 6x = x(x + 6)$

TIP
The letter x is common to both terms.

Write the highest common factor outside the bracket. → (b) $3c^3 - 6c^2 = 3c^2(c - 2)$

TIP
3 and c^2 are factors of $3c^3$ and $6c^2$, so the *highest* common factor is $3c^2$

Start by looking for two numbers whose product is ac and whose sum is b. → (c) Here $ac = 1 \times -15 = -15$ and $b = 2$
$-3 \times 5 = -15$ and $-3 + 5 = 2$
So $x^2 + 2x - 15 = (x - 3)(x + 5)$

TIP
You can check your answer by expanding:
$(x - 3)(x + 5) = x^2 + 2x - 15$

Use the rule for the difference of two squares. → (d) $4t^2 - 9w^4 = (2t - 3w^2)(2t + 3w^2)$

TIP
How to spot a 'difference of two squares':
- ✓ only two terms
- ✓ each term is a perfect square
- ✓ minus sign between the two terms.

Here, $4t^2 = (2t)^2$ and $9w^4 = (3w^2)^2$

- In general, to add two **algebraic fractions** $\frac{1}{n} + \frac{1}{m}$, change them to equivalent fractions with denominator $m \times n$.

$$\frac{1}{n} + \frac{1}{m} = \frac{m + n}{mn}$$

Key words

algebraic fraction ☐
lowest common multiple (LCM) ☐

Example

Grade A*

Simplify $\frac{4}{4 - 3k} - \frac{1}{1 + 2k}$

TIP

The LCM is $(4 - 3k)(1 + 2k)$
For more on adding and subtracting fractions see page 14–15.

You need to work this out as a single fraction. Write each term as an equivalent fraction with the same denominator.

$$\frac{4}{4 - 3k} - \frac{1}{1 + 2k} = \frac{4(1 + 2k)}{(4 - 3k)(1 + 2k)} - \frac{1(4 - 3k)}{(4 - 3k)(1 + 2k)}$$

Expand the bracket in the numerator.

$$= \frac{4 + 8k - (4 - 3k)}{(4 - 3k)(1 + 2k)}$$

$$= \frac{4 + 8k - 4 + 3k}{(4 - 3k)(1 + 2k)}$$

Collect like terms.

$$= \frac{11k}{(4 - 3k)(1 + 2k)}$$

WATCH OUT!

You cannot cancel the k terms here.

Practice

1 Factorise

(a) $7x - 21y$ Grade D

(b) $8t^2 + 4t$ Grade C

2 Factorise

(a) $x^2 - 36$ Grade B

(b) $9x^2 - 25y^2$ Grade A

3 Factorise (Grade A)

(a) $x^2 - xy - 12y^2$

(b) $10x^2 - x - 21$

4 Simplify (Grade A*)

(a) $\frac{x^2 - 3x}{x^2 - 9}$

(b) $\frac{1}{y(y - 3)} + \frac{1}{y(y + 2)}$

Check your answers on page 172. For full worked solutions see the CD.
See the Student Book on the CD if you need more help.

Question	1a	1b	2a	2b	3	4a	4b
Grade	D	C	B	A	A	A	A
Student Book pages	U2 47–48, 51–52		U2 53–54		U2 51–52	U2 54–55	U2 54–55

Manipulative algebra: topic test

Check how well you know this topic by answering these questions.
First cover the answers on the facing page.

Test questions

STAGE 1 AND 2

1 Simplify $4a + 3c - 2a + 4c$

2 Simplify $t \times t \times t \times t$

3 Expand $3(2a - 5)$

4 Expand $p(p - 2)$

5 Factorise $5a - 10$

6 Simplify $g^2 + g^2 + g^2$

7 Simplify $k^4 \times k^5$

8 Expand $3a(a + c)$

9 Factorise $d^2 - 4d$

10 Expand and simplify $(x + 4)(x - 7)$

11 Factorise completely $3d^2 - 12de$

12 Factorise $x^2 + 5x + 6$

2

13 Simplify $\dfrac{3(x + 3)^3}{(x + 3)}$

1&2

14 Expand and simplify $(2x + 5)(4x - 3)$

15 Simplify $(2x^2y)^3$

2

16 Factorise $9x^2 - 16y^2$

17 Simplify $\dfrac{x^2 - 16}{x^2 - x - 12}$

18 Simplify $\dfrac{2}{f - 3} - \dfrac{3}{f - 2}$

Now check your answers – see the facing page.

Cover this page while you answer the test questions opposite.

Worked answers

Revise this on...

1&2

1 $4a + 3c - 2a + 4c = 4a - 2a + 3c + 4c = 2a + 7c$ — page 55 — D

2 $t \times t \times t \times t = t^4$

3 $3(2a - 5) = 6a - 15$ — page 55 — D

4 $p(p - 2) = p^2 - 2p$

5 $5a - 10 = 5(a - 2)$ — pages 55–55 — D

6 $g^2 + g^2 + g^2 = 3g^2$

7 $k^4 \times k^5 = k^9$ — page 54 — D

8 $3a(a + c) = 3a^2 + 3ac$

9 $d^2 - 4d = d(d - 4)$ — pages 54–56 — C

10 $(x + 4)(x - 7) = x^2 - 7x + 4x - 28 = x^2 - 3x - 28$ — pages 54–55 — B

11 $3d^2 - 12de = 3d(d - 4e)$ — page 56 — B

12 $x^2 + 5x + 6 = (x + 2)(x + 3)$ — page 56 — B

2

13 $\frac{3(x + 3)^2}{(x + 3)} = \frac{3(x + 3)\cancel{(x + 3)}}{\cancel{(x + 3)}} = 3(x + 3)$ — page 57 — B

1&2

14 $(2x + 5)(4x - 3) = 8x^2 - 6x + 20x - 15 = 8x^2 + 14x - 15$ — page 55 — A

15 $(2x^2y)^3 = 2 \times 2 \times 2 \times x^2 \times x^2 \times x^2 \times y \times y \times y = 8x^6y^3$ — page 54 — A

2

16 $9x^2 - 16y^2 = (3x - 4y)(3x + 4y)$ — page 56 — A

17 $\frac{x^2 - 16}{x^2 - x - 12} = \frac{(x + 4)(x - 4)}{(x + 3)(x - 4)} = \frac{x + 4}{x + 3}$ — page 57 — A*

18 $\frac{2}{f - 3} - \frac{3}{f - 2} = \frac{2(f - 2)}{(f - 3)(f - 2)} - \frac{3(f - 3)}{(f - 2)(f - 3)}$ — page 57 — A*

$= \frac{2f - 4 - (3f - 9)}{(f - 3)(f - 2)} = \frac{5 - f}{(f - 3)(f - 2)}$

Tick the questions you got right.

Question	1	2	3	4	5	6	7	8	9	10	11	12	13	14	15	16	17	18
Grade	D	D	D	D	D	D	D	C	C	B	B	B	B	A	A	A	A*	A*

Mark the grade you are working at on your revision planner on page vii.

Patterns and sequences

STAGE 1

- When you know a formula for the **nth term** of a **sequence** you can calculate any term in the sequence by substituting a value for n in the formula. n must be a positive **integer** ($n = 1, 2, 3 \ldots$).
- A sequence in which the **differences** between successive terms are equal is called an **arithmetic sequence**.
- The **general term** or the **rule** for an arithmetic sequence is of the form $an + b$. The value of a is the difference between successive terms in the sequence.

Key words

- pattern
- sequence
- term
- nth term
- integer
- difference
- general term
- rule

1

Example

Grade C

Here is a **pattern** made from sticks:

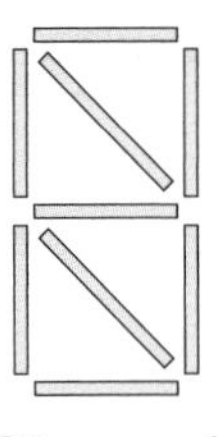
Diagram 1

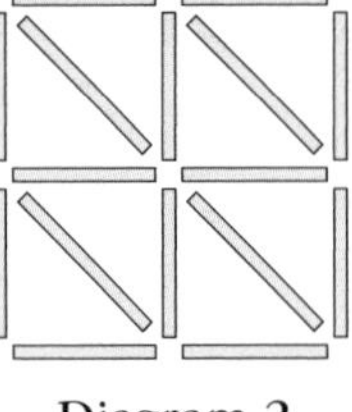
Diagram 2

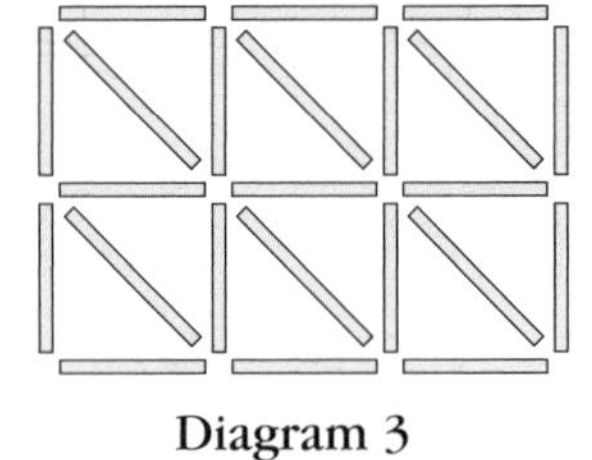
Diagram 3

Find a rule to work out the number of sticks, in terms of n, in the nth diagram.

Write down the numbers of sticks and their differences. → 9 16 23 (+7 +7)

Use the number differences to write the first part of the general rule. → The differences are +7, so the first part of the rule is $7n$

Look at the first term, when $n = 1$ → $7n = 7 \times 1 = 7$
You need to add 2 to get the first term: $7 + 2 = 9$

Write the rule. → $7n + 2$

WATCH OUT!
Don't forget to look back at the numbers in the sequence. A common error is to use the number difference to write the rule as $n + 7$ or $7n$

EXAMINER'S TIP
Do not spoil the rule by writing it as $n = 7n + 2$
The rule is just $7n + 2$
If the question asks for a *formula*, it needs an equals sign, for example $S = 7n + 2$

Example

This is part of an arithmetic sequence of numbers: 13, 6, −1, −8, −15

Grade B

(a) Write down a formula, in terms of n, for the nth term in the sequence, u_n.

(b) Find the 20th term in this sequence.

Find the number differences. → (a) 13 6 −1 −8 −15

differences: −7 −7 −7 −7

Use the number differences to write the first part of the formula. → The differences are −7, so the first part of the rule is $-7n$

Look at the first term, when $n = 1$ → $-7n = -7 \times 1 = -7$
You need to add 20 to get the first term: $-7 + 20 = 13$

Write the formula. → $u_n = -7n + 20 = 20 - 7n$

TIP

Check that your formula gives the next terms correctly:
$u_2 = 20 - 2 \times 7 = 20 - 14 = 6$
$u_3 = 20 - 3 \times 7 = 20 - 21 = -1$

Use your formula. → (b) $n = 20$
$u_{20} = 20 - (7 \times 20) = 20 - 140 = -120$

Practice

1 Here is a pattern made from dots:

Grade C

Diagram 1 Diagram 2 Diagram 3

Find a rule to work out the number of dots, in terms of n, for the nth diagram.

2 Here is an arithmetic sequence of numbers: 3, 10, 17, 24, 31

Grade C

(a) Find a rule, in terms of n, for the nth term in this sequence.

(b) Use your rule to find the 40th term in this sequence.

3 Here is another arithmetic sequence of numbers: 7, 5, 3, 1, −1

Grade B

(a) Find a rule, in terms of n, for the nth term in this sequence.

(b) Use your rule to find the 50th term in this sequence.

Check your answers on page 172. For full worked solutions see the CD.
See the Student Book on the CD if you need more help.

Question	1	2	3
Grade	C	C	B
Student Book pages	U2 76–81	U2 76–81	U2 76–81

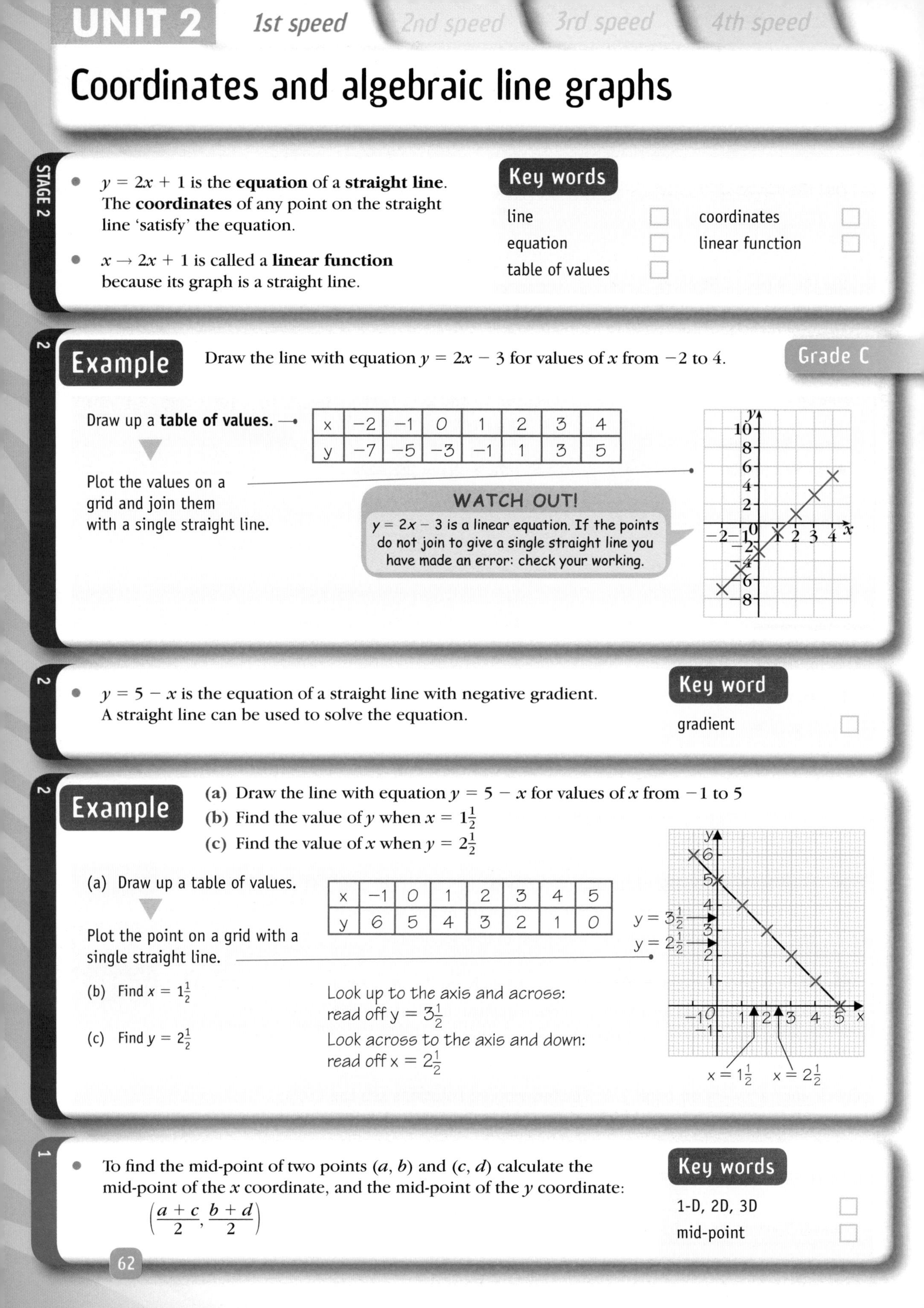

Coordinates and algebraic line graphs

STAGE 2

- $y = 2x + 1$ is the **equation** of a **straight line**. The **coordinates** of any point on the straight line 'satisfy' the equation.
- $x \rightarrow 2x + 1$ is called a **linear function** because its graph is a straight line.

Key words

line ☐ coordinates ☐
equation ☐ linear function ☐
table of values ☐

2

Example

Draw the line with equation $y = 2x - 3$ for values of x from -2 to 4.

Grade C

Draw up a **table of values.**

x	−2	−1	0	1	2	3	4
y	−7	−5	−3	−1	1	3	5

Plot the values on a grid and join them with a single straight line.

WATCH OUT!
$y = 2x - 3$ is a linear equation. If the points do not join to give a single straight line you have made an error: check your working.

2

- $y = 5 - x$ is the equation of a straight line with negative gradient. A straight line can be used to solve the equation.

Key word

gradient ☐

2

Example

(a) Draw the line with equation $y = 5 - x$ for values of x from -1 to 5
(b) Find the value of y when $x = 1\frac{1}{2}$
(c) Find the value of x when $y = 2\frac{1}{2}$

(a) Draw up a table of values.

x	−1	0	1	2	3	4	5
y	6	5	4	3	2	1	0

Plot the point on a grid with a single straight line.

(b) Find $x = 1\frac{1}{2}$

Look up to the axis and across: read off $y = 3\frac{1}{2}$

(c) Find $y = 2\frac{1}{2}$

Look across to the axis and down: read off $x = 2\frac{1}{2}$

1

- To find the mid-point of two points (a, b) and (c, d) calculate the mid-point of the x coordinate, and the mid-point of the y coordinate:

$$\left(\frac{a + c}{2}, \frac{b + d}{2}\right)$$

Key words

1-D, 2D, 3D ☐
mid-point ☐

Example

A is the point $(3, -1)$, B is the point $(5, -3)$.
Calculate the coordinates of the mid-point of the line segment AB.

Mid-point is $\left(\frac{3+5}{2}, \frac{-1+-3}{2}\right) = \left(\frac{8}{2}, \frac{-4}{2}\right) = (4, -2)$

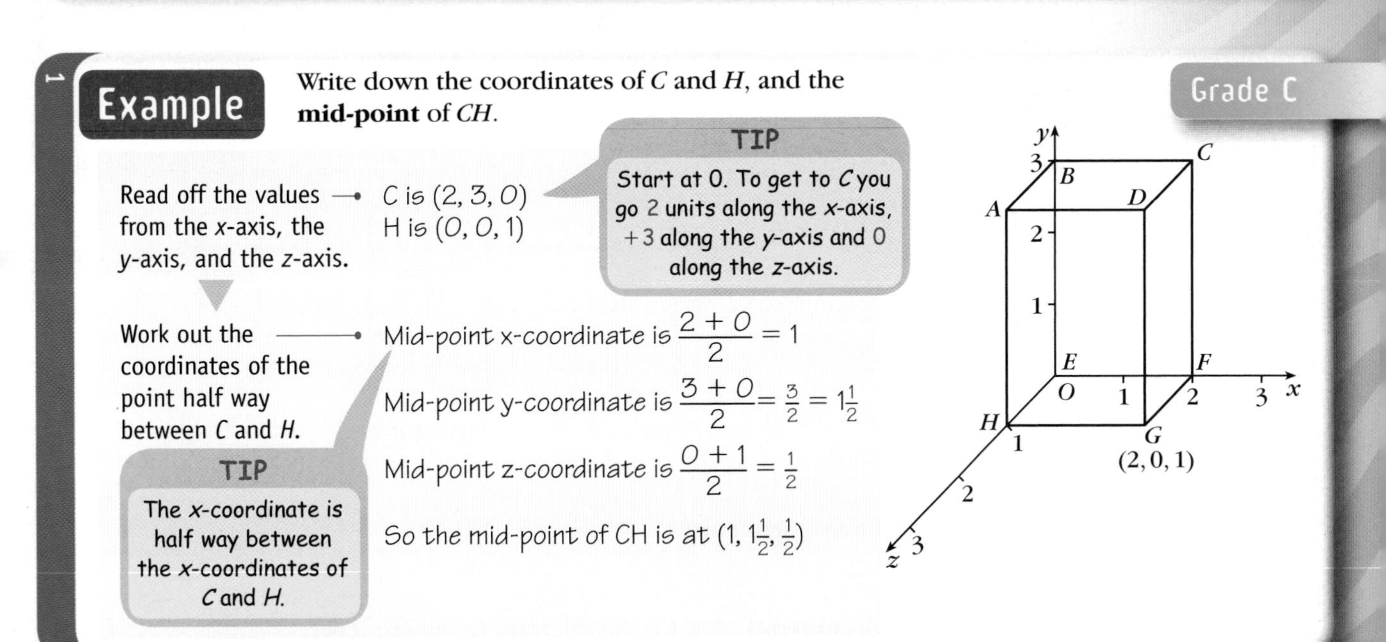

Example

Write down the coordinates of C and H, and the **mid-point** of CH.

Grade C

Read off the values from the x-axis, the y-axis, and the z-axis. → C is (2, 3, 0)
H is (0, 0, 1)

TIP Start at 0. To get to C you go 2 units along the x-axis, +3 along the y-axis and 0 along the z-axis.

Work out the coordinates of the point half way between C and H. →

Mid-point x-coordinate is $\frac{2+0}{2} = 1$

Mid-point y-coordinate is $\frac{3+0}{2} = \frac{3}{2} = 1\frac{1}{2}$

Mid-point z-coordinate is $\frac{0+1}{2} = \frac{1}{2}$

So the mid-point of CH is at $(1, 1\frac{1}{2}, \frac{1}{2})$

TIP The x-coordinate is half way between the x-coordinates of C and H.

Practice

1 (a) Write down the coordinates of points A, D and F. **Grade D**

(b) Find the coordinates of the mid-point of HC. **Grade C**

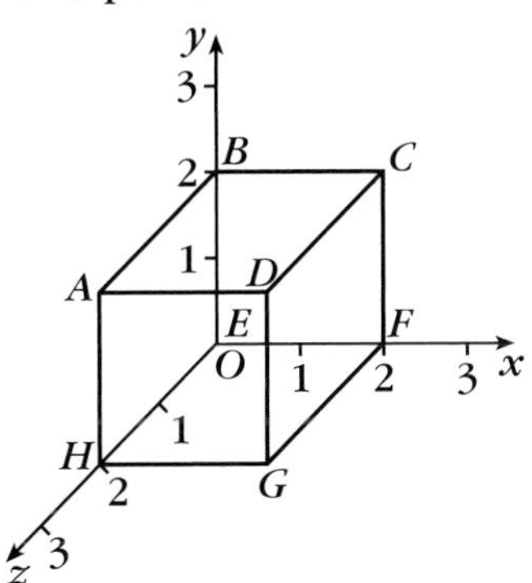

2 (a) Draw the line with equation $y = x - 2$ for values of x from 0 to 5. **Grade C**

(b) Find the value of y when $x = \frac{1}{2}$

(c) Find the value of x when $y = 2\frac{1}{2}$

3 A line passes though the points $A(2, 4)$ and $B(4, -7)$. Find the coordinates of the mid-point of AB. **Grade B**

Check your answers on pages 172–173. For full worked solutions see the CD.
See the Student Book on the CD if you need more help.

Question	1a	1b	2	3
Grade	D	C	C	B
Student Book pages	U2 63–65		U2 69–71	U2 63–64

Patterns, sequences, coordinates and graphs: topic test

Check how well you know this topic by answering these questions.
First cover the answers on the facing page.

Test questions

STAGE 1

1 This is a series of diagrams made from sticks.

(a) Write down a formula for T_d, the total number of sticks in each diagram, in terms of d, where d is the diagram number.

(b) Use your rule to find the number of sticks in the 50th diagram.

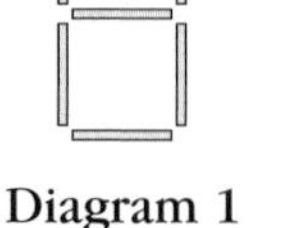

Diagram 1

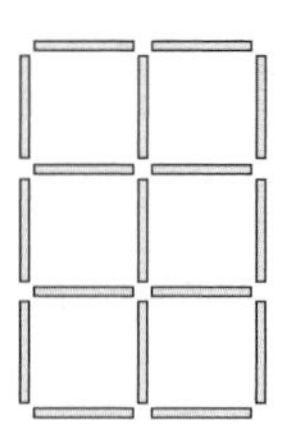

Diagram 2

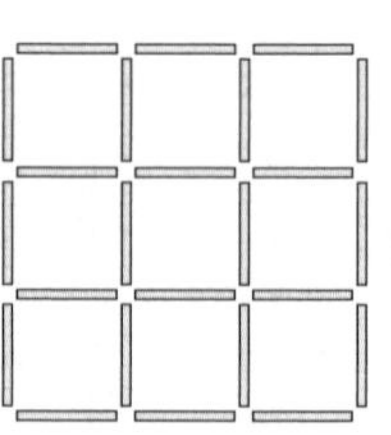

Diagram 3

2 These numbers are part of a number sequence:

$-1,\quad 3,\quad 7\quad 11,\quad 15$

(a) Write down the general rule, in terms of n, for the nth term in the sequence.

(b) Use your rule to find the 20th term in the sequence.

3 Write down the coordinates of the mid-point of DE.

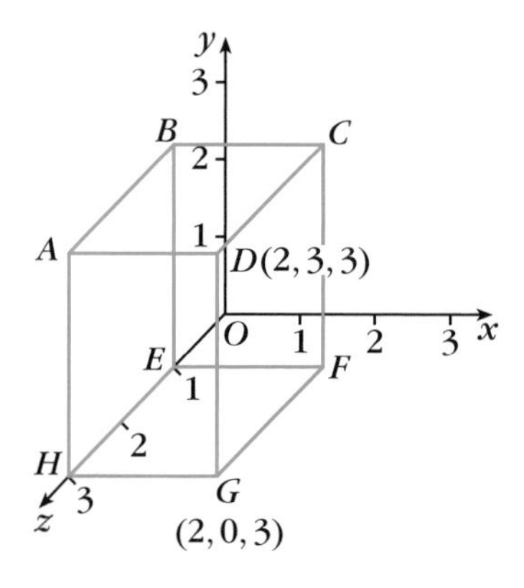

4 Here is part of a sequence of numbers:

$-7,\quad -10,\quad -13,\quad -16,\quad -19$

(a) Write down a formula, in terms of n, for the nth term in the sequence u_n.

(b) Use your formula to find the 20th term in the sequence.

(c) Which is the first term less than -100.

Now check your answers – see the facing page.

Cover this page while you answer the test questions opposite.

Worked answers

Revise this on...

1 (a)

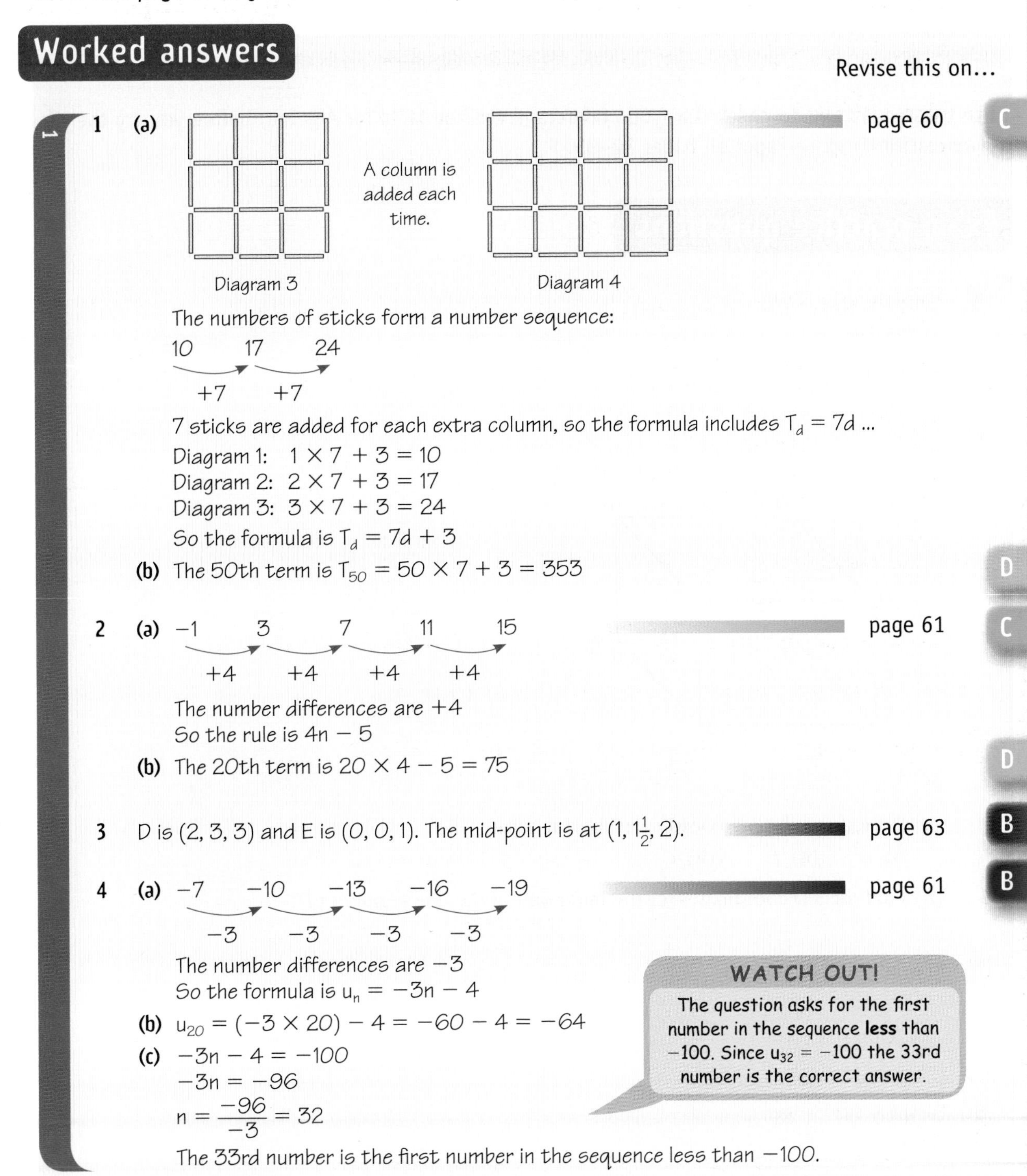

page 60 C

The numbers of sticks form a number sequence:

10 → 17 → 24 (+7, +7)

7 sticks are added for each extra column, so the formula includes $T_d = 7d$ ...

Diagram 1: $1 \times 7 + 3 = 10$

Diagram 2: $2 \times 7 + 3 = 17$

Diagram 3: $3 \times 7 + 3 = 24$

So the formula is $T_d = 7d + 3$

(b) The 50th term is $T_{50} = 50 \times 7 + 3 = 353$

D

2 (a) −1 → 3 → 7 → 11 → 15 (+4, +4, +4, +4)

page 61 C

The number differences are +4

So the rule is $4n - 5$

(b) The 20th term is $20 \times 4 - 5 = 75$

D

3 D is (2, 3, 3) and E is (0, 0, 1). The mid-point is at $(1, 1\frac{1}{2}, 2)$.

page 63 B

4 (a) −7 → −10 → −13 → −16 → −19 (−3, −3, −3, −3)

page 61 B

The number differences are −3

So the formula is $u_n = -3n - 4$

(b) $u_{20} = (-3 \times 20) - 4 = -60 - 4 = -64$

(c) $-3n - 4 = -100$

$-3n = -96$

$n = \frac{-96}{-3} = 32$

The 33rd number is the first number in the sequence less than −100.

WATCH OUT!

The question asks for the first number in the sequence **less** than −100. Since $u_{32} = -100$ the 33rd number is the correct answer.

Tick the questions you got right.

Question	1a	1b	2a	2b	3	4
Grade	C	D	C	D	B	B

Mark the grade you are working at on your revision planner on page viii.

Algebra: subject test

Use these questions to check that you understand the key facts for Algebra, before you try the Examination Practice Paper on pages 82–85.

Exam practice questions

1 Simplify $4(3a + 2b)$

2 Factorise $y^3 - 2y$

3

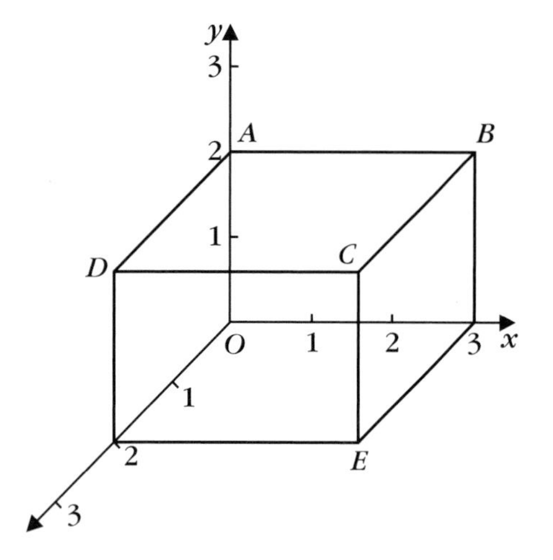

(a) Write down the 3-D coordinates for the point

(i) B (ii) D (iii) C

(b) Find the 3-D coordinates for the mid-point of the line segment CD.

4 Simplify $5(2x + 3) - 3(x - 2)$

5 Expand $t(2t - 3q)$

6 Simplify $12x^5 \div 3x^3$

7 Factorise $k^2 - 3k$

8 These are the five numbers in an arithmetic sequence.

3, 7, 11, 15, 19

Write down the general rule, in terms of n, for the nth term in the sequence.

2

9 Draw the line $y = 3x - 2$ for values of x from $x = -2$ to $x = 3$.
Use your graph to find the value of x when $y = 0$.

1

10 These are the five numbers in an arithmetic sequence.

17, 14, 11, 8, 5

(a) Write down the general rule, in terms of n, for the nth term in the sequence.

(b) Use your rule to find the 20th term in the sequence.

1&2

11 Expand and simplify $(2x - 3)(x + 4)$

12 Expand and simplify $(2x + y)(x - 3y)$

13 Factorise $x^2 - x - 6$

14 C is the point $(5, -7)$, and D is the point $(-2, 5)$.
Find the coordinates of the mid-point of the line segment CD.

15 Factorise $4x^2 - 9y^2$

16 Simplify $(2xy^2)^3$

17 Factorise $2x^2 - x - 15$

18 Simplify $\frac{2(x - 5)^2}{(x - 5)}$

2

19 Simplify $\frac{1}{x + 2} + \frac{2}{2x + 3}$

20 Simplify $\frac{x^2 - 9}{x^2 + 6x + 9}$

Check your answers on page 173. For full worked solutions see the CD.
Tick the questions you got right.

Question	1	2	3	4	5	6	7	8	9	10	11	12	13	14	15	16	17	18	19	20
Grade	D	D	D	C	C	C	C	C	B	B	B	B	B	B	A	A	A	A	A*	A*
Revise this on page	54	56–57	54–55	54–55	54	56–57	56	60–61	62–63	60–61	54–55	56–57	56–57	62–63	56–57	54	56	54–55	57	56–57

Mark the grade you are working at on your revision planner on page viii.
Go to the pages shown to revise for the ones you got wrong.

Working with angles

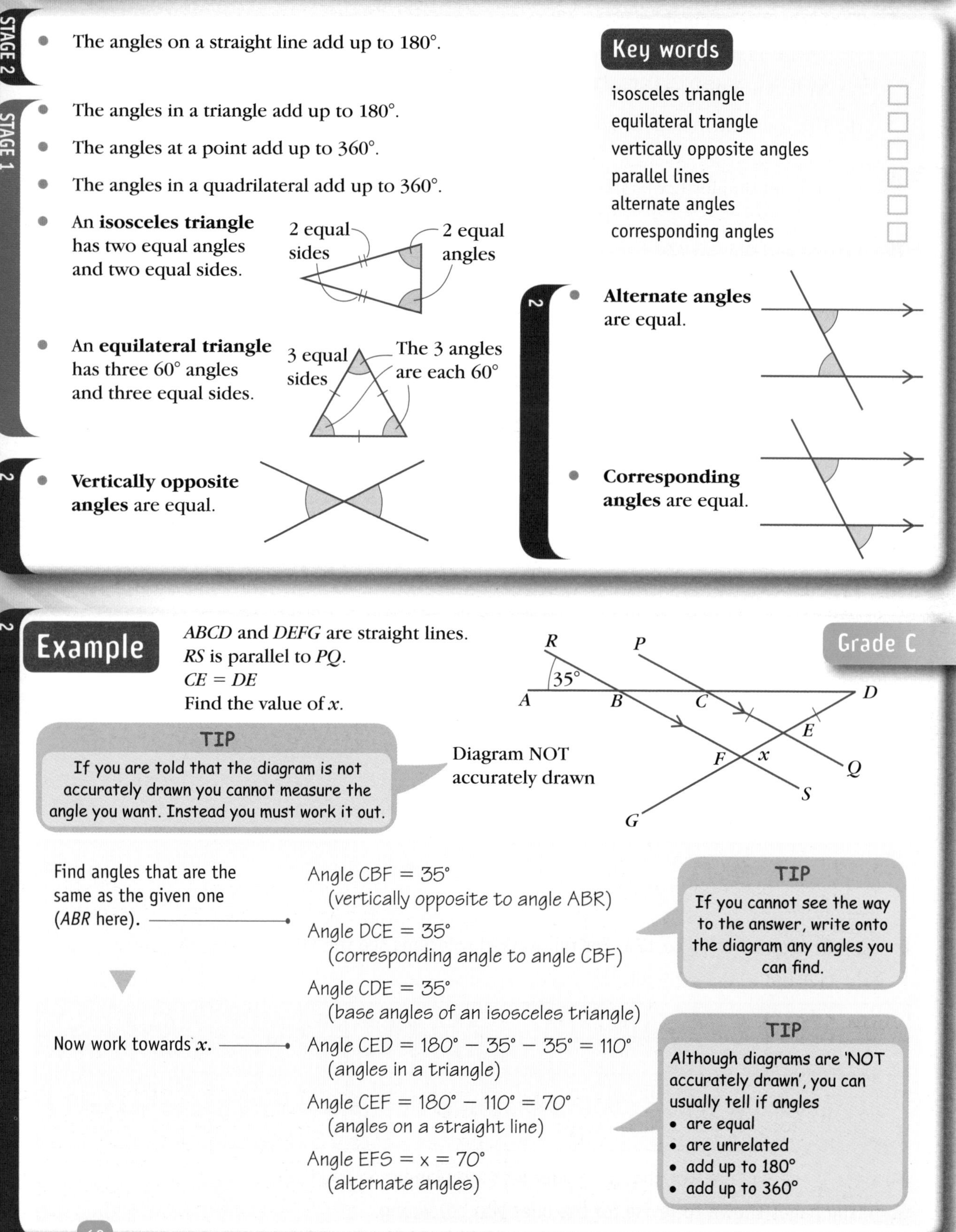

STAGE 2

- The angles on a straight line add up to 180°.

STAGE 1

- The angles in a triangle add up to 180°.
- The angles at a point add up to 360°.
- The angles in a quadrilateral add up to 360°.
- An **isosceles triangle** has two equal angles and two equal sides.

- An **equilateral triangle** has three 60° angles and three equal sides.

2

- **Vertically opposite angles** are equal.

Key words

- isosceles triangle ☐
- equilateral triangle ☐
- vertically opposite angles ☐
- parallel lines ☐
- alternate angles ☐
- corresponding angles ☐

2

- **Alternate angles** are equal.
- **Corresponding angles** are equal.

2

Example

Grade C

ABCD and *DEFG* are straight lines.
RS is parallel to *PQ*.
CE = *DE*
Find the value of x.

Diagram NOT accurately drawn

TIP
If you are told that the diagram is not accurately drawn you cannot measure the angle you want. Instead you must work it out.

Find angles that are the same as the given one (*ABR* here).

Angle CBF = 35°
(vertically opposite to angle ABR)

Angle DCE = 35°
(corresponding angle to angle CBF)

Angle CDE = 35°
(base angles of an isosceles triangle)

Now work towards x.

Angle CED = 180° − 35° − 35° = 110°
(angles in a triangle)

Angle CEF = 180° − 110° = 70°
(angles on a straight line)

Angle EFS = x = 70°
(alternate angles)

TIP
If you cannot see the way to the answer, write onto the diagram any angles you can find.

TIP
Although diagrams are 'NOT accurately drawn', you can usually tell if angles
- are equal
- are unrelated
- add up to 180°
- add up to 360°

- A **bearing** is the angle measured from facing North and turning clockwise. It is always a three-figure number.

- The angle between a **tangent** and a radius is 90°.
- The lengths of the two tangents from a point to a circle are equal.

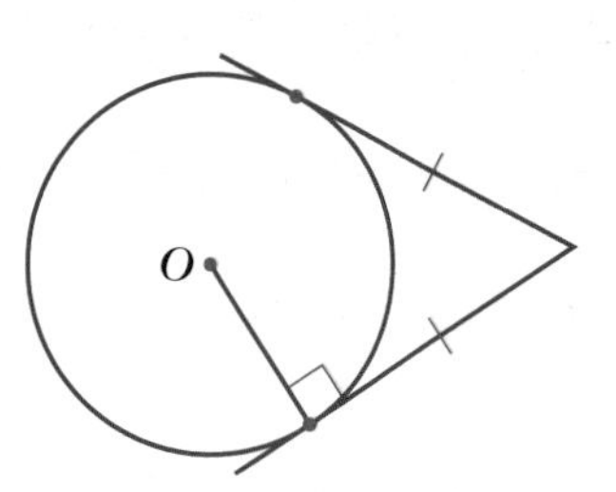

Example The bearing of P from Q is 127°. Find the bearing of Q from P.

As this is the opposite direction a half-turn is needed.
This can be $+180°$ or $-180°$.
The bearing of Q from P is $127° + 180° = 307°$.

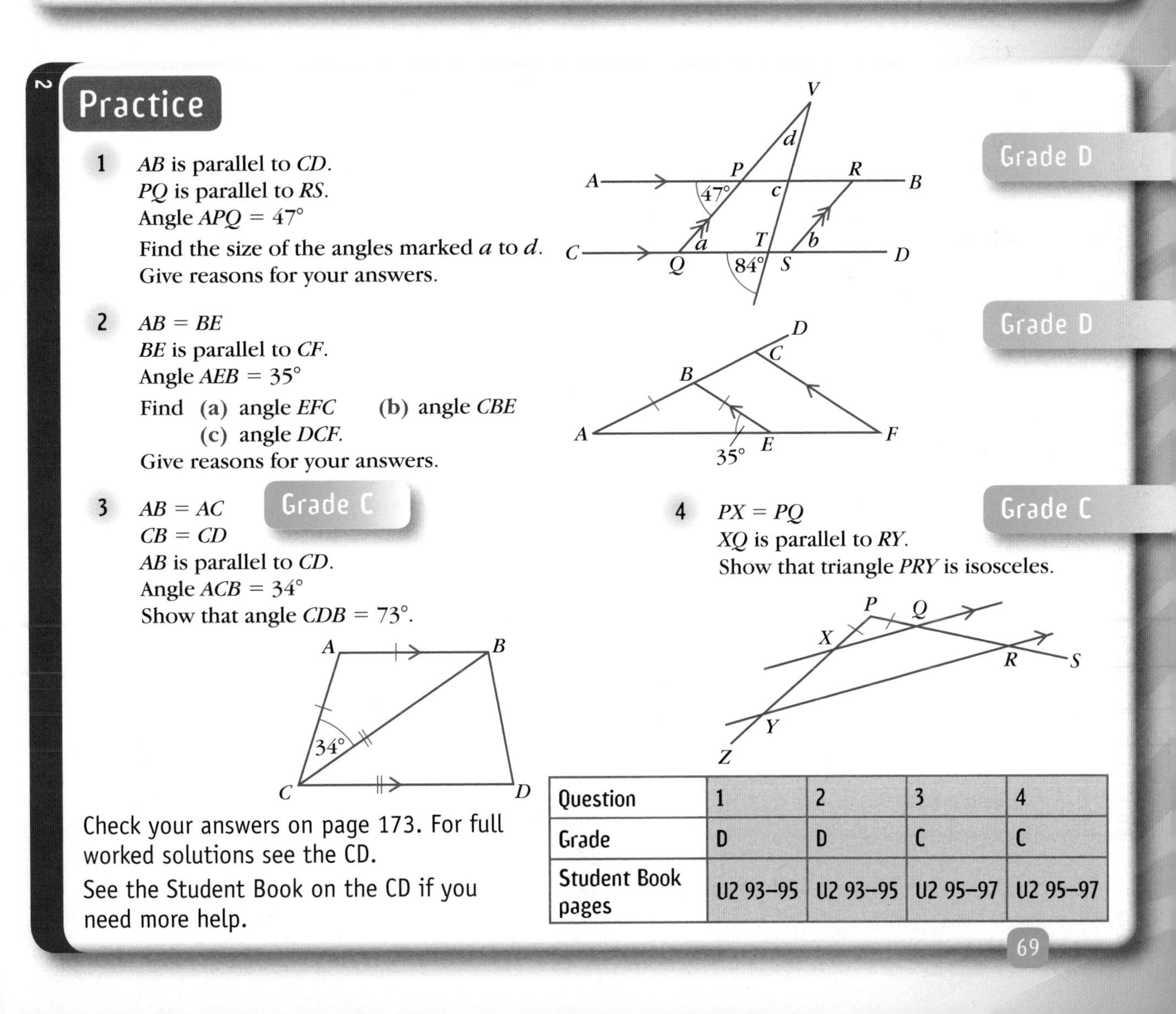

Practice

1 AB is parallel to CD.
PQ is parallel to RS.
Angle $APQ = 47°$
Find the size of the angles marked a to d.
Give reasons for your answers.

Grade D

2 $AB = BE$
BE is parallel to CF.
Angle $AEB = 35°$
Find **(a)** angle EFC **(b)** angle CBE
(c) angle DCF.
Give reasons for your answers.

Grade D

3 $AB = AC$
$CB = CD$
AB is parallel to CD.
Angle $ACB = 34°$
Show that angle $CDB = 73°$.

Grade C

4 $PX = PQ$
XQ is parallel to RY.
Show that triangle PRY is isosceles.

Grade C

Check your answers on page 173. For full worked solutions see the CD.
See the Student Book on the CD if you need more help.

Question	1	2	3	4
Grade	D	D	C	C
Student Book pages	U2 93–95	U2 93–95	U2 95–97	U2 95–97

Angles: topic test

Check how well you know this topic by answering these questions.
First cover the answers on the facing page.

Test questions

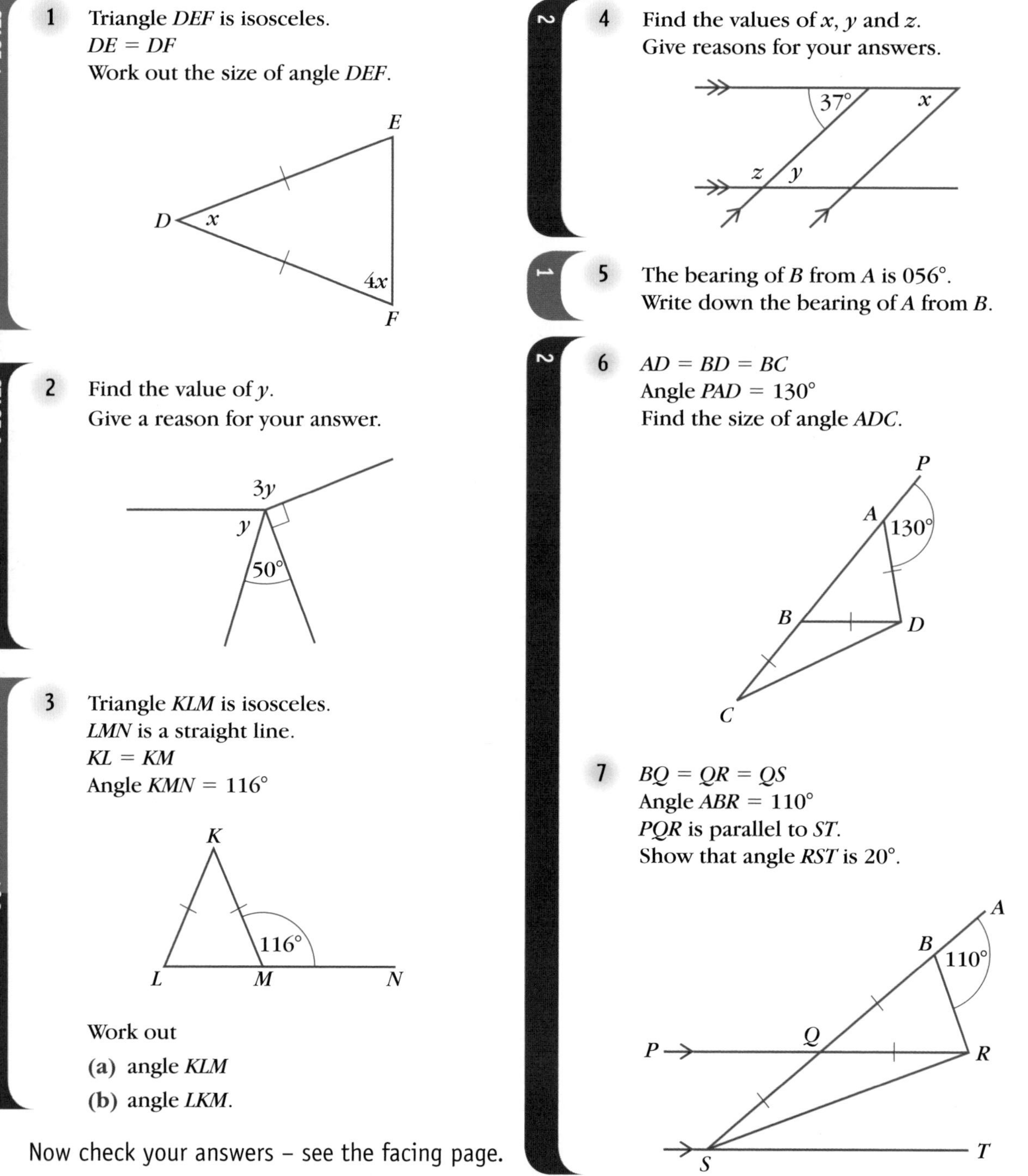

1 Triangle DEF is isosceles.
$DE = DF$
Work out the size of angle DEF.

2 Find the value of y.
Give a reason for your answer.

3 Triangle KLM is isosceles.
LMN is a straight line.
$KL = KM$
Angle $KMN = 116°$

Work out
(a) angle KLM
(b) angle LKM.

4 Find the values of x, y and z.
Give reasons for your answers.

5 The bearing of B from A is 056°.
Write down the bearing of A from B.

6 $AD = BD = BC$
Angle $PAD = 130°$
Find the size of angle ADC.

7 $BQ = QR = QS$
Angle $ABR = 110°$
PQR is parallel to ST.
Show that angle RST is 20°.

Now check your answers – see the facing page.

Cover this page while you answer the test questions opposite.

Worked answers

Revise this on...

1 Triangle DEF is isosceles so angle DEF = angle DFE = $4x$ — page 68 — D

$4x + 4x + x = 9x = 180°$ (angles in a triangle add up to 180°)

$x = 20°$

Angle DEF = $4x = 80°$

2 $y + 3y + 90° + 50° = 4y + 140° = 360°$ (angles at a point add up to 360°) — page 68 — D

$4y = 220°$ and $y = 55°$

3 **(a)** Angle KLM = angle KML = $180° - 116° = 64°$ — page 68 — D

(b) Angle LKM = $180° - 2 \times 64° = 180° - 128° = 52°$

4 $x = 37°$ (corresponding angles) — page 68 — D

$y = 37°$ (alternate angles)

$z = 180° - 37° = 143°$ (angles on a straight line)

5 $056° + 180° = 236°$ (The bearing is in the opposite direction so a half turn of $\pm180°$ is used.) — page 69 — D

6 Angle BAD = $180° - 130° = 50°$ (angles on a straight line) — page 68 — C

Angle ABD = angle BAD = $50°$ (base angle in an isosceles triangle)

Angle ADB = $180° - 50° - 50° = 80°$ (angles in a triangle add up to 180°)

Angle DBC = $180° - 50° = 130°$ (angles on a straight line)

Angle BDC = $\frac{1}{2}(180° - 130°) = 25°$ (base angle in an isosceles triangle)

Angle ADC = angle ADB + angle BDC = $80° + 25° = 105°$

7 Angle QBR = $180° - 110° = 70°$ (angles on a straight line) — page 68 — C

Angle BRQ = $70°$ (base angle in an isosceles triangle)

Angle BQR = $180° - 70° - 70° = 40°$ (angles in a triangle add up to 180°)

Angle RQS = $180° - 40° = 140°$ (angles on a straight line)

Angle QRS = $\frac{1}{2}(180° - 140°) = 20°$ (base angle in an isosceles triangle)

Angle RST = angle QRS = $20°$ (alternate angles)

Tick the questions you got right.

Question	1	2	3	4	5	6	7
Grade	D	D	D	D	D	C	C

Mark the grade you are working at on your revision planner on page viii.

Perimeter, area, volume and measures

STAGE 1

- Approximate **conversions** between units:

Metric	Imperial
8 km	5 miles
1 kg	2.2 pounds
25 g	1 ounce
1 *l*	$1\frac{3}{4}$ pints
4.5 *l*	1 gallon
1 m	39 inches
30 cm	1 foot
2.5 cm	1 inch

- **Speed** $= \dfrac{\text{distance}}{\text{time}}$
- **Average speed** $= \dfrac{\text{total distance}}{\text{total time}}$
- Units of speed are miles per hour (mph), kilometres per hour (km/h) and metres per second (m/s).
- The **perimeter** of a shape is the distance all around the outside.
- The **area** of a 2-D shape is a measure of the amount of space it covers. Typical units of area are mm^2, cm^2, m^2 and km^2.

Key words

metric	☐	area	☐
imperial	☐	rectangle	☐
conversion	☐	triangle	☐
speed	☐	parallelogram	☐
average speed	☐	trapezium	☐
perimeter	☐		

- Area of a **rectangle** $= \text{length} \times \text{width} = l \times w$
- Area of a **triangle** $= \frac{1}{2}\text{base} \times \text{height} = \frac{1}{2} \times b \times h$
- Area of a **parallelogram** $= \text{base} \times \text{vertical height} = b \times h$
- Area of a **trapezium** $= \frac{1}{2} \times$ sum of parallel sides $\times$ distance between parallels $= \frac{1}{2}(a + b)h$

Example

Grade D

Find the area of this shape.

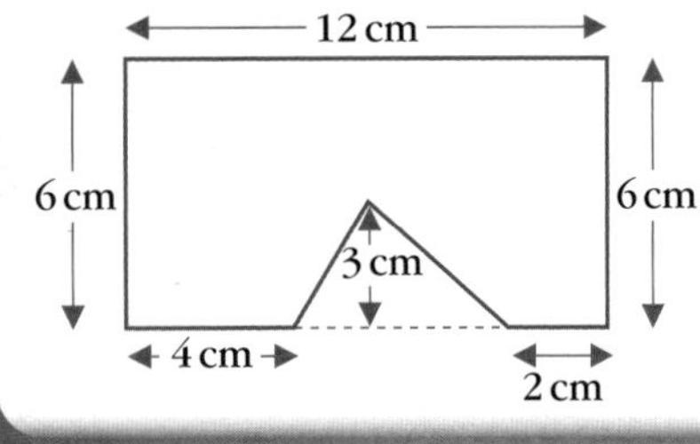

Divide the shape into pieces with areas that are easy to calculate.

Work out the areas and subtract.

Area of rectangle $= 12 \times 6 = 72\,cm^2$

Base of triangle $= 12 - 4 - 2 = 6\,cm$

Area of triangle $= \frac{1}{2} \times 6 \times 3 = 9\,cm^2$

Area of shape $= 72 - 9 = 63\,cm^2$

STAGE 2

- The **volume** is the amount of space occupied by a 3-D shape. Typical units of volume are mm^3, cm^3 and m^3.
- Volume of a **cuboid** $= \text{length} \times \text{width} \times \text{height} = l \times w \times h$

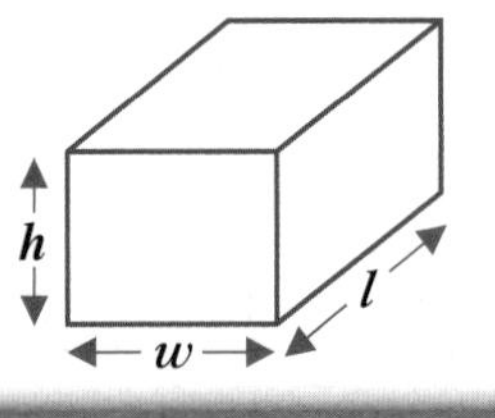

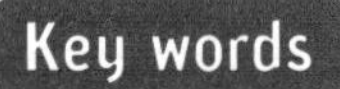

Key words

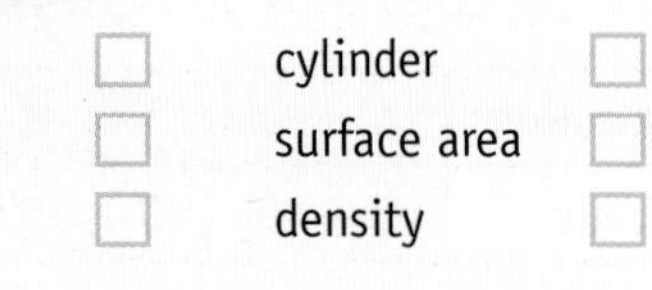

volume	☐	cylinder	☐
cuboid	☐	surface area	☐
prism	☐	density	☐

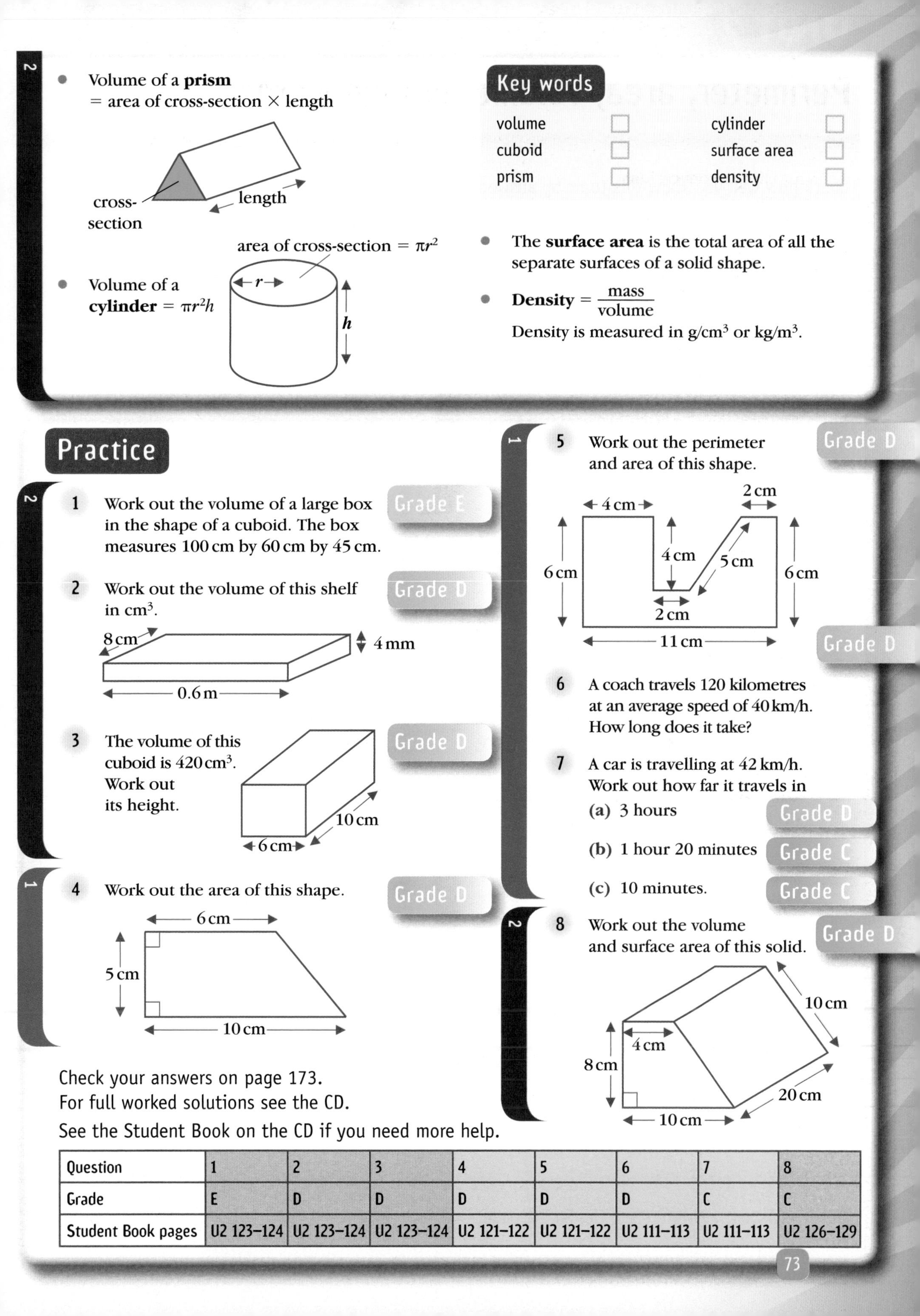

- Volume of a **prism** = area of cross-section × length

- Volume of a **cylinder** = $\pi r^2 h$

Key words

volume	☐	cylinder	☐
cuboid	☐	surface area	☐
prism	☐	density	☐

- The **surface area** is the total area of all the separate surfaces of a solid shape.
- **Density** = $\frac{\text{mass}}{\text{volume}}$
  Density is measured in g/cm³ or kg/m³.

Practice

1 Work out the volume of a large box in the shape of a cuboid. The box measures 100 cm by 60 cm by 45 cm. *Grade E*

2 Work out the volume of this shelf in cm³. *Grade D*

3 The volume of this cuboid is 420 cm³. Work out its height. *Grade D*

4 Work out the area of this shape. *Grade D*

5 Work out the perimeter and area of this shape. *Grade D*

6 A coach travels 120 kilometres at an average speed of 40 km/h. How long does it take? *Grade D*

7 A car is travelling at 42 km/h. Work out how far it travels in

(a) 3 hours *Grade D*

(b) 1 hour 20 minutes *Grade C*

(c) 10 minutes. *Grade C*

8 Work out the volume and surface area of this solid. *Grade D*

Check your answers on page 173.
For full worked solutions see the CD.
See the Student Book on the CD if you need more help.

Question	1	2	3	4	5	6	7	8
Grade	E	D	D	D	D	D	C	C
Student Book pages	U2 123–124	U2 123–124	U2 123–124	U2 121–122	U2 121–122	U2 111–113	U2 111–113	U2 126–129

Perimeter, area, volume and measures: topic test

Check how well you know this topic by answering these questions.
First cover the answers on the facing page.

Test questions

STAGE 1

1 Change 50 litres to gallons.

STAGE 2

2 Packets measuring 5 cm by 3 cm by 10 cm are packed into a box which measures 50 cm by 60 cm by 30 cm.
Work out how many packets will exactly fill the box.

3 Work out the volume of
(a) a box which measures 15 cm by 10 cm by 8 cm
(b) a plank of wood which measures 10 cm by 25 mm by 2 m.

4 The cross-section of a plank of wood is 8 cm by 40 mm. Its volume is 5760 cm^3.
Work out the length of the plank in metres.

1

5 Alan runs 20 km in 1 hour 40 minutes.
Work out his average speed in km/h.

6 Jane drives the 95 kilometres from London to Brighton at an average speed of 42 km/h.
How long does she take?

2

7 Angela cycles 3600 metres in 12 minutes.
Work out her average speed in kilometres per hour (km/h).

1

8 Work out the area of this shape.

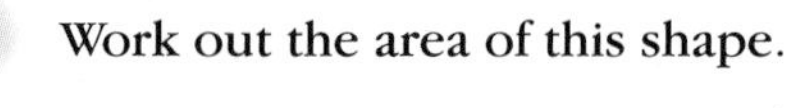
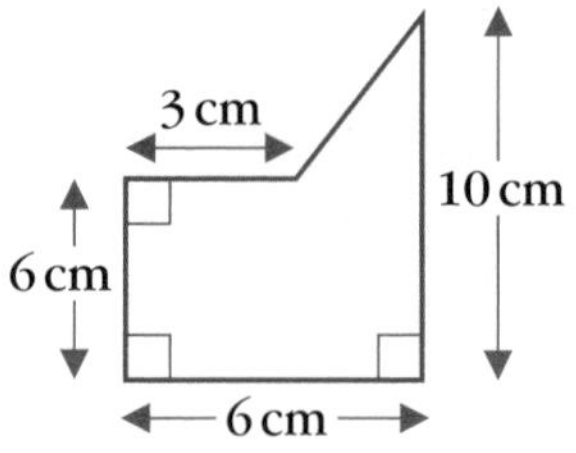

9 The face *ABCD* of this prism is a trapezium.
AB = 14 cm, *AC* = 12 cm and *CD* = 23 cm
Angle *ACD* = angle *CAB* = 90°
The length is 40 cm.

2 **(a)** Work out the volume of the prism.

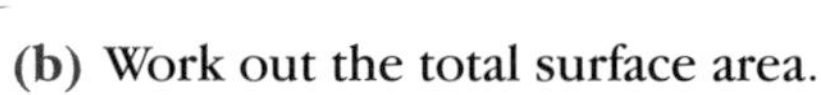

1 **(b)** Work out the total surface area.

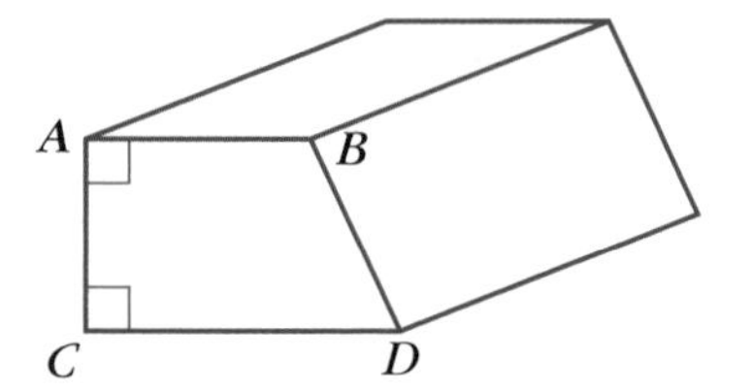

2 **10** Work out the volume and the surface area of a cylinder with base radius 12 cm and height 20 cm.

Now check your answers – see the facing page.

Cover this page while you answer the test questions opposite.

Worked answers

Revise this on...

1 $4.5l$ is 1 gallon
$1l$ is $\frac{1}{4.5}$ gallons
$50l$ is $50 \times \frac{1}{4.5} = 11.1$ gallons.
page 72 E

2 The 50 cm side will take a row of ten 5 cm sides.
The 60 cm side will take a layer of 20 rows using the 3 cm side.
The 30 cm side (height) will take 3 layers of 10 cm.
Total number of packets $= 10 \times 20 \times 3 = 600$
page 72 E

3 (a) Volume $= 15 \times 10 \times 8 = 1200\,\text{cm}^3$ page 72 D
(b) Change lengths to centimetres first: 25 mm = 2.5 cm and 2 m = 200 cm
Volume $= 10 \times 2.5 \times 200 = 5000\,\text{cm}^3$

4 40 mm = 4 cm → Cross-section $= 4 \times 8 = 32\,\text{cm}^2$
Length $= 5760 \div 32 = 180$ cm $= 1.8$ m
page 73 D

5 Average speed $= 20 \div$ 1 hour 40 minutes
$= 20 \div 1\frac{2}{3} = 20 \div \frac{5}{3} = 20 \times \frac{3}{5} = 12$ km/h
page 72 D

6 Time $= 95 \div 42 = 2.2619$ hours
0.2619 hours $= 0.2619 \times 60 = 16$ minutes
Jane takes 2 hours 16 minutes.
page 72 D

7 12 minutes $= \frac{12}{60} = \frac{1}{5}$ hour
3600 metres = 3.6 kilometres
Speed $= \frac{\text{distance}}{\text{time}}$
$= \frac{3.6\text{ km}}{\frac{1}{5}\text{ hour}} = 3.6 \times 5 = 18$ km/h
C

8 Height of triangle $= 10 - 6 = 4$ cm
Base of triangle $= 6 - 3 = 3$ cm
Area $= 6 \times 6 + \frac{1}{2} \times 3 \times 4 = 36 + 6 = 42\,\text{cm}^2$
page 72 C

9 (a) Area of trapezium (cross-section) $= \frac{1}{2}(14 + 23) \times 12 = 222\,\text{cm}^2$ page 73 C
Volume $= 222 \times 40 = 8880\,\text{cm}^2$
(b) $a^2 = 12^2 + 9^2 = 144 + 81 = 225$ so $a = 15$ cm
Perimeter of trapezium $= 14 + 12 + 23 + 15 = 64$ cm
Surface area $= 64 \times 40 + 2 \times 222$
$= 2560 + 444 = 3004\,\text{cm}^2$

10 Volume $= \pi r^2 h = \pi \times 12^2 \times 20 = 2880\pi = 9048\,\text{cm}^3$ page 73 B
Surface area $= 2\pi rh + 2\pi r^2$ (curved surface + two ends)
$= 2\pi(12 \times 20 + 12^2) = 2413\,\text{cm}^2$

Tick the questions you got right.

Question	1	2	3	4	5	6	7	8	9a	9b	10
Grade	E	E	D	D	D	D	C	C	C	C	B

Mark the grade you are working at on your revision planner on page viii.

Shape, space and measure: subject test

Exam practice questions

1 Work out the size of the angles marked with letters.
Give reasons for your answers.

(a)

(b)

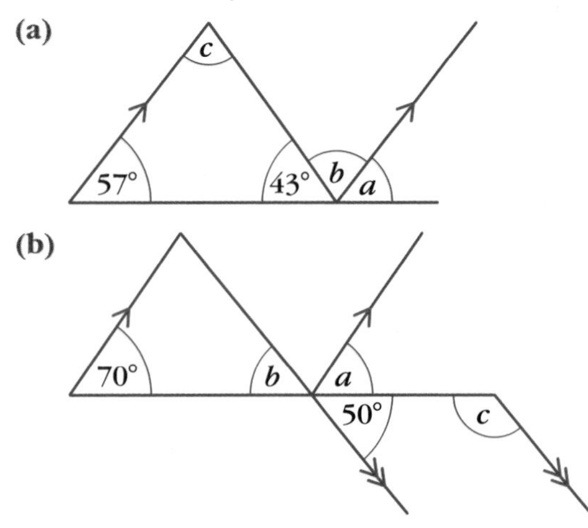

2 The diagram shows the cross-section of a barn.
The barn is 3.5 metres high and 3.5 metres wide.
The sides are 2 metres in height.
Work out the surface area of the cross-section.

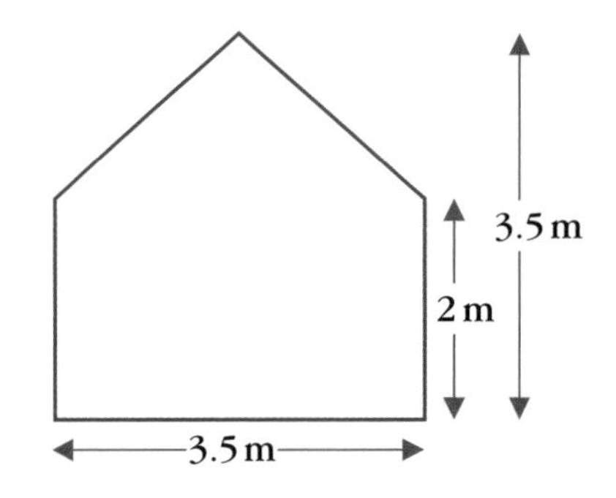

3 The diagram shows a prism.
The cross-section is a triangle with area $35\,\text{cm}^2$. The length is 20 cm.

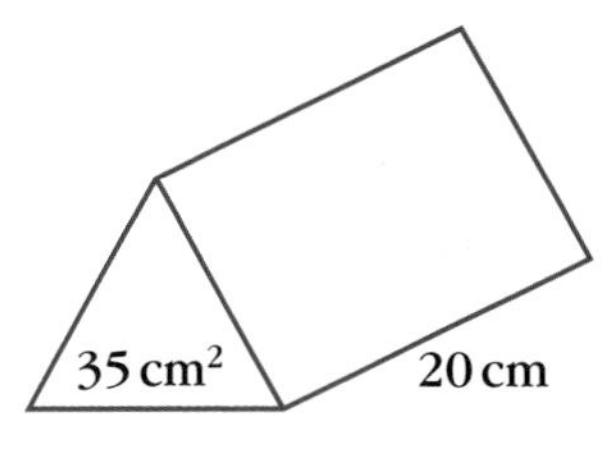

(a) Work out the volume of the prism.

(b) The weight of this solid prism is 6.5 kg.
Work out its density.

4 Work out the area of this shape.

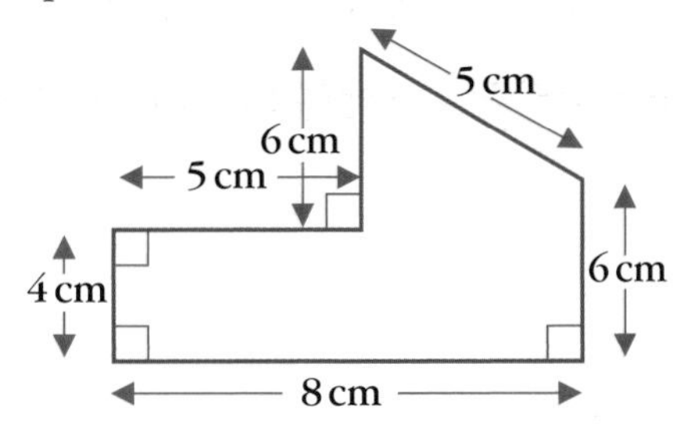

5 Each lap of a Grand prix circuit is 7.2 kilometres.
The race has 62 laps.
The winner takes 1 hour 30 minutes.
Work out the average speed.

6 A container is in the shape of a cuboid and measures 6 metres by 2 metres by 1.6 metres.
Cartons containing electrical equipment measure 80 cm × 50 cm × 30 cm.
Work out how many cartons can be packed into the container.

7 Which speed is the faster?
50 metres per second or 175 kilometres per hour?
You must show working to support your answer.

8 The density of a piece of glass is 2.6 g/cm^3.
Its weight is 5.1 kg.
The glass is melted down and made into cubes with side 0.9 cm from which dice will be made.
Assuming no glass is lost in the process, work out how many cubes can be made.

9 The bearing of *A* from *B* is 252°.
Angle *ACB* is 30°.
Angle *BAC* is 38°.
Find the bearing of *C* from *B*.

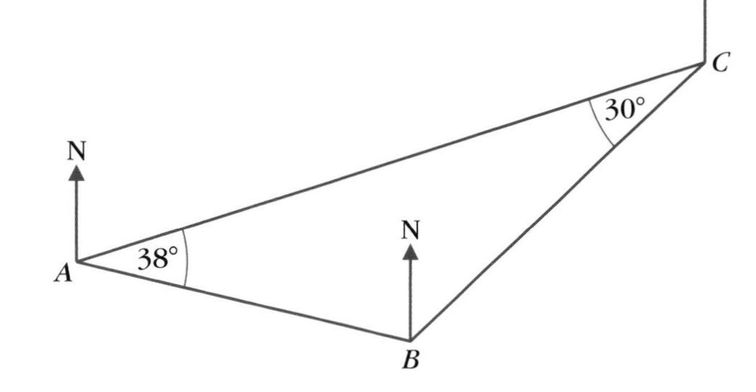

Check your answers on page 173.
For full worked solutions see the CD.
Tick the questions you got right.

Question	1	2	3	4	5	6	7	8	9
Grade	D	D	C	C	C	C	C	B	B
Revise this on page	68	72	73	72	72	72	72	73	69

Mark the grade you are working at on your revision planner on page viii.
Go to the pages shown to revise for the ones you got wrong.

Number

Integers

STAGE 1

- Adding a **negative** number has the same effect as subtracting the **positive** number:
  $4 + -1 = 3$
- Subtracting a negative number has the same effect as adding the positive number:
  $2 - -3 = 5$
- When multiplying or dividing, two **like** signs give a + and two **unlike** signs give a −.
- A **factor** of a number is a whole number that divides exactly into the number.
  1, 2, 3, 4, 6 and 12 are factors of 12.
- The **multiple** of a number is the result of multiplying the number by a positive whole number.
  3, 6 and 9 are multiples of 3.
- A **prime number** is a number greater than 1 which only has 2 factors, itself and 1.
  1 is not a prime number as it can only be divided by one number (itself).
- A **power** or **index** tells how many times a number is multiplied by itself:
  $2 \times 2 \times 2 \times 2 = 2^4$

STAGE 2

- To **multiply** powers of the same number, add the indices:
  $2^3 \times 2^4 = 2^{3+4} = 2^7$
- To **divide** powers of the same number subtract the indices:
  $5^6 \div 5^4 = 5^{6-4} = 5^2$

1

- Use **BIDMAS** to help you remember the order of mathematical operations.
  Brackets
  Indices
  Division
  Multiplication
  Addition
  Subtraction

Decimals and rounding

1

- To **round** to a given number of **decimal places (d.p.)**, count the number of decimal places from the decimal point and look at the next digit. If it is 5 or more, round up; if it is less than 5 round down.
- To **round** to a given number of **significant figures (s.f.)**, count the number of digits from the first non-zero digit, starting from the *left*. Look at the next digit to decide whether to round up or down.

1&2

- When you **multiply decimals**, work out the multiplication without the decimal points and put in the decimal point at the end.
- When **dividing decimals** make sure that you always divide by a whole number. Do this by multiplying *both* numbers by 10, 100 or 1000.
- A number is in **standard form** when

$7.2 \times 10^6 = 7\,200\,000$

$7.2 \times 10^{-6} = 0.000\,007\,2$

This part is written as a number between 1 and 10

This part is written as a **power of 10**

Fractions, decimals and percentages

1

- **Addition and Subtraction**
  - You can only add and subtract fractions that have the same bottom number (denominator).
  - Start by dealing with any whole numbers.
  - Then find equivalent fractions with the same denominator for the fractions.
  - You can then add or subtract.
- **Multiplication and Division**

  When you multiply fractions you write any mixed numbers as improper fractions. You then multiply the numerators and the denominators.

  When you divide fractions you write any mixed numbers as improper fractions. Then write down the first fraction, invert the second fraction and multiply.

2

- Any terminating decimal with n decimal places can be converted into a fraction by multiplying the decimal by 10^n and dividing the result by 10^n.
- Any recurring decimal can be converted into a fraction by multiplying the decimal by 10^m (where m is the number of decimal places in the recurring pattern) and then subtracting.
- To compare **fractions**, **decimals** and **percentages** you can change them all to percentages.
- To find a **percentage** of an amount you can: change the percentage to a fraction and multiply *or* change the percentage to a decimal and multiply *or* work from 10%.

Algebra

Manipulative algebra

1&2

- Use the **index laws** to **simplify** expressions with indices:

  – $x^m \times x^n = x^{m+n}$

  – $(x^m)^n = x^{m \times n} = x^{mn}$

  – $x^m \div x^n = x^{m-n}$

  – $x^0 = 1$ for all non-zero values of x

- **Expanding the brackets** means multiplying to remove the brackets:
  $4(3a + b) = 12a + 4b$.

  In general – $(x + a)^2 = a^2 + 2ax + a^2$
  – $(x - b)^2 = x^2 - 2bx + b^2$

- **Factorising** is the opposite of removing brackets. To factorise an expression completely, the **highest common factor (HCF)** must appear outside the brackets.

Patterns, sequences, coordinates and graphs

1

- To find the **rule** for the nth term of a number pattern, use a table of values.

Term number	1	2	3	4
Term	5	8	11	14
Difference		+3	+3	+3

The rule is $3n + 2$

Shape, space and measure

Angles

- The angles at a point add up to 360° and on a line add up to 180°.
- The angles in a triangle add up to 180°.
- The angles in a quadrilateral add up to 360°.
- **Alternate angles** are equal.

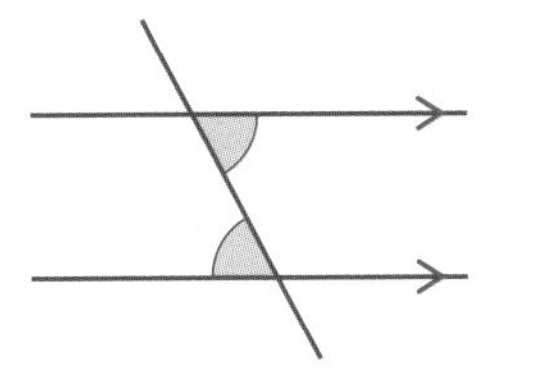

- **Corresponding angles** are equal.

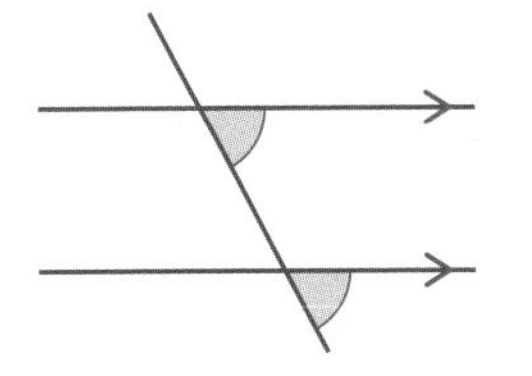

- A **bearing** is the angle measured from facing North and turning clockwise.

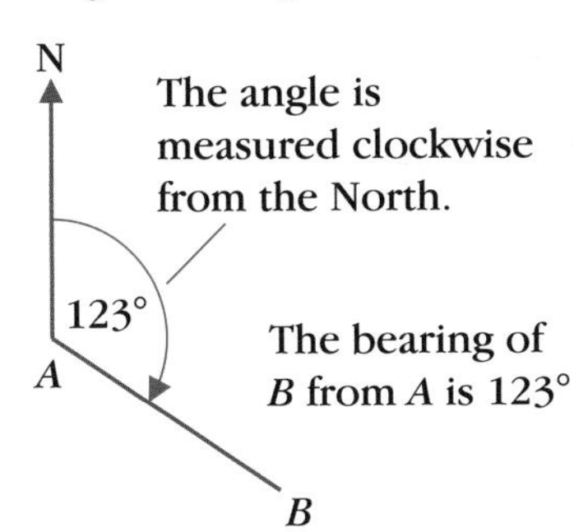

Perimeter, area, volume and measures

- **Average speed** $= \frac{\text{total distance}}{\text{total time}}$ (take care with units)
- You should know that 8 kilometres = 5 miles
  1 kilogram = 2.2 pounds
- **Area** of a **triangle** $= \frac{1}{2}$ base $\times$ height.
- Area of a **parallelogram** = base $\times$ vertical height.
- Area of a **trapezium** $= \frac{1}{2}$ sum of parallel sides $\times$ distance between parallel sides.
- You should know that **volume** is a measure of space occupied and is measured in mm^3, cm^3, m^3.
- Volume of a **cuboid** = length $\times$ width $\times$ height
- The **surface area** is the total area of all the separate surfaces of a solid shape.

Unit 2 Examination practice paper

A formulae sheet can be found on page 167.

Stage 1 (multiple choice)

Non-calculator

1

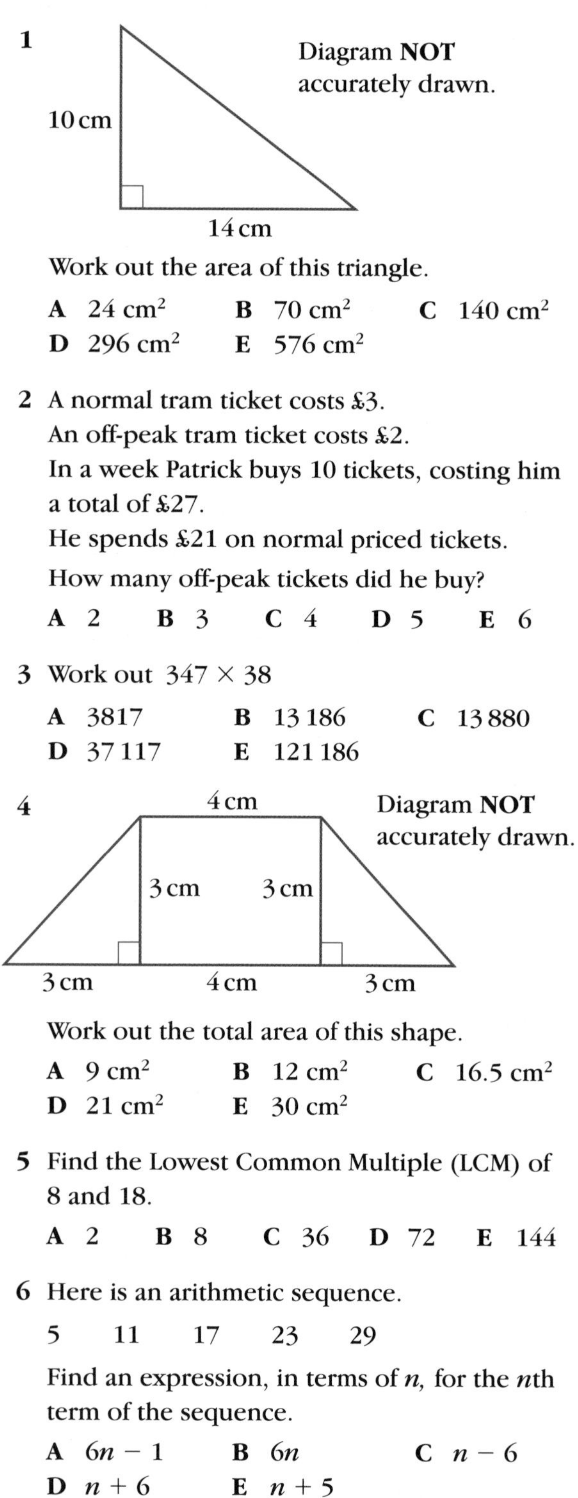

Work out the area of this triangle.

A 24 cm^2 **B** 70 cm^2 **C** 140 cm^2
D 296 cm^2 **E** 576 cm^2

2 A normal tram ticket costs £3.
An off-peak tram ticket costs £2.
In a week Patrick buys 10 tickets, costing him a total of £27.
He spends £21 on normal priced tickets.
How many off-peak tickets did he buy?

A 2 **B** 3 **C** 4 **D** 5 **E** 6

3 Work out 347×38

A 3817 **B** 13 186 **C** 13 880
D 37 117 **E** 121 186

4

Work out the total area of this shape.

A 9 cm^2 **B** 12 cm^2 **C** 16.5 cm^2
D 21 cm^2 **E** 30 cm^2

5 Find the Lowest Common Multiple (LCM) of 8 and 18.

A 2 **B** 8 **C** 36 **D** 72 **E** 144

6 Here is an arithmetic sequence.

5 11 17 23 29

Find an expression, in terms of n, for the nth term of the sequence.

A $6n - 1$ **B** $6n$ **C** $n - 6$
D $n + 6$ **E** $n + 5$

7 A cuboid is shown on a 3-dimensional grid.
The point F has the coordinates (4,6,7).
Work out the coordinates of the point Q.

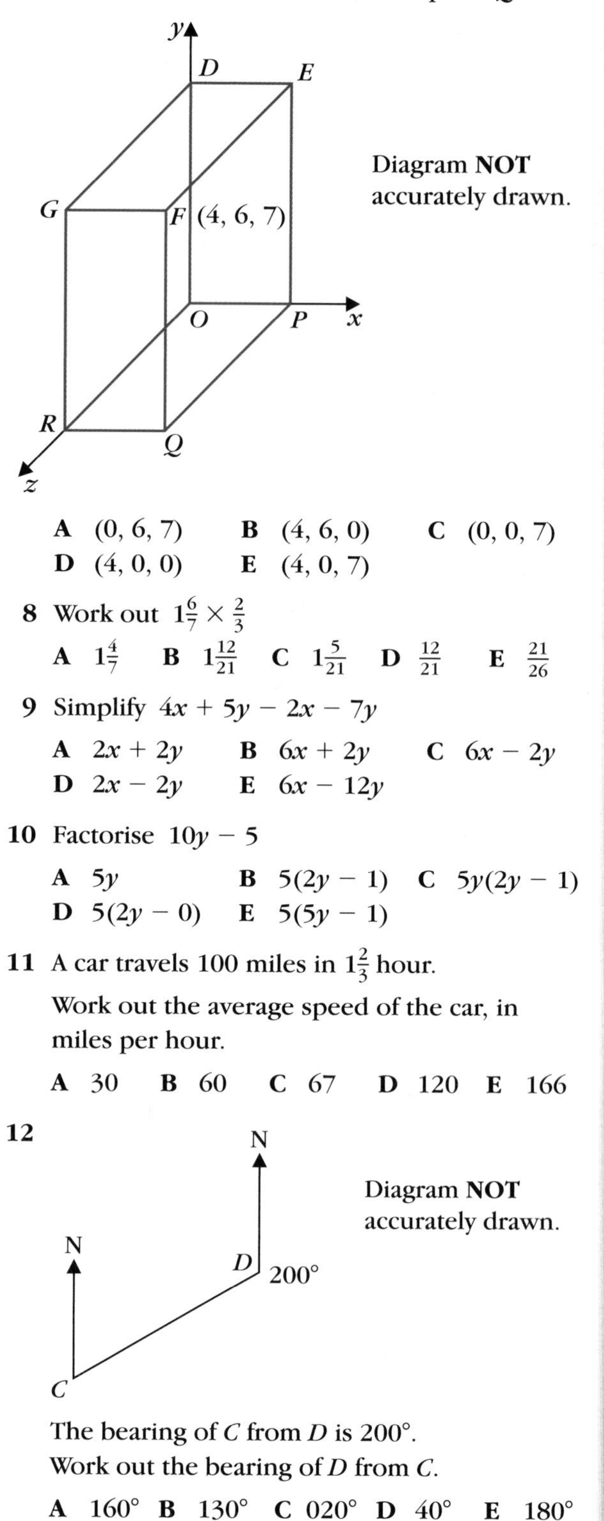

A (0, 6, 7) **B** (4, 6, 0) **C** (0, 0, 7)
D (4, 0, 0) **E** (4, 0, 7)

8 Work out $1\frac{6}{7} \times \frac{2}{3}$

A $1\frac{4}{7}$ **B** $1\frac{12}{21}$ **C** $1\frac{5}{21}$ **D** $\frac{12}{21}$ **E** $\frac{21}{26}$

9 Simplify $4x + 5y - 2x - 7y$

A $2x + 2y$ **B** $6x + 2y$ **C** $6x - 2y$
D $2x - 2y$ **E** $6x - 12y$

10 Factorise $10y - 5$

A $5y$ **B** $5(2y - 1)$ **C** $5y(2y - 1)$
D $5(2y - 0)$ **E** $5(5y - 1)$

11 A car travels 100 miles in $1\frac{2}{3}$ hour.
Work out the average speed of the car, in miles per hour.

A 30 **B** 60 **C** 67 **D** 120 **E** 166

12

The bearing of C from D is 200°.
Work out the bearing of D from C.

A 160° **B** 130° **C** 020° **D** 40° **E** 180°

13 Expand and simplify $(x + 3)(x - 7)$

A $x^2 + 10x + 21$ **B** $2x + 4$
C $x^2 - 4x - 21$ **D** $x^2 + 4x - 21$
E $x^2 - 21$

14 Express 490 as a product of its prime factors.

A $7^4 \times 2 \times 5$ **B** $2 \times 5 \times 5 \times 7$
C 10×49 **D** $2 \times 2 \times 49$
E $2 \times 5 \times 7 \times 7$

15

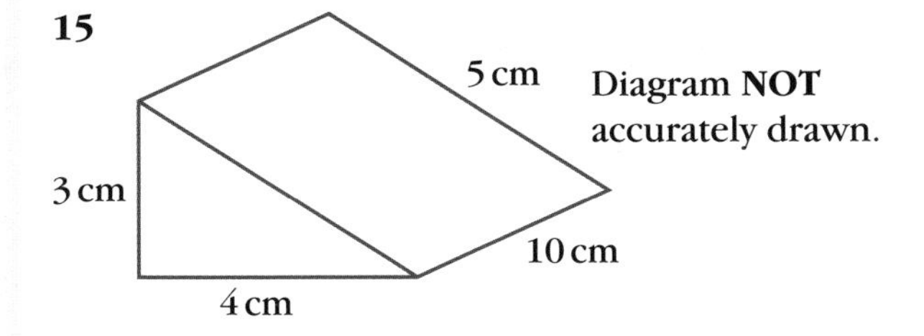

Work out the surface area of this prism.

A 62 cm² **B** 92 cm² **C** 104 cm²
D 132 cm² **E** 144 cm²

16 Factorise $x^2 + x - 30$

A $(x - 5)(x + 6)$ **B** $(x + 5)(x - 6)$
C $(x - 1)(x + 30)$ **D** $(x - 5)(x - 6)$
E $(x + 1)(x - 30)$

17 Find the Highest Common Factor (HCF) of 48 and 72.

A 6 **B** 8 **C** 24 **D** 144 **E** 3456

18

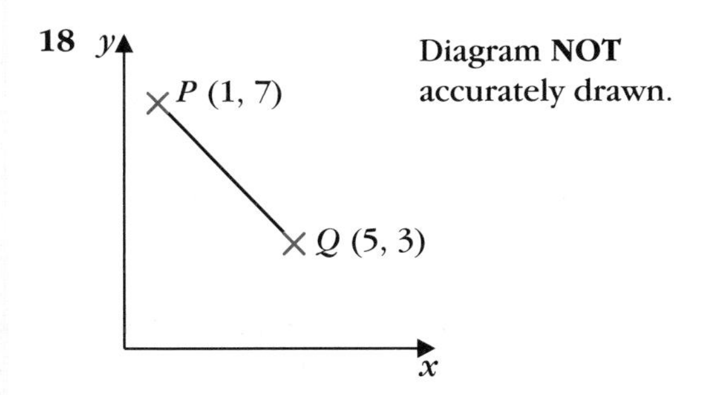

Work out the coordinates of the mid-point of the line *PQ*.

A (4, 4) **B** (3, 4) **C** (4, 5) **D** (3, 3) **E** (3, 5)

19 Factorise completely $6p^2 + 3pq$

A $3(2p^2 + pq)$ **B** $3p(2 + 3q)$
C $3p(2p + q)$ **D** $3p(2p + 3q)$
E $p(6p + 3q)$

20 Write 0.0521 in standard form.

A 521×10^2 **B** 5.21×10^2 **C** 521×10^{-2}
D 5.21×10^{-2} **E** 5.21×10^{-1}

21 A time is given as 55 seconds, to the nearest second.

What is the minimum the time could be?

A 54 seconds **B** 54.4 seconds
C 54.5 seconds **D** 54.9 seconds
E 55.5 seconds

22 A conveyer moves grain at the rate of 0.6 kg per second.

How much grain will be moved every hour?

Give your answer in tonnes.

A 2.16 **B** 1.66 **C** 2160 **D** 1666 **E** 36

23 *C* and *D* are two points on a 3-D coordinate grid.
Point *C* is (−3, −5, 7).
Point *D* is (−4, 3, 2).

Work out the coordinates of the mid-point of the line *CD*.

A $(-3\frac{1}{2}, -4, 4\frac{1}{2})$ **B** $(3\frac{1}{2}, 1, 2\frac{1}{2})$
C $(-3\frac{1}{2}, -4, 2\frac{1}{2})$ **D** $(-3\frac{1}{2}, -1, 4\frac{1}{2})$
E $(-3\frac{1}{2}, -2, 4\frac{1}{2})$

24 Expand and simplify $(3x - 2y)(5x - 2y)$

A $6x^2 - 16xy + 4y^2$ **B** $15x^2 - 16xy + 4y$
C $15x^2 - 4xy - 4y^2$ **D** $15x^2 - 16xy - 4y^2$
E $6x^2 + 4x + 4y^2$

25 Factorise $20x^2 + 23x - 21$

A $(20x - 3)(x + 7)$ **B** $(10x + 7)(2x - 3)$
C $(5x + 3)(4x - 7)$ **D** $(10x - 7)(2x + 3)$
E $(5x - 3)(4x + 7)$

Check your answers on page 173. For full worked solutions see the CD.

Unit 2 Examination practice paper

Stage 2

Calculator

1 **(a)** Work out the value of x.

(b) Give reasons for your answer.

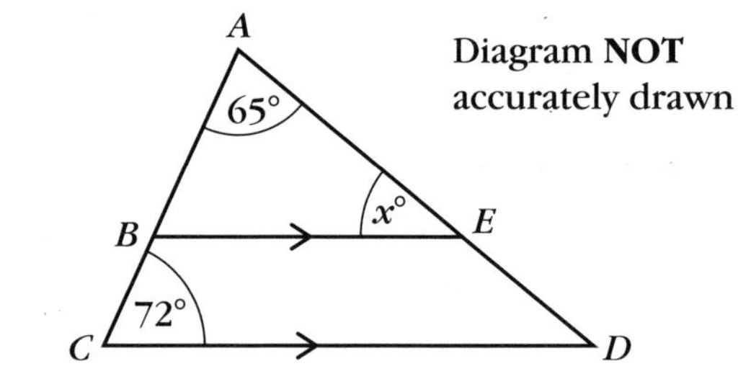

(3 marks)

2 The length of a pen is measured as 9 cm to the nearest centimetre.

Write down the greatest length the pen could be. **(1 mark)**

3 Expand and simplify $(x + 4)(x - 2)$ **(2 marks)**

4 **(a)** Work out $4^5 \times 4^2$

(b) Work out $\dfrac{5^5 \times 5^2}{5^6}$ **(2 marks)**

5 Use your calculator to work out

$4.5 \times 10^6 \div (7.2 \times 10^{-5})$

Give your answer in standard form. **(2 marks)**

6 **(a)** Complete this table of values for $y = 2x + 1$

x	−2	−1	0	1	2	3
y		−1			5	

(b) On the grid, draw the graph of $y = 2x + 1$

10
8
6
4
2
−3 −2 −1 0 1 2 3 4
−2
−4

(4 marks)

7 **(a)** Factorise $p^2 - 8p - 20$

(b) Write $\dfrac{3x}{3x - 2} - \dfrac{2}{x(3x - 2)}$ as a single fraction in its simplest form.

(4 marks)

8

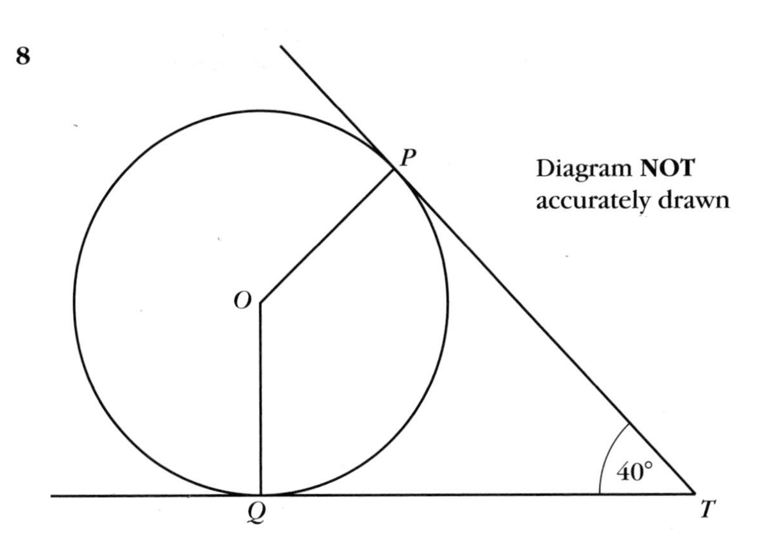

P and Q are two points on a circle centre O.

The tangents to the circle at P and Q intersect at the point T.

(a) Write down the size of angle OQT.

(b) Calculate the size of the obtuse angle POQ.

(c) Give reasons why angle PQT is 70°.

(5 marks)

9 Write the recurring decimal $0.\dot{1}\dot{4}$ as a fraction.

(2 marks)

Check your answers on page 174. For full worked solutions see the CD.

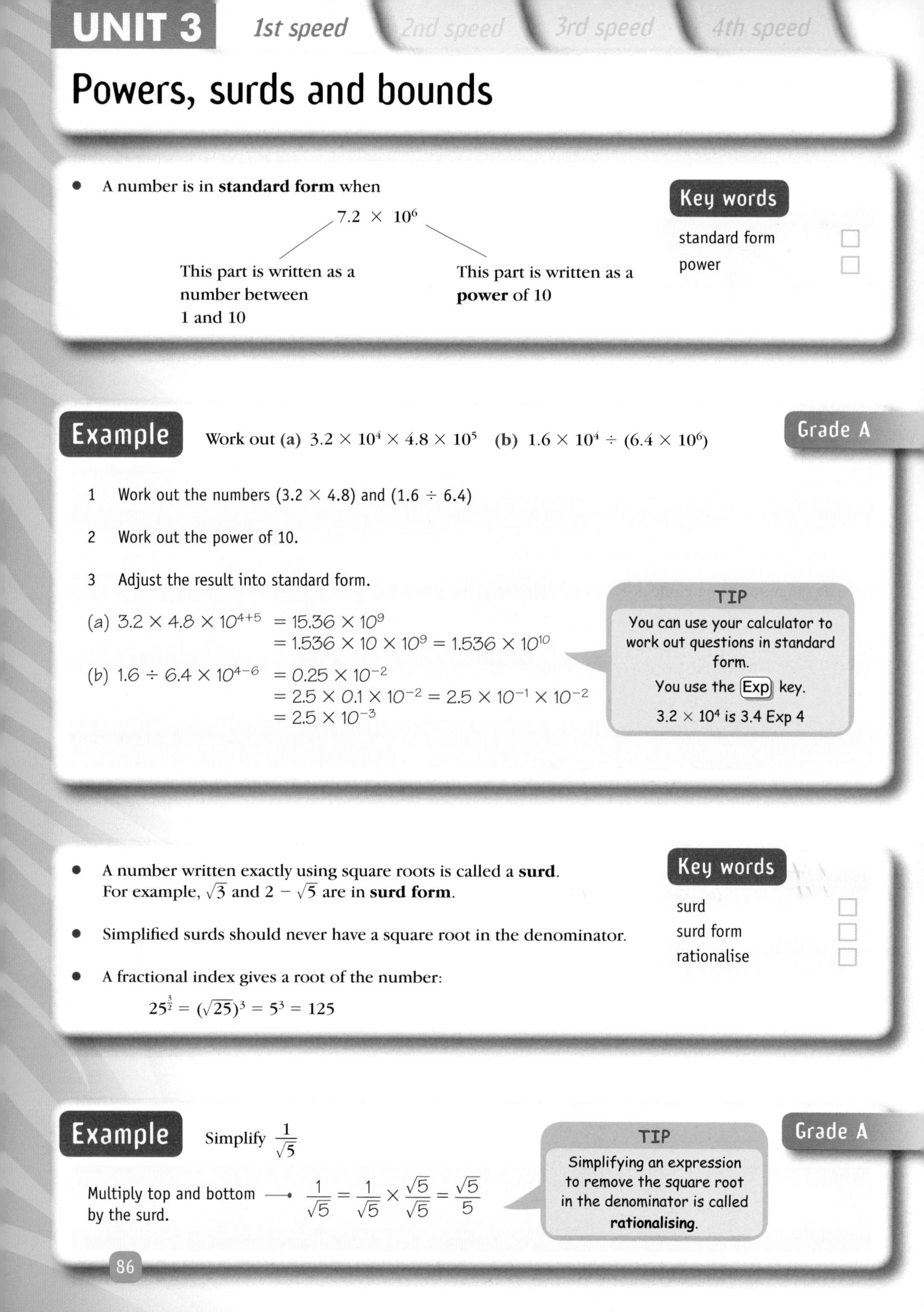

Powers, surds and bounds

- A number is in **standard form** when

$$7.2 \times 10^6$$

This part (7.2) is written as a number between 1 and 10

This part (10^6) is written as a **power** of 10

Key words
- standard form
- power

Example

Grade A

Work out (a) $3.2 \times 10^4 \times 4.8 \times 10^5$ (b) $1.6 \times 10^4 \div (6.4 \times 10^6)$

1. Work out the numbers (3.2×4.8) and $(1.6 \div 6.4)$
2. Work out the power of 10.
3. Adjust the result into standard form.

(a) $3.2 \times 4.8 \times 10^{4+5} = 15.36 \times 10^9$
$= 1.536 \times 10 \times 10^9 = 1.536 \times 10^{10}$

(b) $1.6 \div 6.4 \times 10^{4-6} = 0.25 \times 10^{-2}$
$= 2.5 \times 0.1 \times 10^{-2} = 2.5 \times 10^{-1} \times 10^{-2}$
$= 2.5 \times 10^{-3}$

TIP
You can use your calculator to work out questions in standard form.
You use the Exp key.
3.2×10^4 is 3.4 Exp 4

- A number written exactly using square roots is called a **surd**.
  For example, $\sqrt{3}$ and $2 - \sqrt{5}$ are in **surd form**.
- Simplified surds should never have a square root in the denominator.
- A fractional index gives a root of the number:

$$25^{\frac{3}{2}} = (\sqrt{25})^3 = 5^3 = 125$$

Key words
- surd
- surd form
- rationalise

Example

Grade A

Simplify $\frac{1}{\sqrt{5}}$

Multiply top and bottom by the surd. $\frac{1}{\sqrt{5}} = \frac{1}{\sqrt{5}} \times \frac{\sqrt{5}}{\sqrt{5}} = \frac{\sqrt{5}}{5}$

TIP
Simplifying an expression to remove the square root in the denominator is called **rationalising**.

- The **upper bound** is the maximum possible value of a measurement or result of a calculation.
- The **lower bound** is the minimum possible value of a measurement or result of a calculation.
- In addition and multiplication calculations:
  – use the two lower bounds to obtain the lower bound of the result
  – use the two upper bounds to obtain the upper bound of the result.
- In subtraction and division calculations, you need to use one upper bound and one lower bound to find each bound of the result.

TIP
Think carefully about which pair you need. If in doubt, do the calculation with all four pairs. Then choose the largest and smallest values for the bounds.

Example

Write down the upper and lower bound for these measurements
(a) 45 mm to the nearest mm **(b)** 5.6 seconds to the nearest tenth of a second.

The upper bound is always the measurement + $\frac{1}{2}$ a unit. → (a) Upper bound $45 + 0.5 = 45.5$ mm
Lower bound $45 - 0.5 = 44.5$ mm

The lower bound is always the measurement − $\frac{1}{2}$ a unit. → (b) Upper bound $5.6 + 0.05 = 5.65$ seconds
Lower bound $5.6 - 0.05 = 5.55$ seconds

Example

Grade A*

David's time for a 100 m race is 12.3 seconds. The 100 m is measured to the nearest metre and the time to the nearest tenth of a second. Find the highest and lowest possible values of David's speed.

Highest speed $= \dfrac{\text{greatest distance}}{\text{smallest time}}$ → Upper bound $= 100.5 \div 12.25 = 8.20408$ m/s

Lowest speed $= \dfrac{\text{smallest distance}}{\text{greatest time}}$ → Lower bound $= 99.5 \div 12.35 = 8.05668$ m/s

Practice

Grade A

1 Rationalise
(a) $\dfrac{2}{\sqrt{7}}$ (b) $\dfrac{\sqrt{5}}{\sqrt{20}}$
(c) $\dfrac{2}{\sqrt{2}}$ (d) $\dfrac{3 + \sqrt{5}}{\sqrt{5}}$

Grade A

Grade A*

2 Work out
(a) $36^{\frac{3}{2}}$ (b) $16^{\frac{3}{4}}$ (c) $64^{\frac{3}{2}}$
(d) $8^{-\frac{2}{3}}$ (e) $27^{-\frac{3}{2}}$ (f) $\dfrac{2}{16^{-\frac{3}{4}}}$

3 Susie ran the 400 m race in 70 seconds. Grade A*
The distance was measured to the nearest centimetre and the time to the nearest hundredth of a second. Calculate the upper and lower bound for her speed in metres per second.

4 Calculate the upper and lower bounds for the area of a rectangle with length 4.5 cm and width 3.7 cm, both measured to the nearest millimetre. Grade A*

Check your answers on page 174. For full worked solutions see the CD.
See the Student Book on the CD if you need more help.

Question	1	2abc	2def	3	4
Grade	A	A	A*	A*	A*
Student Book pages	U3 46–48	U2 37		U3 4	U3 49–51

Percentages

- To **increase** or **decrease** an amount by a percentage, you find the percentage of that amount and then add or subtract it from the starting amount.
- **Simple interest** is when interest is paid just on the original amount.
- **Compound interest** is when interest is paid on the original amount *and* on the interest already earned.

Key words

- increase and decrease ☐
- simple interest ☐
- compound interest ☐
- interest rate ☐

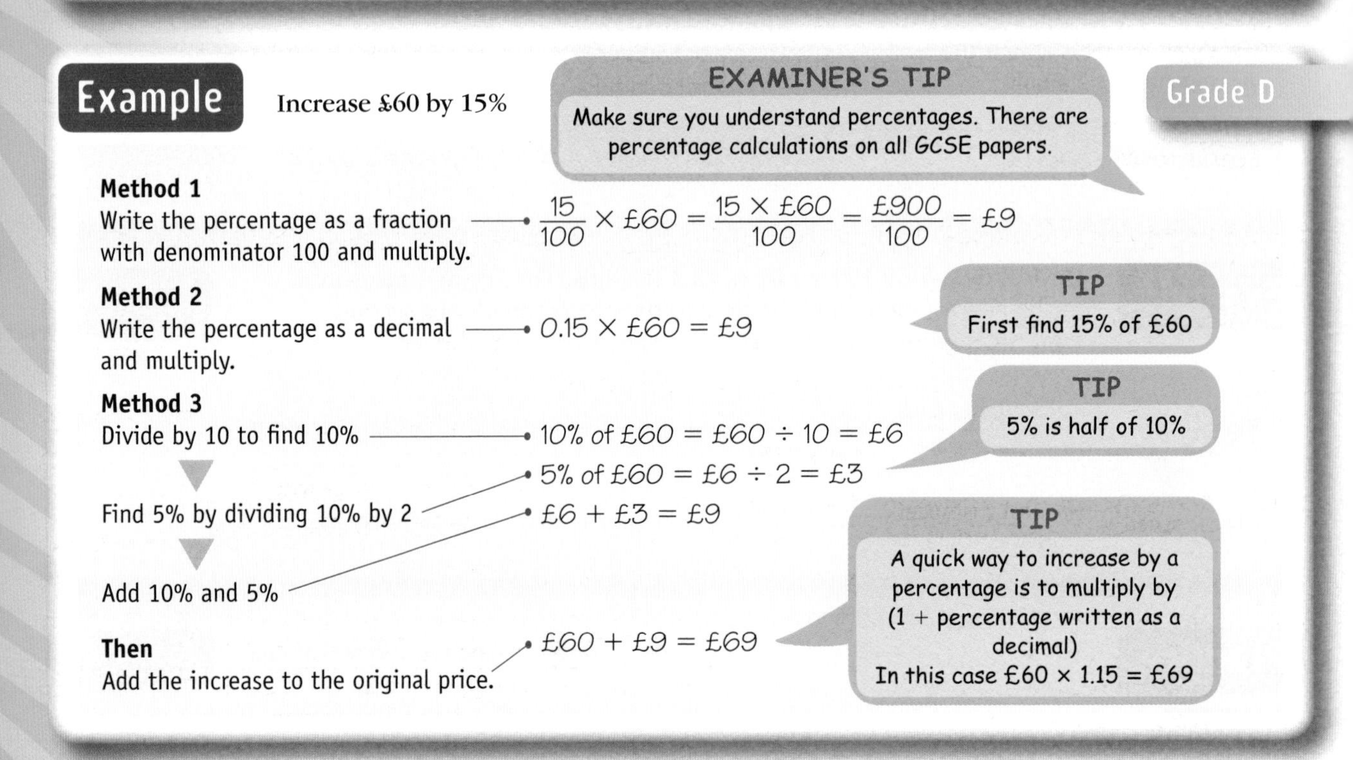

Example Increase £60 by 15% **Grade D**

EXAMINER'S TIP
Make sure you understand percentages. There are percentage calculations on all GCSE papers.

Method 1
Write the percentage as a fraction with denominator 100 and multiply. → $\frac{15}{100} \times £60 = \frac{15 \times £60}{100} = \frac{£900}{100} = £9$

Method 2
Write the percentage as a decimal and multiply. → $0.15 \times £60 = £9$

TIP
First find 15% of £60

Method 3
Divide by 10 to find 10% → 10% of £60 = £60 ÷ 10 = £6

Find 5% by dividing 10% by 2 → 5% of £60 = £6 ÷ 2 = £3

Add 10% and 5% → £6 + £3 = £9

TIP
5% is half of 10%

Then
Add the increase to the original price. → £60 + £9 = £69

TIP
A quick way to increase by a percentage is to multiply by (1 + percentage written as a decimal)
In this case £60 × 1.15 = £69

- Sometimes you are given the end result after a percentage increase or decrease and asked to find the original amount.
  - Write down the end percentage (compared with the original 100%).
  - Divide to find 1%, then multiply by 100 to find the original amount.

Example A watch is reduced by 20% in a sale. Bill buys it for £40
What was the pre-sale price? **Grade B**

Write down the end percentage and put this equal to the sale price. → 100% − 20% = 80%
80% = £40

Find 1% and then 100% → So 1% = 40 ÷ 80 = £0.50
So 100% = £50
The original price was £50

- An **index number** shows how a quantity changes over time.
- A **price index** shows how the price of something changes over time.
  - The index always starts at 100.
  - An index greater than 100 shows a price rise.
  - An index less than 100 shows a price fall.

Key words

index number ☐
price index ☐

Example

Grade C

The table shows the index numbers for the prices of houses in London over a 50-year period.

1960	1970	1980	1990	2000	2010
100	220	300	250	300	350

(a) A house cost £50 000 in 1960. How much would it have cost in 2000?

(b) What can you say about house prices between 1960 and 2000?

Divide the 1960 price by 100 to find 1%. Then multiply by the index number for 2000.

(a) $\frac{£50\,000}{100} \times 300 = £150\,000$

(b) Prices went up until 1980, fell between 1980 and 1990, and then rose again after 1990.

Practice

1 A shop reduces all its prices by 15%. Find the new cost of (Grade D)

(a) a TV that normally costs £90

(b) a DAB radio that normally costs £75

2 Rashmi bought a TV for £240 plus VAT at 17.5% (Grade D)
Work out the total cost of the TV.

TIP

To find $17\frac{1}{2}\%$ of an amount:
find 10%
find 5% ($\frac{1}{2}$ of 10%)
find $2\frac{1}{2}\%$ ($\frac{1}{2}$ of 5%)
Total = $17\frac{1}{2}\%$

3 The table shows the index numbers for computer prices. (Grade C)
In which period was there the biggest reduction in prices?

1960	1970	1980	1990	2000
100	90	80	65	55

4 Jude had a pay increase from £150 to £180. (Grade C)
What percentage increase is this, compared with her original pay?

5 Jade has £400 to invest. The interest rate is 10%

(a) How much interest will she receive after 3 years if simple interest is paid? (Grade D)

(b) How much interest will she receive after 3 years if compound interest is paid? (Grade B)

6 In a '30% off' sale, a coat is reduced to £49. What was the original price? (Grade B)

Check your answers on page 174. For full worked solutions see the CD.
See the Student Book on the CD if you need more help.

Question	1	2	3	4	5a	5b	6
Grade	D	D	C	C	D	B	B
Student Book pages	U3 12–15	U3 12–15	–	U3 17–18	U3 12–15	U3 19–20	U3 15–17

Powers, surds, bounds and percentages: topic test

Check how well you know this topic by answering these questions.
First cover the answers on the facing page.

Test questions

1 Josh buys a games console for £180 plus VAT at $17\frac{1}{2}\%$
Work out the total cost of the games console.

2 A shop reduces all its prices by 15%. Find the new price of
(a) a radio that normally costs £40
(b) a DVD player that normally costs £110

3 The table shows the index numbers for the prices of new cars.

1960	1970	1980	1990	2000	2005
100	350	500	750	900	1000

(a) In which period was there the biggest increase in prices?
(b) A new car cost £1000 in 1960. What would a similar car have cost in 2000?

4 Evan has £300 to invest. The interest rate is 5%
(a) How much interest will he receive after 3 years if simple interest is paid?
(b) How much interest will he receive after 3 years if compound interest is paid?

5 In a '30% off' sale, a pair of trainers are reduced to £56
What was the original price?

6 Rationalise **(a)** $\frac{5}{\sqrt{11}}$ **(b)** $\frac{\sqrt{7}}{\sqrt{21}}$

7 Find the value of **(a)** $16^{\frac{3}{4}}$ **(b)** $16^{-\frac{3}{4}}$ **(c)** $\frac{3^{-2}}{36^{-\frac{3}{2}}}$

8 The squares have sides of 7 cm and 5 cm, each correct to 1 s.f.
Find the upper bound and the lower bound for the shaded area.
Find the shaded area to an appropriate degree of accuracy.

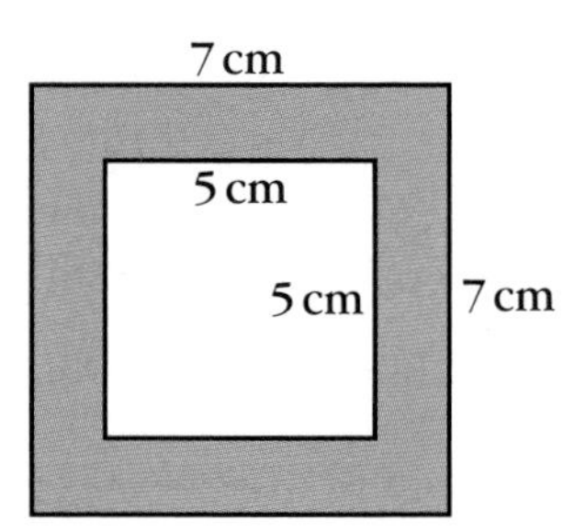

9 The length of the side of a cube is 3.4 cm to 2 s.f.
Calculate the upper bound and the lower bound for
(a) the surface area of the cube
(b) the volume of the cube.

Now check your answers – see the facing page.

Cover this page while you answer the test questions opposite.

Worked answers

Revise this on...

D 1 $17\frac{1}{2}\% = 10\% + 5\% + 2\frac{1}{2}\% = 18 + 9 + 4.50 = £31.50$ page 88

Total price $= 180 + 31.50 = £211.50$

D 2 (a) $15\% = 10\% + 5\% = 4 + 2 = £6$ New price $= 40 - 6 = £34$ page 88

(b) $15\% = £11 + £5.50 = £16.50$ New price $= 110 - 16.50 = £93.50$

C 3 (a) 1960–70 or 1980–90 (b) $\frac{£1000}{100} \times 900 = £9000$ page 89

B 4 (a) 5% of £300 = £15 Total interest = £15 × 3 = £45 page 88

(b) Year 1: Interest = 5% of £300 = £15

New amount = £300 + £15 = £315

Year 2: Interest = 5% of £315 = £15.75

New amount = £315 + £15.75 = £330.75

Year 3: Interest = 5% of £330.75 = £16.54

New amount = £330.75 + £16.54 = £347.29

Total interest = £347.29 − £300 = £47.29

B 5 $70\% = £56 \rightarrow 1\% = £56 \div 70 = £0.80 \rightarrow 100\% = £0.80 \times 100 = £80$ page 88

The original price was £80

A 6 (a) $\frac{5}{\sqrt{11}} \times \frac{\sqrt{11}}{\sqrt{11}} = \frac{5\sqrt{11}}{11}$ page 86

(b) $\frac{\sqrt{7}}{\sqrt{21}} \times \frac{\sqrt{21}}{\sqrt{21}} = \frac{\sqrt{7} \times \sqrt{3 \times 7}}{21} = \frac{7 \times \sqrt{3}}{3 \times 7} = \frac{\sqrt{3}}{3}$

or $\frac{\sqrt{7}}{\sqrt{21}} = \sqrt{\frac{7}{21}} = \sqrt{\frac{1}{3}} = \frac{1}{\sqrt{3}} = \frac{1}{\sqrt{3}} \times \frac{\sqrt{3}}{\sqrt{3}} = \frac{\sqrt{3}}{3}$

A/A* 7 (a) 8 (b) $\frac{1}{8}$ (c) $\frac{3^{-2}}{36^{-\frac{3}{2}}} = \frac{36^{\frac{3}{2}}}{3^2} = \frac{6^3}{9} = \frac{216}{9} = 24$ page 36

A* 8 Upper bound = maximum possible value page 87

= upper bound of area of outside square

− lower bound of area of inside square

$= 7.5 \times 7.5 - 4.5 \times 4.5 = 36\text{ cm}^2$

Lower bound = minimum possible value

= lower bound of area of outside square

− upper bound of area of inside square

$= 6.5 \times 6.5 - 5.5 \times 5.5 = 12\text{ cm}^2$

There is no suitable appropriate value that can be given, not even to 1 s.f.

A* 9 (a) Upper bound $= 6 \times 3.45^2 = 71.415\text{ cm}^2$ page 87

Lower bound $= 6 \times 3.35^2 = 67.335\text{ cm}^2$

(b) Upper bound $= 3.45^3 = 41.063\,625\text{ cm}^3$

Lower bound $= 3.35^3 = 37.595\,375\text{ cm}^3$

Tick the questions you got right.

Question	1	2	3	4	5	6	7a	7bc	7	8	9
Grade	D	D	C	B	B	A	A	A*	B	A*	A8

Mark the grade you are working at on your revision planner on page viii.

Ratio and proportion

- Two quantities are in **direct proportion** if their **ratio** stays the same as the quantities increase or decrease.
- In the **unitary method**, you find the value of *one* item first.

Key words
proportion ☐
ratio ☐
unitary method ☐

Example Grade D

Here is a list of ingredients for making 12 cakes:

200 g sugar	200 g butter	800 g flour
2 eggs	100 g dried fruit	

(a) How many eggs would you need to make 18 cakes?
(b) How many grams of sugar would you need to make 9 cakes?

Look at the ratio of the numbers of cakes. → (a) 12 : 18 ($\times 1\frac{1}{2}$)

12 cakes take 2 eggs
18 cakes take $2 \times 1\frac{1}{2} = 3$ eggs

Use the unitary method. → (b) 12 cakes take 200 grams

1 cake takes $\frac{200}{12}$ grams

9 cakes take $9 \times \frac{200}{12} = 3 \times \frac{200}{4}$
$= 3 \times 50$
$= 150$ grams

TIP
Cancel before multiplying

EXAMINER'S TIP
Changing a recipe for a different number of people is often tested in GCSE maths papers.

- Ratios called **scales** are used to show the relationship between distances on a map and distances on the ground.

Key word
scale ☐

Example Grade D

Two towns are 8.5 cm apart on a map.
The scale of the map is 1 : 50 000
How far apart are the towns in real life?

Write down the scale of the map. → 1 cm represents 50 000 cm

Multiply by the map distance. → 8.5 cm represents $8.5 \times 50\,000$ cm $= 425\,000$ cm

Change to suitable units. → 425 000 cm = 4250 m
4250 m = 4.25 km

TIP
Divide by 100 to change cm to metres.

TIP
Divide by 1000 to change metres to km.

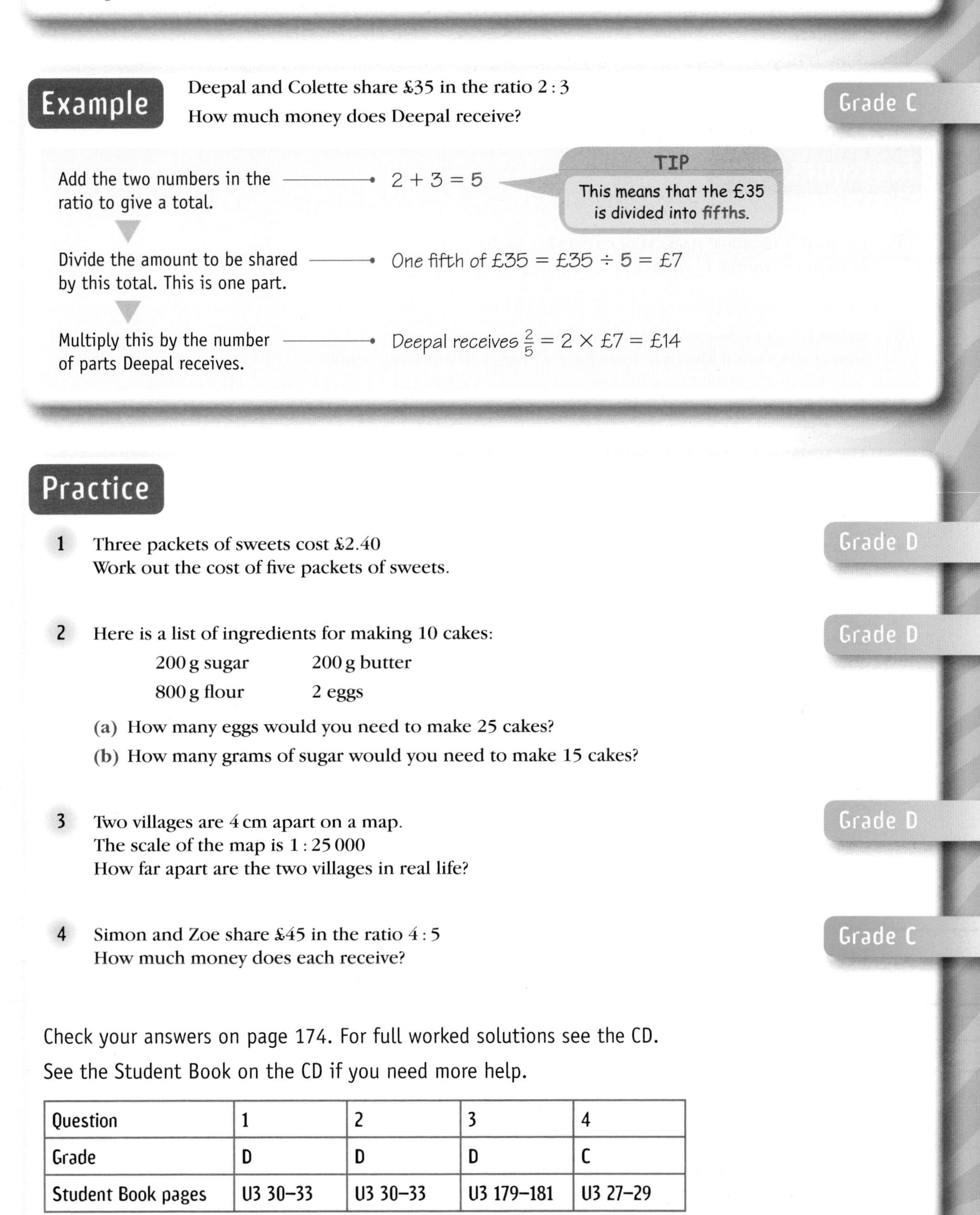

- To share an amount in a given ratio:
  - find the total of all the numbers in the ratio
  - split the amount into fractions with that total as denominator.

Example

Grade C

Deepal and Colette share £35 in the ratio 2 : 3
How much money does Deepal receive?

Add the two numbers in the ratio to give a total. → $2 + 3 = 5$

TIP
This means that the £35 is divided into fifths.

Divide the amount to be shared by this total. This is one part. → One fifth of £35 = £35 ÷ 5 = £7

Multiply this by the number of parts Deepal receives. → Deepal receives $\frac{2}{5} = 2 \times £7 = £14$

Practice

1 Three packets of sweets cost £2.40
Work out the cost of five packets of sweets. — Grade D

2 Here is a list of ingredients for making 10 cakes: — Grade D

200 g sugar	200 g butter
800 g flour	2 eggs

(a) How many eggs would you need to make 25 cakes?

(b) How many grams of sugar would you need to make 15 cakes?

3 Two villages are 4 cm apart on a map.
The scale of the map is 1 : 25 000
How far apart are the two villages in real life? — Grade D

4 Simon and Zoe share £45 in the ratio 4 : 5
How much money does each receive? — Grade C

Check your answers on page 174. For full worked solutions see the CD.
See the Student Book on the CD if you need more help.

Question	1	2	3	4
Grade	D	D	D	C
Student Book pages	U3 30–33	U3 30–33	U3 179–181	U3 27–29

Ratio and proportion: topic test

Check how well you know this topic by answering these questions.
First cover the answers on the facing page.

Test questions

1 Joe buys 7 identical packets of sweets for £9.10
How much would 11 of these packets of sweets cost?

2 Helen bought 6 bottles of drink for £15.60
How much would she have to pay for 9 bottles of the same drink?

3 Two towns are 7.5 cm apart on a map.
The scale of the map is 1 : 50 000
How far apart are the towns in real life?

4 Tom and Shamonti share £3.50 in the ratio 5 : 2
How much money does each of them receive?

5 Rosa prepares the ingredients for pizzas.
She uses cheese, topping and dough in the ratio 2 : 3 : 5
Rosa uses 70 grams of dough.
Work out how many grams of cheese and how many grams of topping she uses.

6 Here is a list of ingredients for making 8 buns:

100 g sugar
100 g butter
300 g flour
2 eggs
60 g dried fruit

(a) How many eggs would you need to make 12 buns?

(b) How many grams of sugar would you need to make 6 buns?

(c) How many buns could you make using 450 g of flour?

7 Two villages are 6 km apart in real life.
How far apart will they be on a map with a scale of 1 : 25 000?

Now check your answers – see the facing page.

Cover this page while you answer the test questions opposite.

Worked answers

Revise this on...

D **1** £9.10 ÷ 7 = £1.30
£1.30 × 11 = £14.30 — page 92

D **2** £15.60 ÷ 6 = £2.60
£2.60 × 9 = £23.40 — page 92

D **3** 7.5 cm × 50 000 = 375 000 cm
375 000 cm = 3750 m = 3.75 km — page 92

C **4** 5 + 2 = 7 parts = £3.50
→ 1 part = £3.50 ÷ 7 = 50p
Tom has 5 parts = 5 × 50p = £2.50
Shamonti has 2 parts = 2 × 50p = £1 — page 93

C **5** Dough is 5 parts so 5 parts = 70 g
→ 1 part = 70 ÷ 5 = 14 g
Cheese = 2 parts = 2 × 14 = 28 g
Topping = 3 parts = 3 × 14 = 42 g — page 93

C **6** (a) 8 : 12 ($\times 1\frac{1}{2}$)
8 buns take 2 eggs, so 12 buns take $2 \times 1\frac{1}{2} = 3$ eggs — page 92

(b) 8 : 6 ($\times \frac{3}{4}$)
8 buns take 100 g, so 6 buns take $100 \times \frac{3}{4} = 75$ g

(c) 300 g : 450 g ($\times 1\frac{1}{2}$)
300 g makes 8 buns, so 450 g will make $8 \times 1\frac{1}{2} = 12$ buns

C **7** 6 km = 6000 m = 600 000 cm
600 000 ÷ 25 000 = 24 cm — page 92

Tick the questions you got right.

Question	1	2	3	4	5	6	7
Grade	D	D	D	C	C	C	C

Mark the grade you are working at on your revision planner on page viii.

Number: subject test

Use these questions to check that you understand the key facts for Number, before you try the Examination Practice Paper on pages 160–163.

Exam practice questions

1 Find 15% of £60

2 Bill's garage bill was £180 plus VAT.
Work out the VAT at $17\frac{1}{2}\%$ on £180

3 Ronni bought an MP3 player for £60 plus VAT at $17\frac{1}{2}\%$.
Work out the total cost of the MP3 player.

4 Jason mixes bags of gravel and cement in the ratio 6 : 1 to make concrete.
Jason uses 5 bags of cement.

(a) How many bags of gravel does he need?

Jason uses 48 bags of gravel.

(b) How many bags of cement does he need?

5 Rosa prepares the ingredients for pizzas.
She uses cheese, topping and dough in the ratio 2 : 3 : 5.
Rosa uses 70 grams of dough.
Work out the number of grams of cheese and the number of grams of topping Rosa uses.

6 Ron and Shanna share £4.50 in the ratio 5 : 4
How much does each of them receive?

7 Derek shares £300 with his three children Amy, Ben and Clive.
He shares the money in the ratio of their ages.
Amy is 3, Ben is 5 and Clive is 7.
Work out the amount he gives each of his children.

8 What is the total amount saved when £400 is invested for two years at a compound interest rate of 5%?

9 Rhiannon buys a second-hand car for £8000.
The car depreciates at 10% each year.
What is the value of the car after 3 years?

10 Jade invests £300 at an interest rate of 6%.

(a) How much interest will Jade receive after 3 years simple interest?

(b) How much interest will Jade receive at compound interest after 3 years

11 A coat is reduced in price to £60 in a sale with 20% off.
What was the original price?

12 Work out

(a) $3.7 \times 10^3 \times 4.9 \times 10^5$

(b) $1.8 \times 10^3 \div (7.2 \times 10^7)$

(c) $3.2 \times 10^{-4} \times 4.8 \times 10^{-3}$

(d) $4.62 \times 10^{-2} \div (6.4 \times 10^{-6})$

13 Find the value of

(a) $81^{\frac{3}{4}}$

(b) $16^{\frac{3}{2}}$

(c) $8^{-\frac{2}{3}}$

(d) $\frac{2}{27^{-\frac{2}{3}}}$

14 Rationalise the denominator of these fractions

(a) $\frac{5}{\sqrt{17}}$

(b) $\frac{\sqrt{7}}{\sqrt{35}}$

15 $x = 4$ m to the nearest metre and $y = 5.60$ m to the nearest 10 cm.
Calculate the upper and lower bound of

(a) $x + y$

(b) $x - y$

(c) xy

(d) $\frac{x}{y}$

16 Leah ran the 100 m race in 12 seconds.
The distance was measured to the mearest centimetre and the time to the nearest thousandth of a second.
Calculate the upper and lower bound for her speed in metres per second.

Check your answers on page 174. For full worked solutions see the CD.
Tick the questions you got right.

Question	1	2	3	4	5	6	7	8	9	10a	10b	11	12	13	14	15	16
Grade	D	D	D	C	C	C	C	C	C	D	B	B	A	A	A	A	A*
Revise this on pages	88	88	88	93	93	93	93	89	89	88		88	86	86–87	88	86	87

Mark the grade you are working at on your revision planner on page viii.
Go to the pages shown to revise for the ones you got wrong.

Graphs

- $y = 2x + 1$ is the **equation** of a **straight line**. The **coordinates** of any point on the straight line 'satisfy' the equation.
- $x \to 2x + 1$ is called a **linear function** because its graph is a straight line.
- An **intercept** is a point at which a line cuts the y-axis or the x-axis.
- **Gradient** $= \dfrac{\text{change in } y\text{-direction}}{\text{change in } x\text{-direction}}$
- The straight line with equation $y = mx + c$ has a gradient of m and its intercept on the y-axis is $(0, c)$.

Key words

line	☐	parallel	☐
intercept	☐	perpendicular	☐
gradient	☐	coordinates	☐

- Lines with equation $y = mx + c$, with the same gradient (m), are **parallel**.
- If a line has a gradient of m, a line **perpendicular** to it has a gradient of $-\frac{1}{m}$.

Example

(a) Draw the line with equation $y = 3x - 2$ for values of x from -2 to 4. Grade C

(b) Another line is parallel to this one, and passes through the point (2, 10). Find the equation of this line. Grade B

(c) A different line is perpendicular to the line $y = 3x - 2$ and crosses the y-axis at (0, 12). Find the equation of this line. Grade A

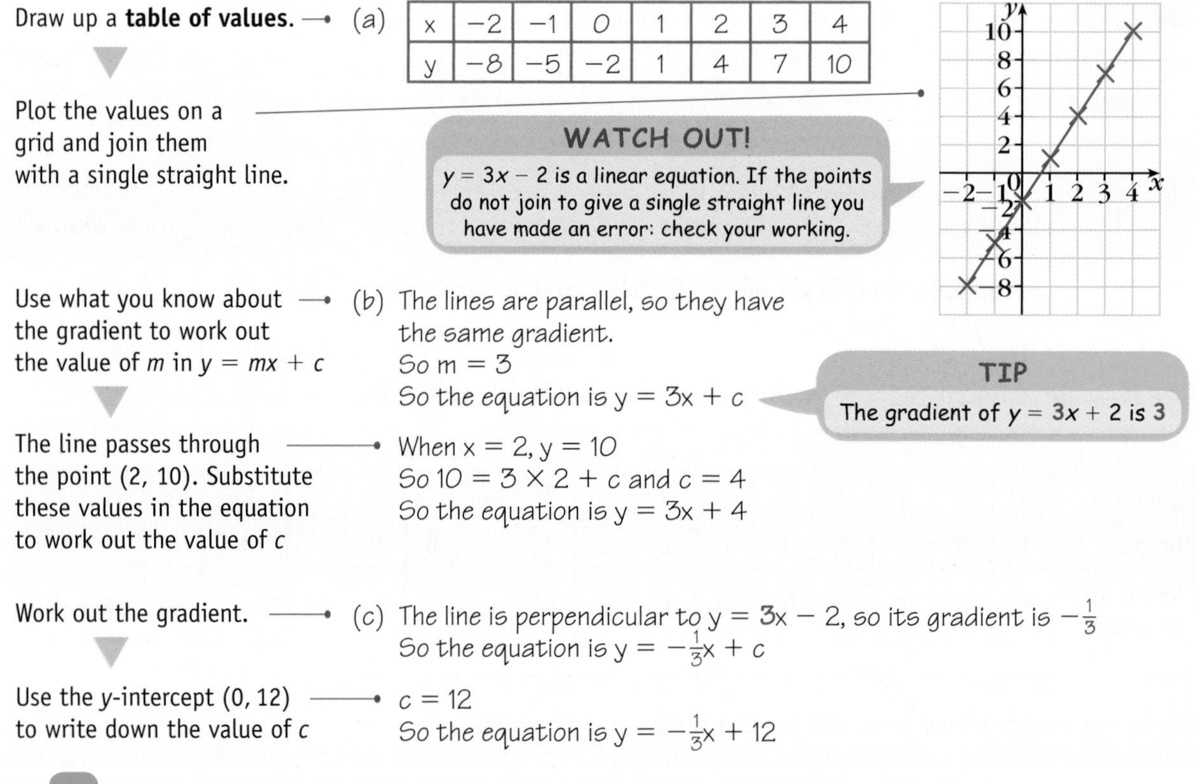

Draw up a **table of values.** → (a)

x	−2	−1	0	1	2	3	4
y	−8	−5	−2	1	4	7	10

Plot the values on a grid and join them with a single straight line.

WATCH OUT!
$y = 3x - 2$ is a linear equation. If the points do not join to give a single straight line you have made an error: check your working.

Use what you know about the gradient to work out the value of m in $y = mx + c$ → (b) The lines are parallel, so they have the same gradient.
So $m = 3$
So the equation is $y = 3x + c$

TIP
The gradient of $y = 3x + 2$ is 3

The line passes through the point (2, 10). Substitute these values in the equation to work out the value of c → When $x = 2, y = 10$
So $10 = 3 \times 2 + c$ and $c = 4$
So the equation is $y = 3x + 4$

Work out the gradient. → (c) The line is perpendicular to $y = 3x - 2$, so its gradient is $-\frac{1}{3}$
So the equation is $y = -\frac{1}{3}x + c$

Use the y-intercept (0, 12) to write down the value of c → $c = 12$
So the equation is $y = -\frac{1}{3}x + 12$

Example

Write down the gradient and y-intercept of the line $2y = 3x - 4$

Grade B

Write the equation in the form $y = mx + c$

$2y = 3x - 4$

$y = \frac{3}{2}x - 2$ Divide both sides by 2

So $m = \frac{3}{2}$ and $c = -2$

The gradient is $\frac{3}{2}$ and the y-intercept is $(0, -2)$

Example

Point A is (1, 1). Point B is (3, 9).
Find the equation of the line that passes through points A and B.

Grade A

Use $y = mx + c$

$x = 1$ and $y = 1$ gives $1 = m + c$

$x = 3$ and $y = 9$ gives $9 = 3m + c$

Solving the simultaneous equations:

$8 = 2m$ so $m = 4$

$1 = 4 + c$ so $c = -3$

The equation is $y = 4x - 3$

Practice

Grade C

1 Draw the graph of $y = 3x + 2$ for values of x from $x = -3$ to $x = 2$.

Grade B

2 A line is parallel to the line $y = 4x - 2$ and passes through the point (3, 17). Find the equation of the line.

Grade B

3 A line passes though the points $A(2, 4)$ and $B(4, -7)$.
Find the equation of the line.

Grade A

4 A line is perpendicular to the line $y = 5x + 2$ and passes through the point (0, 4). Find the equation of the line.

Check your answers on page 174. For full worked solutions see the CD.
See the Student Book on the CD if you need more help.

Question	1	2	3	4
Grade	C	B	B	A
Student Book pages	U3 81–83	U3 81–83	U3 81–83	U3 83–85

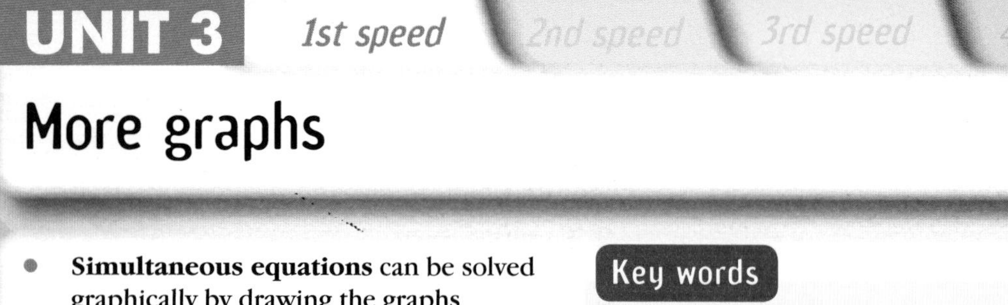

More graphs

- **Simultaneous equations** can be solved graphically by drawing the graphs for the two equations and finding the **coordinates** of their **point of intersection**.

Key words

simultaneous equations ☐ point of intersection ☐
coordinates ☐ solution ☐

Example

Grade B

Use the diagram to solve the simultaneous equations

$$y = 3x - 2$$
$$x + y = 6$$

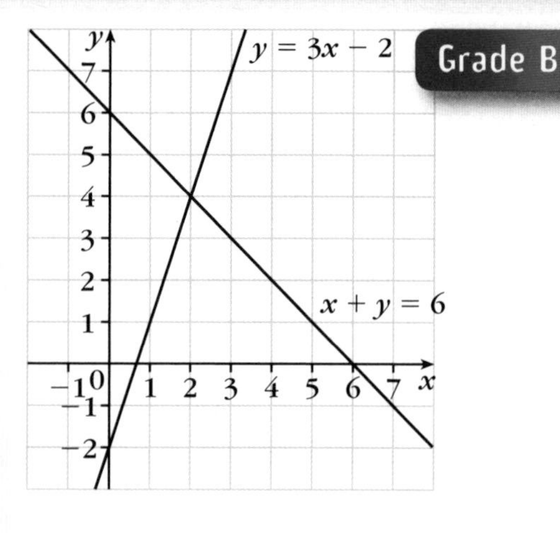

Read off the coordinates from the point of intersection. → The point of intersection is at (2, 4). So the **solution** is $x = 2$, $y = 4$

EXAMINER'S TIP

If you are given a diagram, do not attempt to solve the simultaneous equations algebraically. This can be done but it takes much longer.

Example

Grade C

The diagram shows an empty bottle.

Water is poured into it at a steady rate.

Draw a sketch graph to show the relationship between the water level and the volume of water added.

Water level

Volume added

The bottom of the bottle is spherical so the part of the graph showing the sphere being filled should be symmetrical. The bottle fills at its slowest when the water reaches the widest part of the sphere.

When the neck is reached the level rises faster and at a constant rate.

Water level

Volume added

TIP

Think about what would happen if you were to do this in real life.

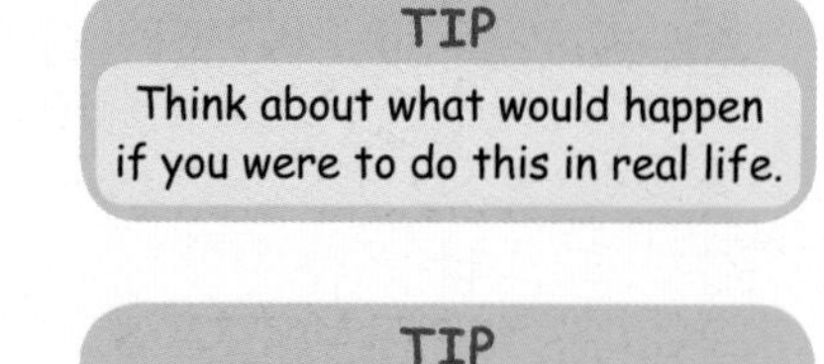

TIP

An increase at a constant rate is represented by a straight line.

- On a **distance–time graph**, the gradient gives the speed.
- On a **speed–time graph**, the gradient gives the acceleration. When the graph is a curve, draw a tangent and find its gradient.

Key words

time ☐ travel graph ☐
distance ☐ average speed ☐

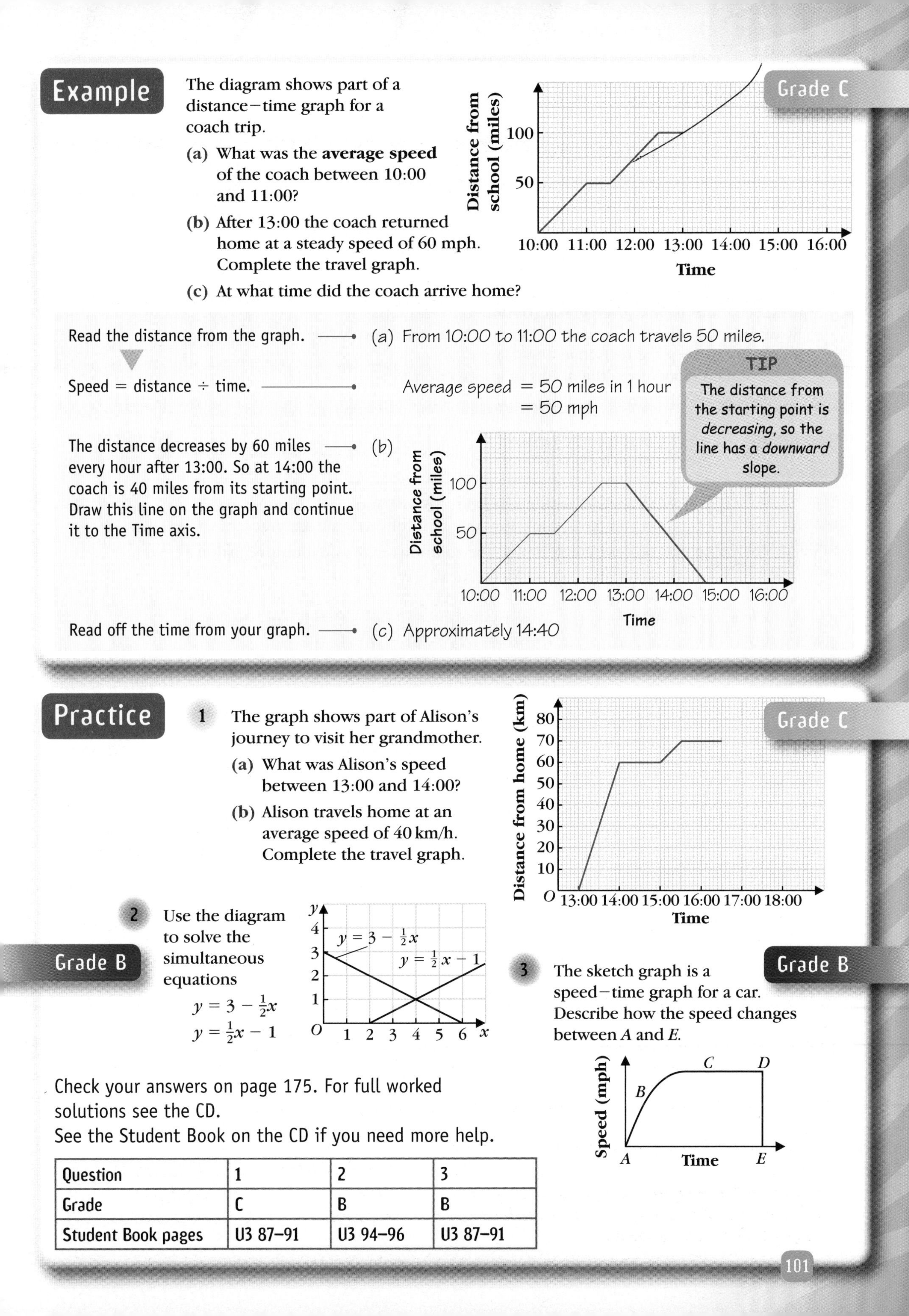

Example

Grade C

The diagram shows part of a distance–time graph for a coach trip.

(a) What was the **average speed** of the coach between 10:00 and 11:00?

(b) After 13:00 the coach returned home at a steady speed of 60 mph. Complete the travel graph.

(c) At what time did the coach arrive home?

Read the distance from the graph. → (a) From 10:00 to 11:00 the coach travels 50 miles.

Speed = distance ÷ time. → Average speed = 50 miles in 1 hour = 50 mph

The distance decreases by 60 miles every hour after 13:00. So at 14:00 the coach is 40 miles from its starting point. Draw this line on the graph and continue it to the Time axis. → (b)

TIP

The distance from the starting point is *decreasing*, so the line has a *downward* slope.

Read off the time from your graph. → (c) Approximately 14:40

Practice

Grade C

1 The graph shows part of Alison's journey to visit her grandmother.

(a) What was Alison's speed between 13:00 and 14:00?

(b) Alison travels home at an average speed of 40 km/h. Complete the travel graph.

Grade B

2 Use the diagram to solve the simultaneous equations

$y = 3 - \frac{1}{2}x$

$y = \frac{1}{2}x - 1$

Grade B

3 The sketch graph is a speed–time graph for a car. Describe how the speed changes between A and E.

Check your answers on page 175. For full worked solutions see the CD.
See the Student Book on the CD if you need more help.

Question	1	2	3
Grade	C	B	B
Student Book pages	U3 87–91	U3 94–96	U3 87–91

Curved graphs

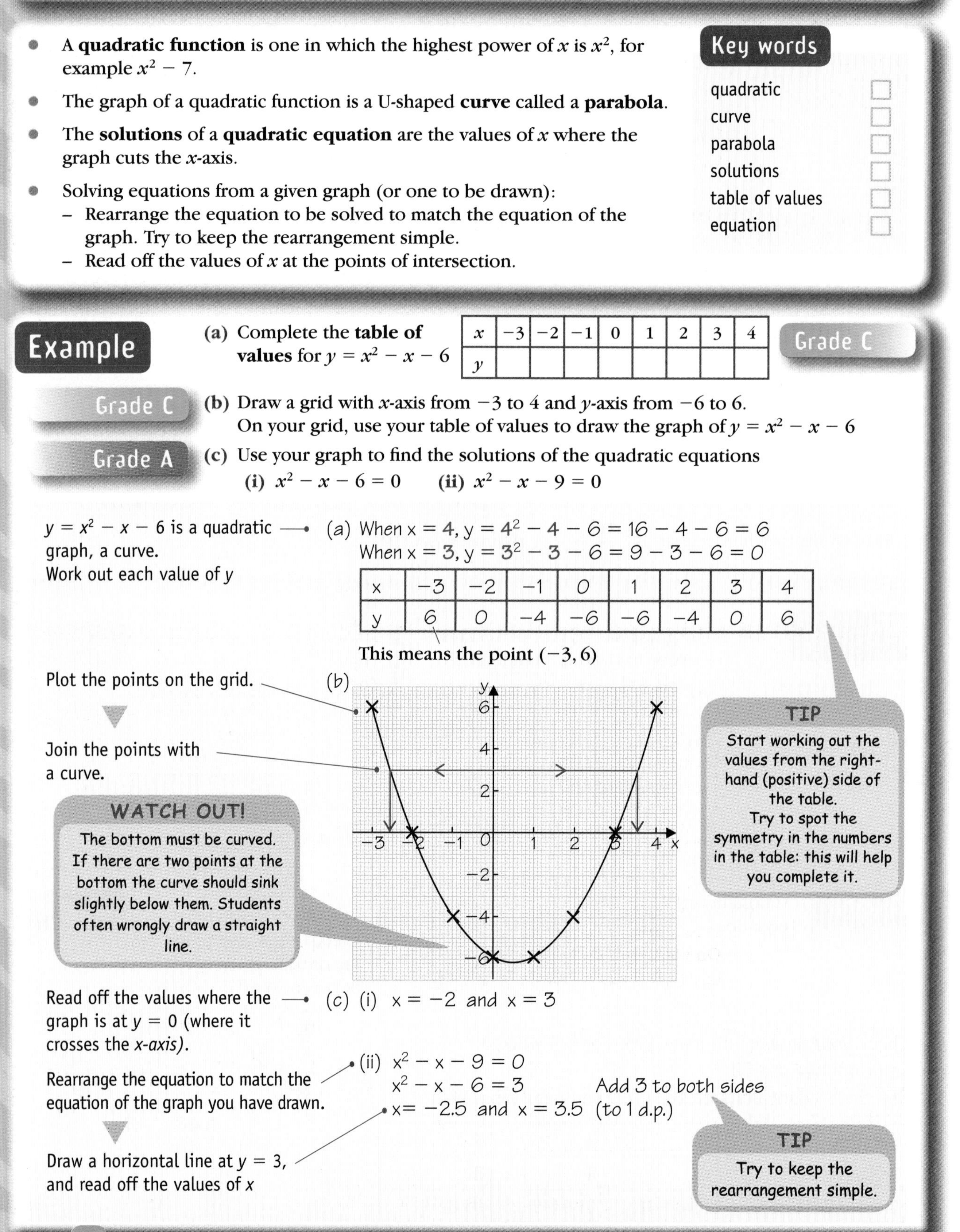

- A **quadratic function** is one in which the highest power of x is x^2, for example $x^2 - 7$.
- The graph of a quadratic function is a U-shaped **curve** called a **parabola**.
- The **solutions** of a **quadratic equation** are the values of x where the graph cuts the x-axis.
- Solving equations from a given graph (or one to be drawn):
  - Rearrange the equation to be solved to match the equation of the graph. Try to keep the rearrangement simple.
  - Read off the values of x at the points of intersection.

Key words

- quadratic
- curve
- parabola
- solutions
- table of values
- equation

Example

Grade C

(a) Complete the **table of values** for $y = x^2 - x - 6$

x	−3	−2	−1	0	1	2	3	4
y								

Grade C

(b) Draw a grid with x-axis from −3 to 4 and y-axis from −6 to 6.
On your grid, use your table of values to draw the graph of $y = x^2 - x - 6$

Grade A

(c) Use your graph to find the solutions of the quadratic equations
(i) $x^2 - x - 6 = 0$ (ii) $x^2 - x - 9 = 0$

$y = x^2 - x - 6$ is a quadratic graph, a curve. Work out each value of y

(a) When $x = 4$, $y = 4^2 - 4 - 6 = 16 - 4 - 6 = 6$
When $x = 3$, $y = 3^2 - 3 - 6 = 9 - 3 - 6 = 0$

x	−3	−2	−1	0	1	2	3	4
y	6	0	−4	−6	−6	−4	0	6

This means the point (−3, 6)

TIP
Start working out the values from the right-hand (positive) side of the table.
Try to spot the symmetry in the numbers in the table: this will help you complete it.

Plot the points on the grid.

Join the points with a curve.

(b)

WATCH OUT!
The bottom must be curved. If there are two points at the bottom the curve should sink slightly below them. Students often wrongly draw a straight line.

Read off the values where the graph is at $y = 0$ (where it crosses the x-axis).

(c) (i) $x = -2$ and $x = 3$

Rearrange the equation to match the equation of the graph you have drawn.

(ii) $x^2 - x - 9 = 0$
$x^2 - x - 6 = 3$ Add 3 to both sides
$x = -2.5$ and $x = 3.5$ (to 1 d.p.)

Draw a horizontal line at $y = 3$, and read off the values of x

TIP
Try to keep the rearrangement simple.

- A **cubic function** is one in which the highest power of x is x^3, for example $x^3 - 3x + 2$.

Key words

cubic ☐

Example

Grade B

(a) Complete the table of values for $y = x^3 - 4x$.

x	-3	-2	-1	0	1	2	3
y							

(b) Draw a grid with x-axis from -3 to 3 and y-axis from -20 to 20.
On your grid, use your table of values to draw the graph of $y = x^3 - 4x$.

$y = x^3 - 4x$ is a cubic graph, a curve. Work out each value of y

(a) When $x = 3$, $y = 3^3 - 4 \times 3 = 27 - 12 = 15$
When $x = 2$, $y = 2^3 - 4 \times 2 = 8 - 8 = 0$

x	−3	−2	−1	0	1	2	3
y	−15	0	3	0	−3	0	15

Plot the table values on the grid.

Join the points with a curve.

(b)

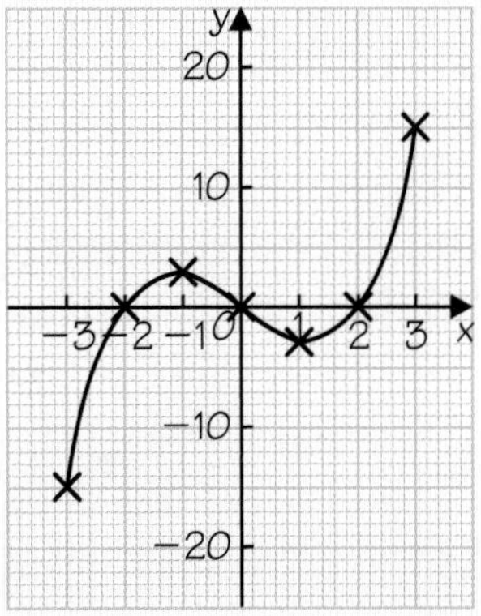

y
20
10
−3 −2 −1 0 2 3 x
−10
−20

Practice

Grade B

1 (a) Complete the table of values for $y = x^3 - 6x^2 + 9x$

x	0	1	2	3	4
y					

(b) Draw a grid with x- and y-axes from -1 to 4.
On your grid, draw the graph of $y = x^3 - 6x^2 + 9x$

2 (a) Complete the table of values for $y = 2x^2 - x - 10$ Grade C

x	-3	-2	-1	0	1	2	3
y							

(b) Draw a grid with x-axis from -3 to 3 and y-axis from -10 to 15. On your grid, draw the graph of $y = 2x^2 - x - 10$ Grade C

(c) Use your graph to find the solutions of the quadratic equations

(i) $2x^2 - x - 10 = 0$ Grade B

(ii) $2x^2 - x - 3 = 0$ Grade A

Check your answers on page 175. For full worked solutions see the CD.
See the Student Book on the CD if you need more help.

Question	1	2ab	2c i	2c ii
Grade	B	C	B	A
Student Book pages	U3 115–117	U3 115–117	U3 117–119	U3 119–124

Transformations of graphs

Key words

function	☐	vertex	☐
translation	☐	sine	☐
reflection	☐	cosine	☐
stretch	☐	solution	☐
scale factor	☐		

- For any **function** f, the graph of $y = f(x + a) + b$ is the graph of $y = f(x)$ **translated** a units horizontally (in the *negative* x-direction if $a > 0$, in the *positive* x-direction if $a < 0$) followed by a translation of b units vertically (*upwards* if $b > 0$, *downwards* if $b < 0$).

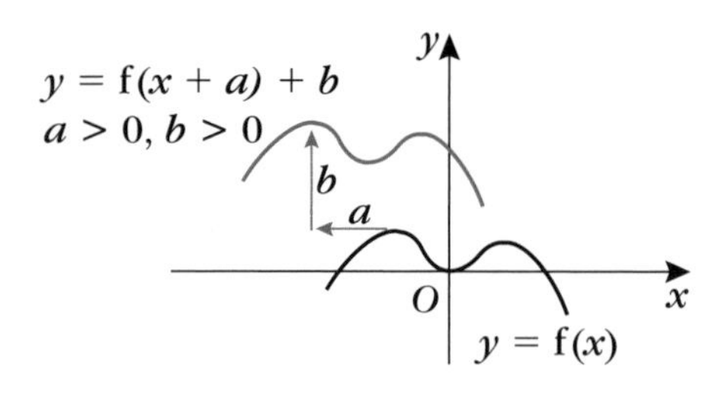

- For any function f, the graph of $y = f(-x)$ is obtained by **reflecting** $y = f(x)$ in the y-axis.

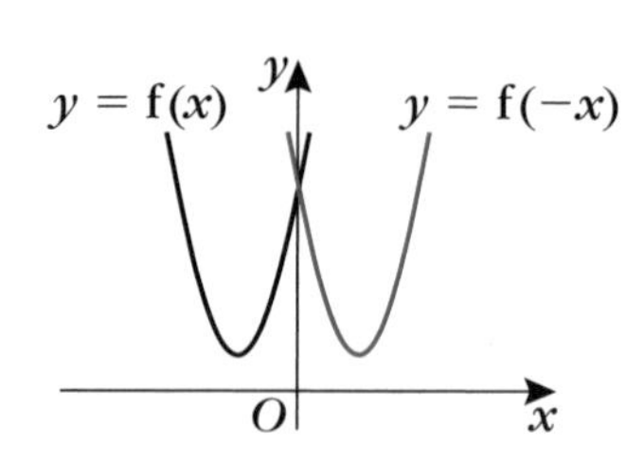

- For any function f, the graph of $y = -f(x)$ is obtained by **reflecting** $y = f(x)$ in the x-axis.

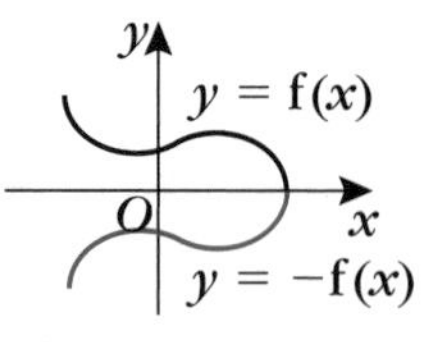

- For any function f, the graph of $y = af(x)$, where a is a positive constant, is obtained from $y = f(x)$ by applying a **stretch** of **scale factor** a parallel to the y-axis.

- For any function f, the graph of $y = f(ax)$, where a is a positive constant, is obtained from $y = f(x)$ by applying a **stretch** of **scale factor** $\frac{1}{a}$ parallel to the x-axis.

Example

Grade A*

The diagram shows a sketch of the function $y = f(x)$, with a **vertex** at the point $(-2, 1)$.

Write down the coordinates of the vertex of

(a) $y = f(x + 4)$ **(b)** $y = -f(x)$ **(c)** $y = 3f(x)$

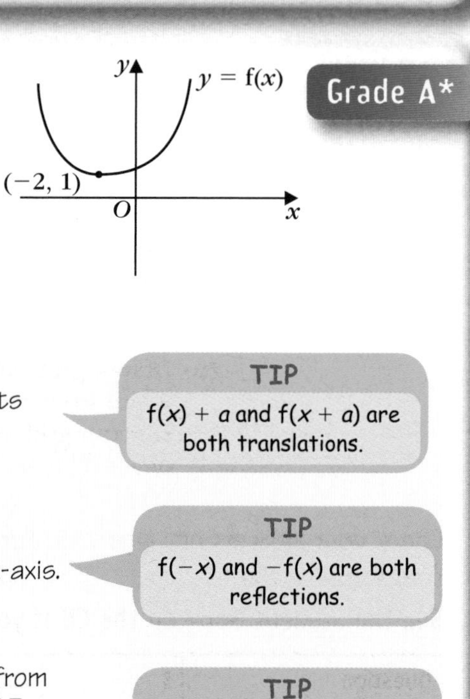

TIP
If you need to, sketch the new graph to help you judge where the vertex moves.

$y = f(x + 4)$ is the graph of $y = f(x)$ translated 4 units horizontally in the negative x-direction.

(a) The vertex is translated 4 units horizontally.
So $(-2, 1)$ becomes $(-6, 1)$

TIP
$f(x) + a$ and $f(x + a)$ are both translations.

$y = -f(x)$ is the graph of $y = f(x)$ reflected in the x-axis.

(b) The vertex is reflected in the x-axis.
So $(-2, 1)$ becomes $(-2, -1)$

TIP
$f(-x)$ and $-f(x)$ are both reflections.

$y = 3f(x)$ is obtained from $y = f(x)$ by applying a stretch of scale factor 3 parallel to the y-axis.

(c) The vertex is stretched away from the x-axis by a scale factor of 3.
So $(-2, 1)$ becomes $(-2, 3)$

TIP
$af(x)$ and $f(ax)$ are both stretches.

Example

Grade A*

This is a sketch graph of $y = f(x)$, where $f(x) = \cos x$, for $0° \leq x \leq 360°$

(a) Sketch the graph of $y = 2\cos x$ for $0° \leq x \leq 360°$

(b) How many **solutions** of the equation $2\cos x = 1$ lie within the range $0° \leq x \leq 360°$?

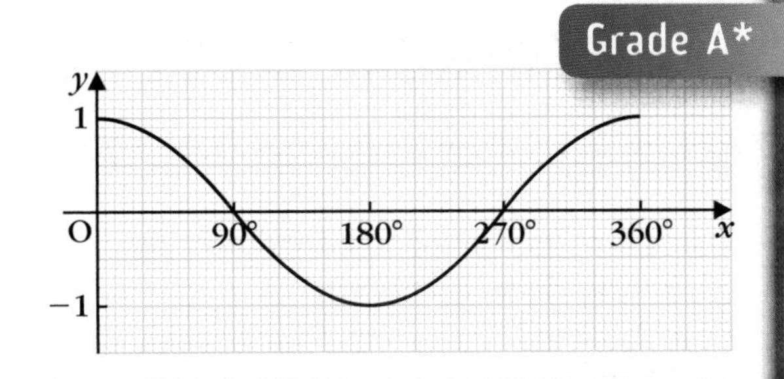

Apply a stretch parallel to the y-axis of scale factor 2. → (a) $y = 2\cos x$ is $y = 2f(x)$

Draw a line on your graph at $y = 1$. How many intersections are there in this range? → (b) $2\cos x = 1$ has two solutions in the range $0° \leq x \leq 360°$

Practice

Grade A*

1 The diagram shows a sketch of the function $y = f(x)$, with a vertex at the point $(2, 3)$. Write down the coordinates of the vertex of

(a) $y = f(x) - 5$ **(b)** $y = f(-x)$

(c) $y = f(4x)$

Grade A*

2 The diagram shows a sketch of $y = f(x)$, where $f(x) = \sin x$, for $-360° \leq x \leq 360°$

Sketch the graph of $y = 2f(\frac{1}{2}x)$

Check your answers on page 175. For full worked solutions see the CD.
See the Student Book on the CD if you need more help.

Question	1	2
Grade	A*	A*
Student Book pages	U3 147–153	U3 147–153

Graphs: topic test

Check how well you know this topic by answering these questions.
First cover the answers on the facing page.

Test questions

1 This diagram shows part of a train journey.

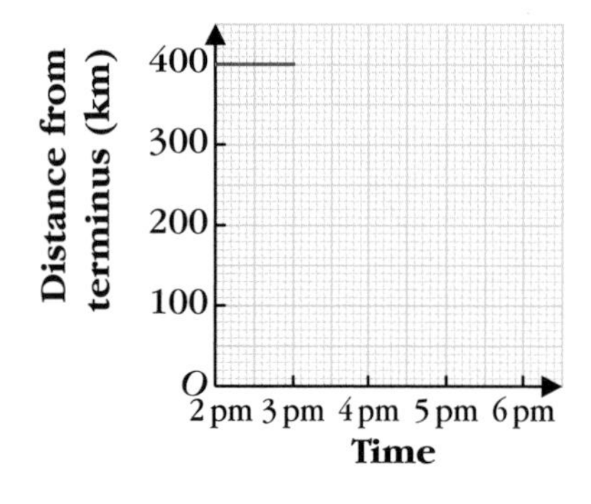

At 3 pm the train is 400 km from the terminus.
It then travels at an average speed of 150 km/h.

(a) Complete the travel graph.

(b) At what time did the train arrive at the terminus?

2 Use the diagram to write down the solution to the simultaneous equations

$2y = x + 2$
$y = -2x + 6$

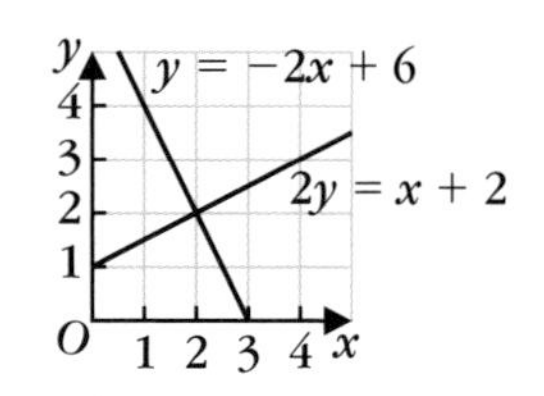

Now check your answers – see the facing page.

3 The line **L** passes through the points (0, 3) and (2, 9).

(a) Find the equation of the line.

(b) Write down the equation of the line that is parallel to **L** and passes through the point (0, 5).

(c) Write down the equation of the line that is perpendicular to **L** and passes through the point (0, 5).

4 (a) Complete the table of values for $y = x^2 - 3x - 2$

x	-1	0	1	2	3	4
y						

(b) Draw a grid with x-axis from -1 to 4 and y-axis from -5 to 2. On your grid, draw the graph of $y = x^2 - 3x - 2$

(c) Use your graph to find the solutions of the equations

(i) $x^2 - 3x - 2 = 0$

(ii) $x^2 - 3x - 4 = 0$

5 This is a sketch graph of $y = f(x)$.

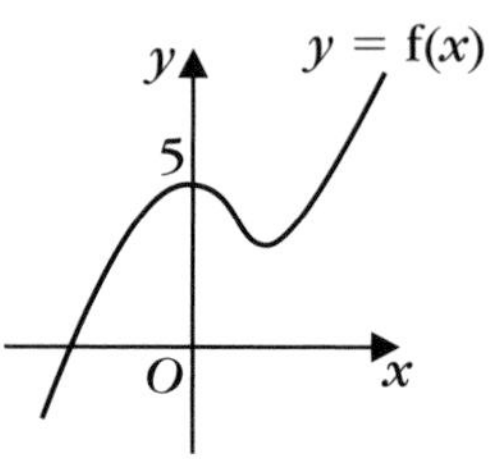

(a) Sketch the graph of $y = f(x) - 6$

(b) Write down the coordinates of the point where $y = f(x) - 6$ crosses the y-axis.

Cover this page while you answer the test questions opposite.

Worked answers

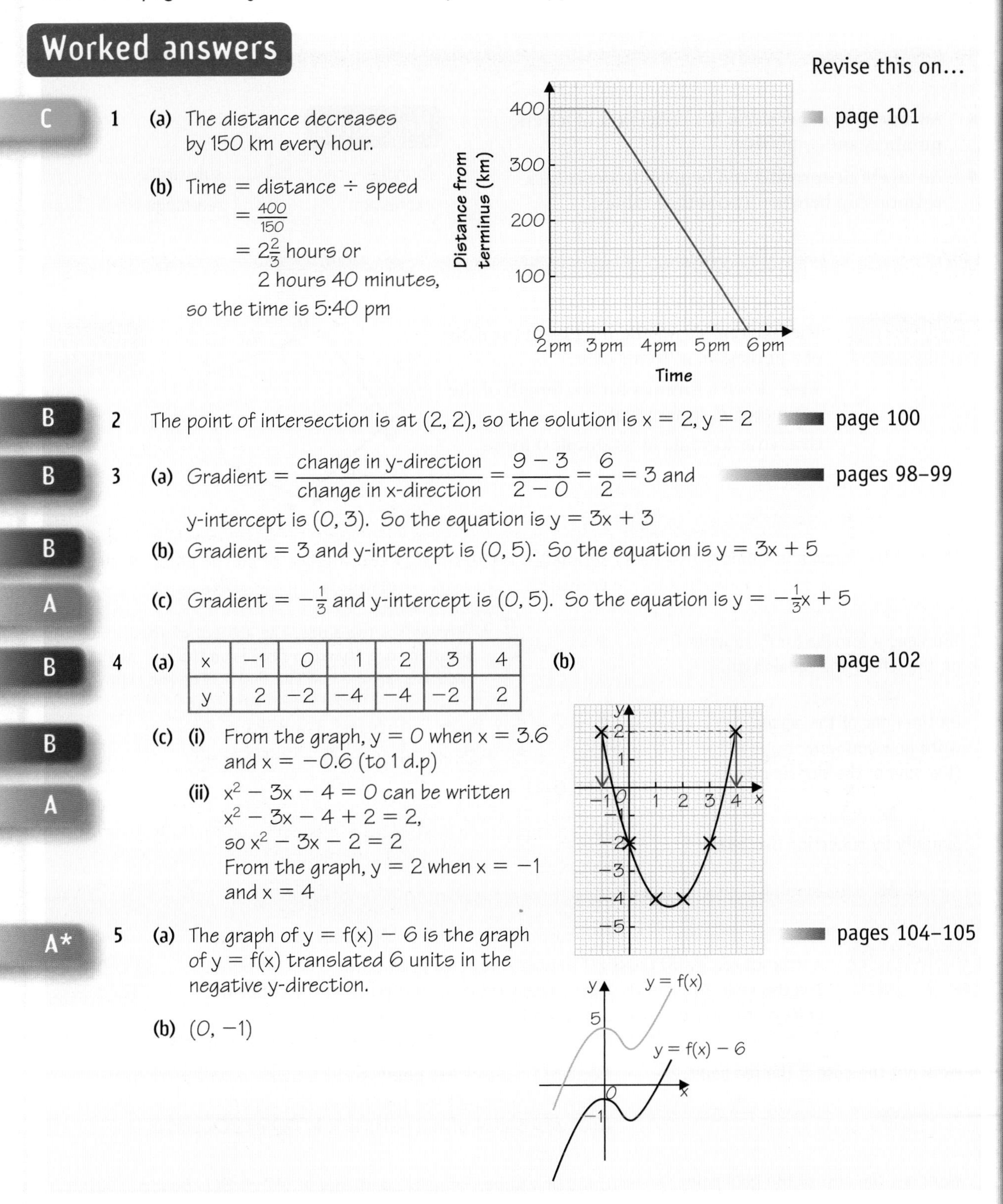

Revise this on...

C

1 **(a)** The distance decreases by 150 km every hour.

(b) Time = distance ÷ speed

$= \frac{400}{150}$

$= 2\frac{2}{3}$ hours or 2 hours 40 minutes,

so the time is 5:40 pm

page 101

B

2 The point of intersection is at (2, 2), so the solution is $x = 2$, $y = 2$ page 100

B

3 **(a)** Gradient $= \frac{\text{change in y-direction}}{\text{change in x-direction}} = \frac{9-3}{2-0} = \frac{6}{2} = 3$ and y-intercept is (0, 3). So the equation is $y = 3x + 3$ pages 98–99

B

(b) Gradient = 3 and y-intercept is (0, 5). So the equation is $y = 3x + 5$

A

(c) Gradient $= -\frac{1}{3}$ and y-intercept is (0, 5). So the equation is $y = -\frac{1}{3}x + 5$

B

4 **(a)**

x	−1	0	1	2	3	4
y	2	−2	−4	−4	−2	2

(b)

page 102

B

(c) **(i)** From the graph, $y = 0$ when $x = 3.6$ and $x = -0.6$ (to 1 d.p)

A

(ii) $x^2 - 3x - 4 = 0$ can be written $x^2 - 3x - 4 + 2 = 2$, so $x^2 - 3x - 2 = 2$

From the graph, $y = 2$ when $x = -1$ and $x = 4$

A*

5 **(a)** The graph of $y = f(x) - 6$ is the graph of $y = f(x)$ translated 6 units in the negative y-direction.

(b) (0, −1)

pages 104–105

Tick the questions you got right.

Question	1	2	3ab	3c	4abc i	4c ii	5
Grade	C	B	B	A	B	A	A*

Mark the grade you are working at on your revision planner on page ix.

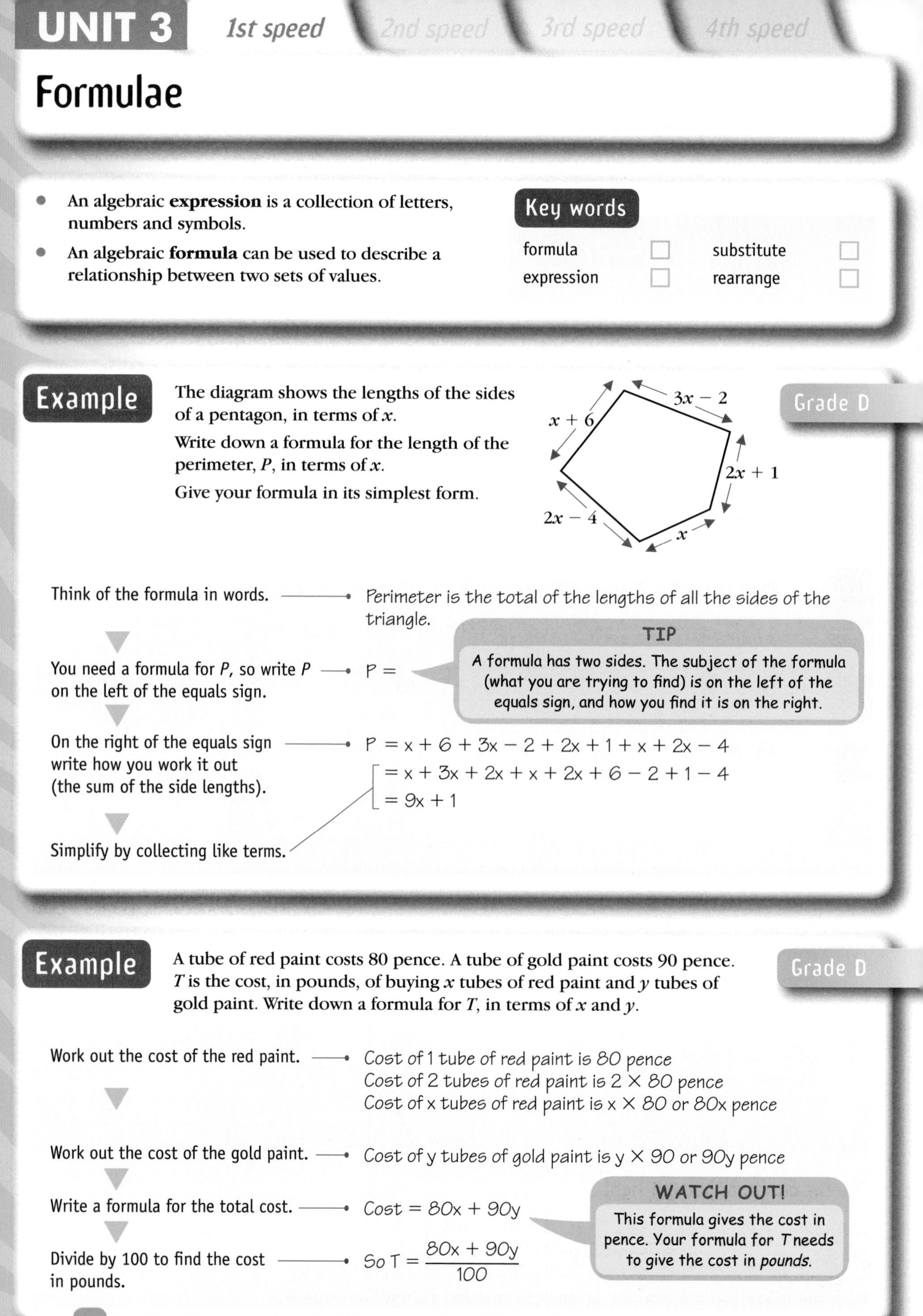

Formulae

- An algebraic **expression** is a collection of letters, numbers and symbols.
- An algebraic **formula** can be used to describe a relationship between two sets of values.

Key words

formula	☐	substitute	☐
expression	☐	rearrange	☐

Example

Grade D

The diagram shows the lengths of the sides of a pentagon, in terms of x.

Write down a formula for the length of the perimeter, P, in terms of x.

Give your formula in its simplest form.

Think of the formula in words. → Perimeter is the total of the lengths of all the sides of the triangle.

You need a formula for P, so write P on the left of the equals sign. → $P =$

TIP
A formula has two sides. The subject of the formula (what you are trying to find) is on the left of the equals sign, and how you find it is on the right.

On the right of the equals sign write how you work it out (the sum of the side lengths). → $P = x + 6 + 3x - 2 + 2x + 1 + x + 2x - 4$

Simplify by collecting like terms. → $= x + 3x + 2x + x + 2x + 6 - 2 + 1 - 4$

$= 9x + 1$

Example

Grade D

A tube of red paint costs 80 pence. A tube of gold paint costs 90 pence. T is the cost, in pounds, of buying x tubes of red paint and y tubes of gold paint. Write down a formula for T, in terms of x and y.

Work out the cost of the red paint. → Cost of 1 tube of red paint is 80 pence
Cost of 2 tubes of red paint is 2×80 pence
Cost of x tubes of red paint is $x \times 80$ or $80x$ pence

Work out the cost of the gold paint. → Cost of y tubes of gold paint is $y \times 90$ or $90y$ pence

Write a formula for the total cost. → Cost $= 80x + 90y$

Divide by 100 to find the cost in pounds. → So $T = \frac{80x + 90y}{100}$

WATCH OUT!
This formula gives the cost in pence. Your formula for T needs to give the cost in *pounds*.

Example

Given the formula $C = ab + 8$

(a) find C when $a = 3$ and $b = -4$

(b) find a when $C = 20$ and $b = 4$

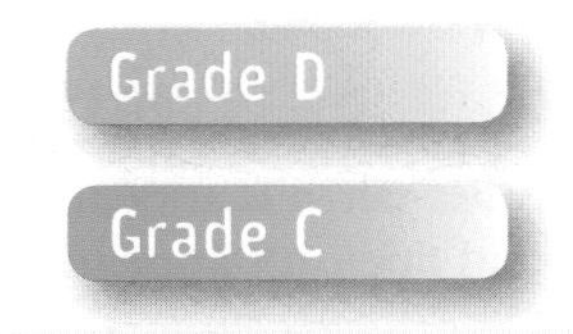

Substitute the numbers for the letters. → (a) $C = ab + 8$

$C = 3 \times -4 + 8$

Work out in the correct order. → $= -12 + 8$

$= -4$

Substitute the numbers for the letters. → (b) $C = ab + 8$

$20 = a \times 4 + 8$

$20 = 4a + 8$

Rearrange to make a the subject. → $4a = 20 - 8$

$a = 12 \div 4 = 3$

Practice

1 The sizes of the angles of a quadrilateral are shown in the diagram.

Write down a formula for A, the sum of the angles of the quadrilateral, in terms of x.

Give your formula in its simplest form.

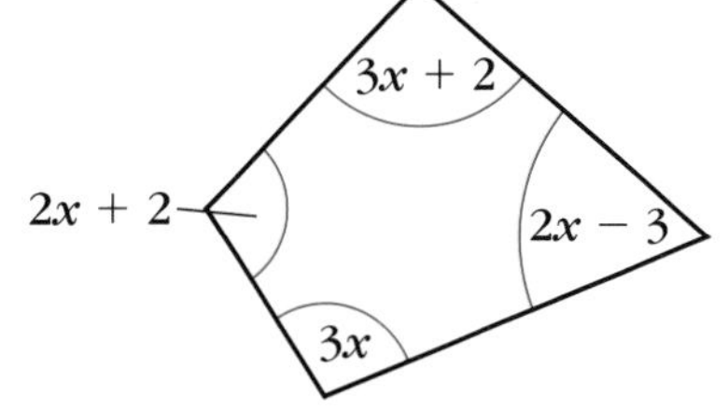

Grade D

2 A café charges x pence for a cup of tea, and y pence for a cup of coffee.

In one day the café sells 53 cups of tea and 65 cups of coffee.

S is the total amount, in pounds, that it takes from selling tea and coffee on this day.

Write down a formula for S, in terms of x and y.

Grade D

3 Given the formula $H = 2b + \frac{c^2}{4}$, calculate the value of H when $b = -\frac{3}{4}$ and $c = 6$

Grade D

4 Given the formula $A = 2\pi r(r + h)$, calculate the value of h, correct to 3 significant figures, when $A = 200$ and $r = 2.5$

Check your answers on page 175. For full worked solutions see the CD.
See the Student Book on the CD if you need more help.

Question	1	2	3	4
Grade	D	D	D	C
Student Book pages	U3 71	U3 71	U3 72–73	U3 74–75

Formulae and proof

- x is the **subject** of a **formula** when it appears on its own on one side of the formula and does not appear on the other side.
- A **fractional index** gives a root of the number: $25^{\frac{3}{2}} = (\sqrt{25})^3 = 5^3 = 125$
- You can use the balancing method to **rearrange** a formula so that it has a different subject.

Key words

rearrange ☐ subject ☐ formula ☐ product ☐ proof ☐

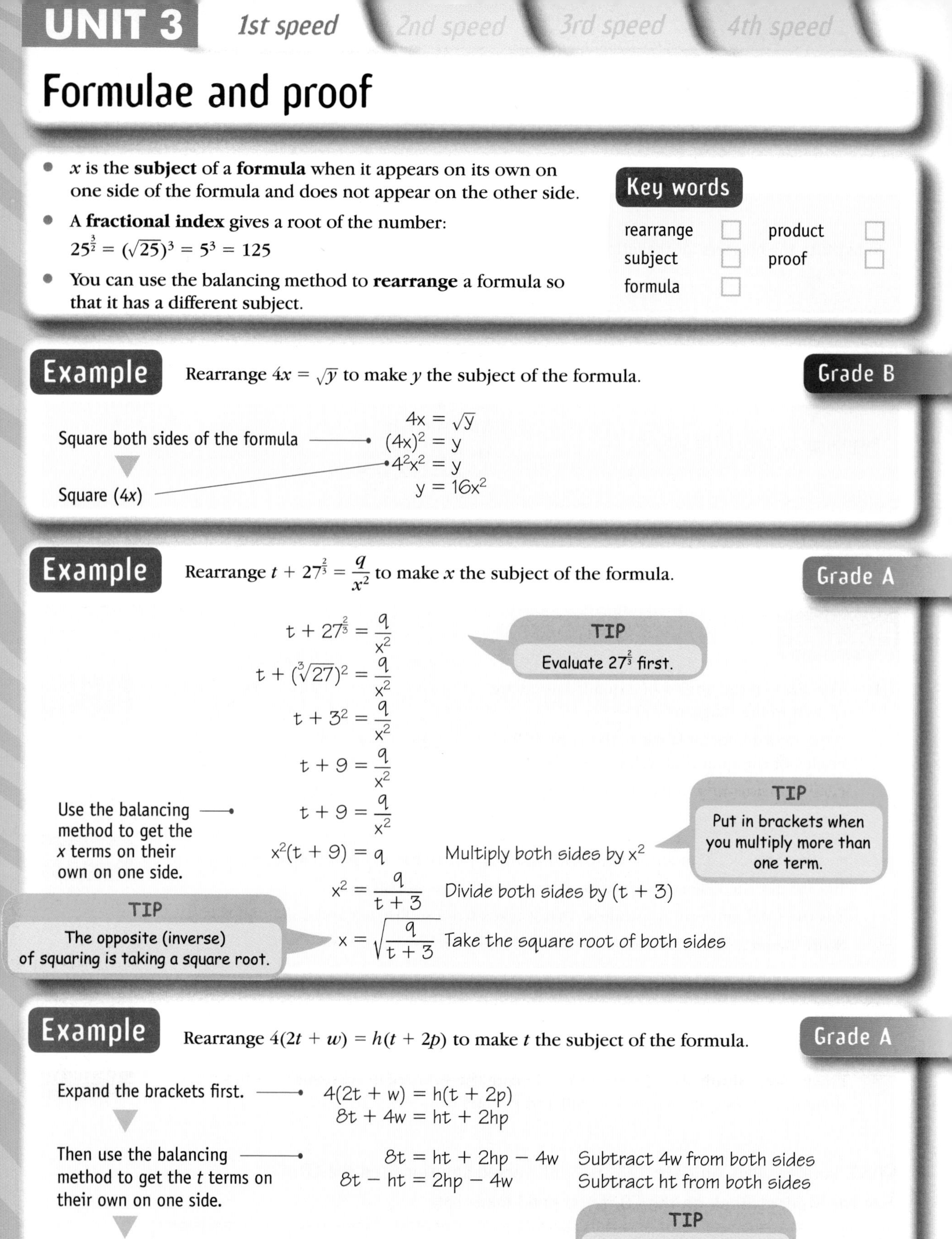

Example Rearrange $4x = \sqrt{y}$ to make y the subject of the formula. **Grade B**

$$4x = \sqrt{y}$$

Square both sides of the formula → $(4x)^2 = y$

Square $(4x)$ → $4^2x^2 = y$

$$y = 16x^2$$

Example Rearrange $t + 27^{\frac{2}{3}} = \frac{q}{x^2}$ to make x the subject of the formula. **Grade A**

$$t + 27^{\frac{2}{3}} = \frac{q}{x^2}$$

$$t + (\sqrt[3]{27})^2 = \frac{q}{x^2}$$

$$t + 3^2 = \frac{q}{x^2}$$

$$t + 9 = \frac{q}{x^2}$$

TIP Evaluate $27^{\frac{2}{3}}$ first.

Use the balancing method to get the x terms on their own on one side. → $t + 9 = \frac{q}{x^2}$

$x^2(t + 9) = q$ Multiply both sides by x^2

TIP Put in brackets when you multiply more than one term.

$x^2 = \frac{q}{t + 3}$ Divide both sides by $(t + 3)$

$x = \sqrt{\frac{q}{t + 3}}$ Take the square root of both sides

TIP The opposite (inverse) of squaring is taking a square root.

Example Rearrange $4(2t + w) = h(t + 2p)$ to make t the subject of the formula. **Grade A**

Expand the brackets first. → $4(2t + w) = h(t + 2p)$

$$8t + 4w = ht + 2hp$$

Then use the balancing method to get the t terms on their own on one side. →

$8t = ht + 2hp - 4w$ Subtract $4w$ from both sides

$8t - ht = 2hp - 4w$ Subtract ht from both sides

TIP When the subject appears more than once, factorise.

Factorise and divide to get t on its own. → $t(8 - h) = 2hp - 4w$

$t = \frac{2hp - 4w}{8 - h}$ Divide both sides by $(8 - h)$

Example

Prove algebraically that the **product** of any three consecutive odd numbers is odd.

Grade A*

Use letters for the numbers. → Call the first (odd) number x
So the next odd number is $x + 2$
and the third odd number is $x + 4$

EXAMINER'S TIP
The question asks for an algebraic **proof** (using letters). A common error is to attempt a proof using lots of different numbers.

Use letters to write down the product.
Expand the brackets. → Product $= x(x + 2)(x + 4)$
$= x^3 + 6x^2 + 8x$

TIP
Product means 'numbers multiplied together'.

Is this expression odd or even?
Look at each part separately. → If you multiply any number by an even number, the result is even.
So $6x^2$ and $8x$ will both be even numbers.
If you have odd × odd × odd, the result is odd.
So x^3 is an odd number.

Then look at how the parts combine. → If you have odd + even + even, the result is odd.
So $x^3 + 6x^2 + 8x$ is odd.

Practice

1 Rearrange $p = 3 + \frac{q}{2}$ to make q the subject of the formula. Grade C

2 Rearrange $\sqrt{xy} = f + e$ to make x the subject of the formula. Grade B

3 Rearrange $k = 3\sqrt{5 - m^2}$ to make m the subject of the formula. Grade A

4 Rearrange $a(2c - d) = 3(c + 4f)$ to make c the subject of the formula. Grade A

5 Prove algebraically that the cube of any odd number is odd. Grade A*

6 Prove algebraically that the sum of any three consecutive numbers is a multiple of 3. Grade A*

Check your answers on page 175.
See the Student Book on the CD if you need more help.

Question	1	2	3	4	5	6
Grade	C	B	A	A	A*	A*
Student Book pages	U3 75–77	U3 75–77	U3 75–77	U3 75–77	U3 78	U3 78

Formulae: topic test

Check how well you know this topic by answering these questions.
First cover the answers on the facing page.

Test questions

1 A packet of crisps costs 24 pence and a chocolate bar costs 18 pence.
T is the total cost, in pence, of buying p packets of crisps and c chocolate bars.
Write down a formula for T, in terms of p and c.

2 Given the formula $P = 5a - 2b$, find the value of P when $a = 2$ and $b = -3$

3 The diagram shows the lengths of the four sides of a quadrilateral, in terms of x.
Write down a formula for the length of the perimeter, P, in terms of x.

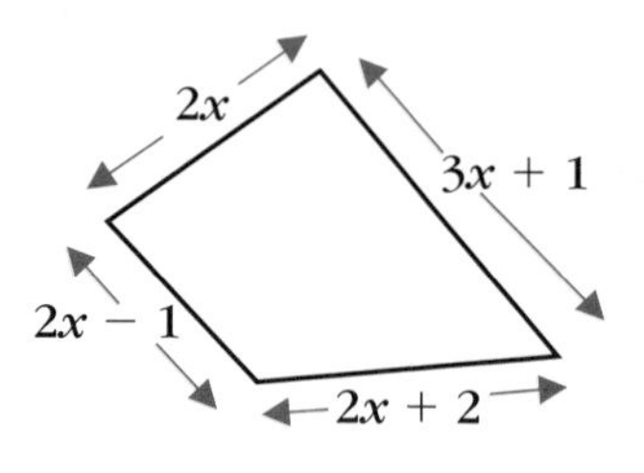

4 Rearrange the formula $y = 4x - 16$ to make x the subject.

5 Rearrange the formula $t = \frac{v}{4} - 5$ to make v the subject.

6 Rearrange the formula $c = \sqrt{de}$ to make d the subject.

7 Rearrange the formula $5(y + 2) = 3w - 4$ to make y the subject.

8 Given the formula $s = ut + \frac{1}{2}at^2$, calculate the value of u when $s = 16$, $t = 4$ and $a = 0.5$

9 Rearrange the formula $3(x - 2) = y(5 - 2x)$ to make x the subject.

10 Prove that, for any whole number n, the expression $4n^2 + 4n + 1$ is always a square number and is odd.

Now check your answers – see the facing page.

Cover this page while you answer the test questions opposite.

Worked answers

Revise this on...

D

1 The cost of buying the crisps is $p \times 24$ or $24p$ pence
The cost of buying the chocolate is $c \times 18$ or $18c$ pence
So the total cost $T = 24p + 18c$ — page 108

D

2 $P = 5 \times 2 - (2 \times -3) = 10 - -6 = 10 + 6 = 16$ — page 109

D

3 Perimeter $P = 2x + (2x - 1) + (2x + 2) + (3x + 1) = 9x + 2$ — page 108

D

4 page 110

$y = 4x - 16$

$4x = y + 16$ Add 16 to both sides

$x = \frac{y + 16}{4}$ Divide both sides by 4

C

5 page 110

$t = \frac{v}{4} - 5$

$\frac{v}{4} = t + 5$ Add 5 to both sides

$v = 4(t + 5)$ Multiply both sides by 4

B

6 page 110

$c = \sqrt{de}$

$c^2 = de$ Square both sides

$d = \frac{c^2}{e}$ Divide both sides by e

B

7 page 110

$5(y + 2) = 3w - 4$

$5y + 10 = 3w - 4$ Expand the bracket

$5y = 3w - 4 - 10$ Take 10 from both sides

$y = \frac{3w - 14}{5}$ Divide both sides by 10

B

8 $16 = u \times 4 + ½(0.5 \times 4^2) \rightarrow 16 = 4u + 4 \rightarrow 12 = 4u \rightarrow u = 3$ — page 109

A

9 page 110

$3(x - 2) = y(5 - 2x)$

$3x - 6 = 5y - 2xy$ Expand the brackets

$3x + 2xy = 5y + 6$ Add $2xy + 6$ to both sides

$x(3 + 2y) = 5y + 6$ Factorise the left-hand side

$x = \frac{5y + 6}{3 + 2y}$ Divide both sides by $3 + 2y$

A*

10 $4n^2 + 4n + 1 = (2n + 1)^2$ — page 111
$2n$ is always even, so $2n + 1$ is an odd number.
$(2n + 1)^2$ is clearly a square number. It is odd $\times$ odd, so it is an odd number.

Tick the questions you got right.

Question	1	2	3	4	5	6	7	8	9	10
Grade	D	D	D	D	C	B	B	B	A	A*

Mark the grade you are working at on your revision planner on page ix.

Solving linear equations

- To remove brackets from an algebraic expression, multiply each term inside the brackets by the term outside.
  This is sometimes called **expanding the brackets**.
- Terms with the same power of the same letter(s) are called **like terms**.
- To **rearrange** an **equation** you can use the **balancing method**:
  - add the same quantity to both sides
  - subtract the same quantity from both sides
  - multiply both sides by the same quantity
  - divide both sides by the same quantity.

Key words

equation	☐	balance	☐
expand	☐	rearrange	☐
brackets	☐	like terms	☐

- Whatever you do to one side of an equation you must also do to the other side.
- For equations with the unknown on both sides, rearrange so the unknowns are on one side of the equation and the numbers are on the other.

Example

Solve (a) $6 + 4y = 2y + 16$ Grade D

(b) $4(4y + 1) = 3(5y + 4)$ Grade C

(c) $\frac{7 + 2x}{5} = x + 6$ Grade C

Use the balancing method to get the letters on their own on one side.

(a)

$6 + 4y = 2y + 16$

$6 + 4y - 2y = 2y - 2y + 16$ Subtract 2y from both sides

$6 - 6 + 2y = 16 - 6$ Subtract 6 from both sides

$2y = 10$

$y = 5$ Divide both sides by 2

WATCH OUT!
Make sure you multiply every term inside the bracket by the number outside. Students often wrongly multiply just the first term.
For more on expanding brackets, see pages 54-55.

Expand the brackets. Then use the balancing method.

(b)

$4(4y + 1) = 3(5y + 4)$

$16y + 4 = 15y + 12$

$16y - 15y + 4 = 15y - 15y + 12$ Subtract 15y from both sides

$y + 4 - 4 = 12 - 4$ Subtract 4 from both sides

$y = 8$

TIP
Put in brackets when you multiply more than one term.

Multiply both sides of the equation to get rid of the fraction. Expand the brackets. Then use the balancing method.

(c)

$\frac{7 + 2x}{5} = x + 6$

$5\left(\frac{7 + 2x}{5}\right) = 5(x + 6)$ Multiply both sides by 5

$7 + 2x = 5x + 30$

$7 + 2x - 2x = 5x - 2x + 30$ Subtract 2x from both sides

$7 - 30 = 3x + 30 - 30$ Subtract 30 from both sides

$-23 = 3x$

$x = \frac{-23}{3} = -7\frac{2}{3}$ Divide both sides by 3

Example

Solve (a) $\frac{3x - 4}{2} = \frac{4x - 3}{4}$

(b) $\frac{2}{x} - \frac{1}{2x} = 1$

Grade A

Grade A*

(a) $\frac{3x - 4}{2} = \frac{4x - 3}{4}$

TIP The LCM of 2 and 4 is 4

Multiply both sides of the equation by the LCM of the denominators.

$4\left(\frac{3x - 4}{2}\right) = 4\left(\frac{4x - 3}{4}\right)$ Multiply both sides by 4

Expand the brackets.

$6x - 8 = 4x - 3$

Then use the balancing method.

$6x - 4x - 8 = 4x - 4x - 3$ Subtract 4x from both sides

$2x - 8 + 8 = -3 + 8$ Add 8 to both sides

$2x = 5$

$x = 2\frac{1}{2}$ Divide both sides by 2

(b) $\frac{2}{x} - \frac{1}{2x} = 1$

TIP The LCM of x and $2x$ is $2x$

Multiply both sides of the equation by the LCM of the denominators.

$2x\left(\frac{2}{x}\right) - 2x\left(\frac{1}{2x}\right) = 2x(1)$ Multiply both sides by 2x

Expand the brackets.

$4 - 1 = 2x$

$3 = 2x$

$x = \frac{3}{2}$ or $1\frac{1}{2}$ Divide both sides by 2

Practice

1 Solve $5b + 8 = 43 - 2b$ Grade D

2 Solve $3(6m - 3) = 4(5m - 4)$ Grade C

3 Solve $\frac{8 + 3x}{2} = 7 + 6x$ Grade C

4 Solve $\frac{2y}{3} + \frac{3y}{4} = 34$ Grade B

5 Solve $\frac{2}{3x + 1} = \frac{1}{2x - 5}$ Grade A*

Check your answers on page 176. For full worked solutions see the CD.
See the Student Book on the CD if you need more help.

Question	1	2	3	4	5
Grade	D	C	C	B	A*
Student Book pages	U3 56–57	U3 56–57	U3 62–63	U3 62–63	U3 63–64

Solving quadratic and cubic equations

- The **quadratic equation** $ax^2 + bx + c = 0$, with $a \neq 0$, has two solutions (or **roots**) which may be equal.
- If $xy = 0$ then either $x = 0$ or $y = 0$.
- The roots of the quadratic equation $ax^2 + bx + c = 0$, where $a \neq 0$, are given by the formula

$$x = \frac{-b \pm \sqrt{b^2 - 4ac}}{2a}$$

Key words

- quadratic equation
- factorising
- roots

Example

Grade B

Solve $x^2 - 4x - 5 = 0$

Factorise the left-hand side.

$x^2 - 4x - 5 = 0$

$(x - 5)(x + 1) = 0$

So either $x - 5 = 0$ or $x + 1 = 0$

So either $x = 5$ or $x = -1$

WATCH OUT!
If the equation is not in the form $ax^2 + bx + c = 0$ you will need to rearrange it first.

TIP
The solutions are sometimes called the **roots** of the equation.

Example

Grade A

Solve $2x^2 - 9x + 5 = 0$, giving your solutions correct to 2 decimal places.

$2x^2 - 9x + 5 = 0$

EXAMINER'S TIP
If a question asks you to round your answer, you can tell that the quadratic equation *cannot* be solved by factorising.

Use the formula $x = \frac{-b \pm \sqrt{b^2 - 4ac}}{2a}$

This equation is of the form $ax^2 + bx + c = 0$ with $a = 2$, $b = -9$ and $c = 5$

TIP
The formula for solving quadratic equations will be on the formulae sheet in your GCSE exam.

$$x = \frac{-(-9) \pm \sqrt{(-9)^2 - 4(2)(5)}}{2(2)}$$

$$= \frac{9 \pm \sqrt{81 - 40}}{4}$$

$$= \frac{9 \pm \sqrt{41}}{4}$$

So $x = \frac{9 + \sqrt{41}}{4}$ or $x = \frac{9 - \sqrt{41}}{4}$

TIP
Work out the numerator first, then divide by 4

So $x = 3.85$ or $x = 0.65$ (to 2 d.p.)

- You can use a **trial and improvement method** to solve an equation. Try a value in the equation and use your result to improve your estimate.
  Repeat, getting closer and closer to the correct value.

Key words

- cubic equation ☐
- trial and improvement ☐

Example

Grade C

The **cubic equation** $3x^3 - 2x^2 = 50$ has a solution between 2 and 3.
Use a trial and improvement method to find this solution.
Give your answer correct to one decimal place.

Work out $3x^3 - 2x^2$ using different values of x, until you get near to 50.
Start with the x values given.

When $x = 2$, $3x^3 - 2x^2 = 16$ Smaller than 50
When $x = 3$, $3x^3 - 2x^2 = 63$ Bigger than 50
When $x = 2.5$, $3x^3 - 2x^2 = 34.375$ Smaller than 50
When $x = 2.7$, $3x^3 - 2x^2 = 44.469$ Smaller than 50
When $x = 2.8$, $3x^3 - 2x^2 = 50.176$ Bigger than 50

TIP
Make sure you show the results of all your trials and improvements.

So x lies between 2.7 and 2.8. Try the half-way value.

When $x = 2.75$, $3x^3 - 2x^2 = 47.2656$ Smaller than 50
The solution is between $x = 2.75$ and $x = 2.80$
Any number in this range rounds to 2.8 (to 1 d.p.).
So $x = 2.8$ (to 1 d.p.)

TIP
Always go to the next decimal place to confirm your solution.

EXAMINER'S TIP
Remember to write down the solution to the equation. This is the value of x (2.8), not the number calculated (50.176)

Practice

1. The equation $4x^3 - x^2 = 50$ has a solution between 2 and 3. Use a trial and improvement method to find this solution. Give your answer to 1 d.p. (Grade C)

2. Solve $x^2 + 2x - 15 = 0$ (Grade B)

3. Solve $3x^2 - 5x - 4 = 0$, giving your answers correct to 2 d.p. (Grade A)

Check your answers on page 176. For full worked solutions see the CD.
See the Student Book on the CD if you need more help.

Question	1	2	3
Grade	C	B	A
Student Book pages	U3 110–111	U3 98–100	U3 102–103

Solving simultaneous equations

- **Simultaneous equations** can be solved algebraically by multiplying one or both of the equations by a number, if necessary, and then adding or subtracting before dividing by the coefficient.

Key words

- simultaneous equations ☐
- solution ☐
- substitute ☐
- eliminate ☐
- rearrange ☐

Example

Solve the simultaneous equations $x + y = 5$
$-x + y = 1$

Grade B

$x + y = 5$ (1)
$-x + y = 1$ (2)

TIP
It helps to number the equations so you can refer to them easily.

Add the two equations to **eliminate** x. → (1) + (2): $2y = 6$
$y = 3$

TIP
You could subtract the equations to eliminate y instead.

Substitute this value of y into one of the original equations and rearrange to find x. → Substitute $y = 3$ into (1):
$x + 3 = 5$
So $x = 2$
The solution is $x = 2$, $y = 3$

TIP
Since this is a long answer it is important you show all your working.

Example

Solve the simultaneous equations $3x + 4y = 1$
$4x + 3y = 6$

Grade B

$3x + 4y = 1$ (1)
$4x + 3y = 6$ (2)

You need equal numbers of x's or y's, so multiply both equations first. →
(1) × 4: $12x + 16y = 4$
(2) × 3: $12x + 9y = 18$

WATCH OUT!
Make sure you multiply *all* terms in the equation.

Subtract to eliminate x. → $7y = -14$
$y = -2$

TIP
If the coefficients have the same sign, subtract.
If they have different signs, add.

Substitute this value of y into one of the original equations. → Substitute $y = -2$ into (1):
$3x + 4(-2) = 1$
$3x - 8 = 1$
So $3x = 9$
$x = 3$
The solution is $x = 3$, $y = -2$

- Solving a **linear equation** ($y = px + q$) and a **quadratic equation** ($y = ax^2 + bx + c$) simultaneously:
  - Find y in terms of x from the linear equation (or x in terms of y).
  - Substitute for y (or x) in the quadratic equation.
  - Solve the resulting quadratic equation for x (or y).
  - Substitute the values of x (or y) into the linear equation to find y (or x).

  If the roots of the quadratic equation are equal, the line will be a tangent to the curve.

Key words

linear equation ☐
quadratic equation ☐

Example

Grade A*

Solve the simultaneous equations $4x + y = 2$ (1)
$4x + y^2 = 8$ (2)

Start by eliminating one of the variables (here, x). → From (1): $(4x) = 2 - y$
Substitute into (2): $(4x) + y^2 = 8$
So $(2 - y) + y^2 = 8$

TIP Use brackets to help you.

Rearrange to form a quadratic equation. → $y^2 - y + 2 - 8 = 0$
$y^2 - y - 6 = 0$

EXAMINER'S TIP Show all your working.

Factorise the left-hand side. → $(y - 3)(y + 2) = 0$

Solve for y → So $y = 3$ or $y = -2$

TIP It is simpler if you substitute into the original linear equation rather than the quadratic equation.

Substitute these values of y into one of the original equations. → Substitute $y = 3$ and $y = -2$ into (1):
$4x = 2 - y$
So $4x = 2 - 3$ or $4x = 2 - (-2)$
$4x = -1$ or $4x = 4$
$x = -\frac{1}{4}$ or $x = 1$

The solutions are $x = -\frac{1}{4}$, $y = 3$ and $x = 1$, $y = -2$

WATCH OUT! Make sure you pair up the x and y values correctly.

Practice

1 Solve the simultaneous equations $3x + y = 8$
$3x + 5y = -2$

Grade B

2 Solve the simultaneous equations $x + y^2 = 10$
$x - 2y = 2$

Grade A*

Check your answers on page 176. For full worked solutions see the CD.
See the Student Book on the CD if you need more help.

Question	1	2
Grade	B	A*
Student Book pages	U3 96–97	U3 124–126

Solving inequalities

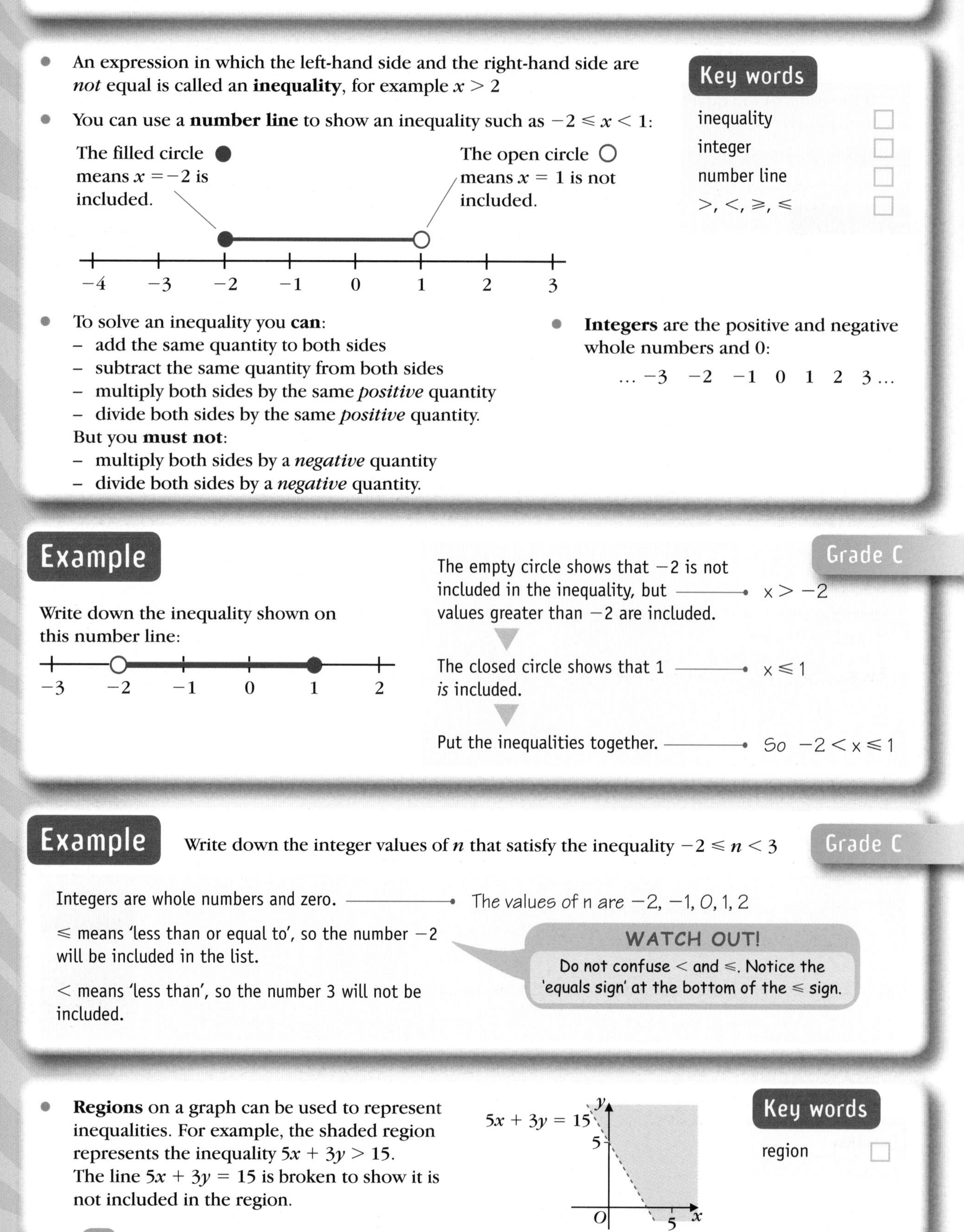

- An expression in which the left-hand side and the right-hand side are *not* equal is called an **inequality**, for example $x > 2$
- You can use a **number line** to show an inequality such as $-2 \leqslant x < 1$:

Key words

- inequality
- integer
- number line
- $>, <, \geqslant, \leqslant$

- To solve an inequality you **can**:
  - add the same quantity to both sides
  - subtract the same quantity from both sides
  - multiply both sides by the same *positive* quantity
  - divide both sides by the same *positive* quantity.

  But you **must not**:
  - multiply both sides by a *negative* quantity
  - divide both sides by a *negative* quantity.

- **Integers** are the positive and negative whole numbers and 0:

  $\ldots -3 \quad -2 \quad -1 \quad 0 \quad 1 \quad 2 \quad 3 \ldots$

Example

Grade C

Write down the inequality shown on this number line:

The empty circle shows that -2 is not included in the inequality, but values greater than -2 are included. → $x > -2$

The closed circle shows that 1 is included. → $x \leqslant 1$

Put the inequalities together. → So $-2 < x \leqslant 1$

Example

Write down the integer values of n that satisfy the inequality $-2 \leqslant n < 3$

Grade C

Integers are whole numbers and zero. → The values of n are $-2, -1, 0, 1, 2$

$\leqslant$ means 'less than or equal to', so the number -2 will be included in the list.

$<$ means 'less than', so the number 3 will not be included.

WATCH OUT!

Do not confuse $<$ and $\leqslant$. Notice the 'equals sign' at the bottom of the $\leqslant$ sign.

- **Regions** on a graph can be used to represent inequalities. For example, the shaded region represents the inequality $5x + 3y > 15$. The line $5x + 3y = 15$ is broken to show it is not included in the region.

Key words

- region

Example

Solve the inequality $4 + 3x > 22 - 3x$

Grade C

Use the balancing method to make x the subject, just like an equation.

$$4 + 3x > 22 - 3x$$
$$4 + 3x + 3x > 22 - 3x + 3x \quad \text{Add } 3x \text{ to both sides}$$
$$4 + 6x > 22$$
$$4 - 4 + 6x > 22 - 4 \quad \text{Subtract 4 from both sides}$$
$$6x > 18$$
$$x > 3 \quad \text{Divide both sides by 6}$$

Example

Draw a diagram and shade the region which satisfies the inequalities

$x > 1, \quad y \geqslant 3, \quad x + y \leqslant 7$

Grade B

First draw the lines $x = 1$, $y = 3$ and $x + y = 7$

TIP

Label each line clearly. Draw the line solid if points on it are included, and dotted if they are not included.

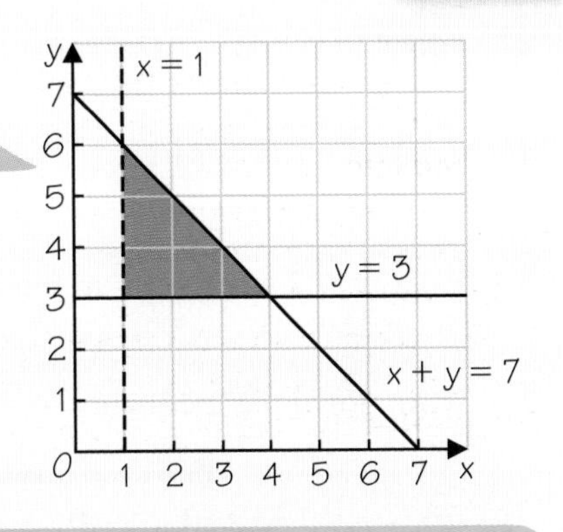

Think about each inequality in turn.

- The region for $x > 1$ is to the *right* of the line $x = 1$. Points on the line $x = 1$ are not included
- The region for $y \leqslant 3$ is *above* the line $y = 3$, and includes the line.
- The region $x + y \leqslant 7$ is *below* the line $x + y = 7$, and includes the line.

Shade the region where *all* the inequalities are satisfied.

WATCH OUT!

Read the question carefully. Sometimes you need to shade the regions that are *not* required.

Practice

1 Write down the integer values of n that satisfy the inequality $-4 < n \leqslant 2$ **Grade C**

2 Write down the inequality shown on this number line: **Grade C**

3 Solve the inequality $4 - 2x < 12 - 3x$ **Grade C**

4 Draw a diagram and shade the region which satisfies the inequalities **Grade B**

$x + y \leqslant 6, \quad y \geqslant 2, \quad x \geqslant 1$

Check your answers on page 176. For full worked solutions see the CD.
See the Student Book on the CD if you need more help.

Question	1	2	3	4
Grade	C	C	C	B
Student Book pages	U3 106–107	U3 104–105	U3 105–106	U3 107–110

Solving equations and inequalities: topic test

Check how well you know this topic by answering these questions.
First cover the answers on the facing page.

Test questions

1 Solve $3x - 6 = 14 + x$

2 Solve $9(x + 1) = 4(2x + 3)$

3 Solve $\frac{15 - x}{3} = 13 + x$

4 $-3 < x \leqslant 2$ where n is an integer. Write down all the possible values of n.

5 Write down the inequality shown on this number line:

−3 −2 −1 0 1 2 3 4

6 The equation $2x^3 - x^2 = 7$ has a solution between 1 and 2.
Use a trial and improvement method to find this solution.
Give your answer correct to one decimal place.

7 Solve the simultaneous equations

$$4x + 2y = 8$$
$$2x - 5y = 10$$

8 Solve $\frac{x + 4}{3} = \frac{x + 1}{2}$

9 Solve the inequality $2x - 14 < 5 + x$

10 Draw a diagram and shade the region which satisfies the inequalities

$$x + y < 10, \quad y < 7, \quad x < 6$$

11 Solve $2x^2 + 5x + 1 = 0$, giving any solutions correct to 2 decimal places.

12 Solve $\frac{5}{4x} - \frac{2}{x} = 3$

Now check your answers – see the facing page.

Cover this page while you answer the test questions opposite.

Worked answers

Revise this on...

D **1**
$$3x - 6 = 14 + x$$
$$3x - x - 6 + 6 = 14 + 6 + x - x$$
$$2x = 20$$
$$x = 10$$

page 114

D **2**
$$9(x + 1) = 4(2x + 3)$$
$$9x + 9 = 8x + 12$$
$$9x - 8x + 9 - 9 = 8x - 8x + 12 - 9$$
$$x = 3$$

C **3**
$$15 - x = 3(13 + x)$$
$$15 - x = 39 + 3x$$
$$-4x = 24$$
$$x = -6$$

page 114

C **4** The integer values are $-2, -1, 0, 1, 2$.

5 $-1 \leqslant x < 3$

page 120

C **6**

When $x = 1$, $2x^3 - x^2 = 0$ Smaller than 7
When $x = 2$, $2x^3 - x^2 = 12$ Bigger
When $x = 1.5$, $2x^3 - x^2 = 4.5$ Smaller
When $x = 1.7$, $2x^3 - x^2 = 6.936$ Smaller
When $x = 1.8$, $2x^3 - x^2 = 8.424$ Bigger
When $x = 1.75$, $2x^3 - x^2 = 7.656\,25$ Bigger

The solution is between 1.7 and 1.75. Any number in this range rounds to 1.7 (to 1 d.p.)
So $x = 1.7$ (to 1 d.p.)

page 117

B **7**
$4x + 2y = 8$ (1)
$2x - 5y = 10$ (2)
$2 \times (2)$: $4x - 10y = 20$
$(1) - 2 \times (2)$: $12y = -12$
$y = -1$
Substitute $y = -1$ into equation (1):
$4x - 2 = 8$
$x = 2\frac{1}{2}$
The solution is $x = 2\frac{1}{2}$, $y = -1$

8
$$\overset{2}{\cancel{6}}\left(\frac{x + 4}{\cancel{3}}\right) = \overset{3}{\cancel{6}}\left(\frac{x + 1}{\cancel{2}}\right)$$
$$2(x + 4) = 3(x + 1)$$
$$2x + 8 = 3x + 3$$
$$x = 5$$

pages 114, 118

9
$$2x - 14 < 5 + x$$
$$2x - x - 14 + 14 < 5 + 14 + x - x$$
$$x < 19$$

page 121

B **10**

page 121

A **11** $a = 2$, $b = 5$ and $c = 1$
$$x = \frac{-5 \pm \sqrt{5^2 - 4(2)(1)}}{2(2)}$$
$$= \frac{-5 \pm \sqrt{25 - 8}}{4}$$
$$= \frac{-5 \pm \sqrt{17}}{4}$$
So $x = \frac{-5 + \sqrt{17}}{4}$ or $x = \frac{-5 - \sqrt{17}}{4}$
$x = -0.22$ or $x = -2.28$ (to 2 d.p.)

A* **12**
$$\frac{5}{4x} - \frac{2}{x} = 3$$
Multiply both sides by $4x$:
$$\cancel{4x}\left(\frac{5}{\cancel{4x}}\right) - 4\cancel{x}\left(\frac{2}{\cancel{x}}\right) = 4x\,(3)$$
$$5 - 8 = 12x$$
$$x = -\tfrac{3}{12} = -\tfrac{1}{4}$$

pages 116, 115

Tick the questions you got right.

Question	1	2	3	4	5	6	7	8	9	10	11	12
Grade	D	D	C	C	C	C	B	B	B	B	A	A*

Mark the grade you are working at on your revision planner on page ix.

Proportion

- The symbol $\propto$ means 'is **proportional** to'.
- $y \propto x$ means 'y is **directly proportional** to x'.
- When a graph of two quantities is a straight line through the origin, one quantity is **directly proportional** to the other.
- When y is directly proportional to x, you can write a proportionality statement and a formula connecting y and x:
  - the proportionality statement is $y \propto x$
  - the proportionality formula is $y = kx$, where k is the **constant of proportionality**.
- To find the value of k, the constant of proportionality, substitute known values of y and x into $y = kx$.
- When y is directly proportional to the **square** of x:
  - the proportionality statement is $y \propto x^2$
  - the proportionality formula is $y = kx^2$, where k is the constant of proportionality.
- When y is directly proportional to the **cube** of x:
  - the proportionality statement is $y \propto x^3$
  - the proportionality formula is $y = kx^3$, where k is the constant of proportionality.

Key words

- proportional ☐
- directly proportional ☐
- constant of proportionality ☐
- square ☐
- cube ☐

Example

Grade A

The cost of paint in a tub is directly proportional to the amount of paint in the tub. A 20-litre tub costs £50

(a) Find a rule connecting the cost of the paint in a tub, £C, and the amount of paint in the tub, A litres.

(b) Find the cost of 15 litres of paint.

(c) Find the amount of paint that costs £20

Use what you are told to write down the proportionality formula.

(a) $C \propto A$
So $C = kA$

TIP Write down the formula first, before finding k.

Find the constant of proportionality by substituting in the values.

$C = 50$ when $A = 20$
So $50 = k \times 20$
$k = \frac{50}{20} = 2.5$ Divide both sides by 20
The rule is $C = 2.5A$

Substitute the numbers for the letters.

(b) $A = 15$,
so $C = 2.5 \times 15 = 37.5$

Give your answer in the correct units.

The cost is £37.50

TIP You are finding a cost, so the answer must be an amount of money.

Substitute the numbers for the letters and rearrange.

(c) $C = 20$, so $20 = 2.5 \times A$
$A = 20 \div 2.5 = 8$ Divide both sides by 2.5
The amount is 8 litres.

- $y \propto \frac{1}{x}$ means 'y is **inversely proportional** to x'.
- When y is inversely proportional to x:
  - the proportionality statement is $y \propto \frac{1}{x}$
  - the proportionality formula can be written as $y = k \times \frac{1}{x}$ or $y = \frac{k}{x}$, where k is the constant of inverse proportionality.

Key words

inversely proportional ☐

Example

Grade A

The electrical resistance of a wire (R) varies inversely as the square of the radius r.
The resistance is 0.4 ohms when the radius is 0.3 cm.

(a) Find a rule connecting R and r.

(b) Find the resistance of a wire with radius 0.5 cm.

(c) Find the radius of a wire with resistance 0.1 ohm.

Write down the proportionality formula.

(a) $R \propto \frac{1}{r^2}$

TIP

'varies inversely as' means the same as 'is inversely proportional to'.

So $R = \frac{k}{r^2}$

Find the constant of proportionality by substituting in the values.

$R = 0.4$ when $r = 0.3$

So $0.4 = \frac{k}{0.3^2}$

$k = 0.4 \times 0.3^2 = 0.4 \times 0.09 = 0.036$ Multiply both sides by 0.3^2

The formula is $R = \frac{0.036}{r^2}$

Substitute the numbers for the letters.

(b) $r = 0.5$, so $R = \frac{0.036}{0.5^2} = 0.036 \div 0.25 = 0.144$

Give your answer in the correct units.

The resistance is 0.144 ohms

Substitute the numbers for letters and rearrange.

(c) $R = 0.1$, so $0.1 = \frac{0.036}{r^2}$

$r^2 = 0.036 \times 10 = 0.36$ Multiply both sides by 10 and by r^2

$r = \sqrt{0.36} = 0.6$

The radius is 0.6 cm

Practice

Grade A

1 The distance (d metres) travelled by a stone falling vertically varies in direct proportion to the square of the time (t seconds) for which it falls.

A stone takes 2 seconds to fall 20 metres.

(a) Write a formula connecting d and t.

(b) How far will the stone fall in 4 seconds?

(c) How long will it take to travel 31.25 metres?

Grade A

2 f is inversely proportional to w.
When $w = 2$, $f = 40$.

(a) Find a rule connecting f and w.

(b) Find f when $w = 0.2$

(c) Find w when $f = 200$

Check your answers on page 176. For full worked solutions see the CD.
See the Student Book on the CD if you need more help.

Question	1	2
Grade	A	A
Student Book pages	U3 30–31	U3 31–33

Proportion: topic test

Check how well you know this topic by answering these questions.
First cover the answers on the facing page.

Test questions

1 The volume of sand in a container is proportional to the cube of its height in the container.

When the height of sand is 3 metres, the volume is $108\,m^3$.

(a) Write down a formula for the volume, $V\,m^3$, in terms of the height, h metres.

(b) Use your rule to find the volume when the height is 7 metres.

(c) Use your rule to find the height when the volume is $500\,m^3$.

2 The cost of a square rug is directly proportional to the length of a side of the rug.

A square rug with side 3 m costs £120.

Work out the cost of a similar rug with side 4 m.

3 d is inversely proportional to f.

When f is 2, d is 4.

(a) Find a formula connecting d and f.

(b) Find the value of d when $f = 0.5$

(c) Find the value of f when $d = 3.2$

4 The time taken in seconds, t, for a full kettle to boil is inversely proportional to the power P of the kettle in kilowatts.

When $P = 3$ kW, $t = 120$ s.

How long would a kettle with power 2 kilowatts take to boil the same amount of water?

Now check your answers – see the facing page.

Cover this page while you answer the test questions opposite.

Worked answers

Revise this on...

A **1** (a) $V \propto h^3$ so $V = kh^3$ page 124

$V = 108$ when $h = 3$, so

$108 = k \times 3^3$

$108 = k \times 27$

$k = 4$

So the proportionality formula is $V = 4h^3$

(b) When $h = 7$, $V = 4 \times 7^3 = 4 \times 343 = 1372\ m^3$

(c) When $V = 500\ m^3$, $500 = 4 \times h^3$

So $h^3 = 500 \div 4 = 125$

$h = \sqrt[3]{125} = 5$ metres

A **2** $C \propto l$ C = cost, l = length page 124

$C = kl$

When $l = 3$, $C = 120$

$120 = 3k$

$k = 40$

When $l = 4$

$C = 40 \times 4 = £160$

A **3** (a) $d \propto \frac{1}{f}$ so $d = \frac{k}{f}$ page 125

When $f = 2$, $d = 4$, so

$4 = \frac{k}{2}$

$k = 4 \times 2 = 8$

So the formula is $d = \frac{8}{f}$

(b) When $f = 0.5$, $d = \frac{8}{0.5} = 16$

(c) When $d = 3.2$, $3.2 = \frac{8}{f}$

So $3.2f = 8$

$f = 8 \div 3.2 = 2.5$

A **4** $t \propto \frac{1}{P}$ page 125

$t = \frac{k}{P}$

When $P = 3$, $t = 120$

$120 = \frac{k}{3}$

$k = 3 \times 120 = 560$

When $P = 2$

$t = \frac{360}{2} = 180$ s

Tick the questions you got right.

Question	1	2	3	4
Grade	A	A	A	A

Mark the grade you are working at on your revision planner on page ix.

Algebra: subject test

Use these questions to check that you understand the key facts for Algebra, before you try the Examination Practice Paper on pages 148–156.

Exam practice questions

1

$3x - 2$

$2x$ $3x + 4$

(a) Write down an expression, in terms of x, for the total of the angles in the triangle.

(b) Write down an equation in x.

(c) Solve your equation to find a value for x.

2 $v^2 = u^2 + 2as$

Find v when $u = 6$, $a = 7$ and $s = 2$

3 This is part of the travel graph for a motorist.
After 2 pm the motorist travels home at an average speed of 80 km/h.
Complete the travel graph.

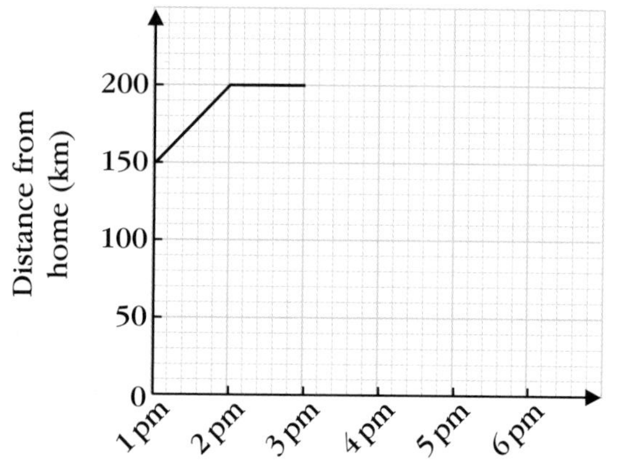

4 Solve $3(x - 2) = 2(x - 1)$

5 $-3 \leqslant n < 4$ where n is an integer. Write down all the possible values of n.

6 The equation $3x^3 - x^2 = 58$ has a solution between 2 and 3. Use a trial and improvement method to find this solution. Give your answer correct to one decimal place.

7 **(a)** Complete the table of values for $y = x^2 - 3x + 2$

x	-1	0	1	2	3	4
y						

(b) Draw the graph of $y = x^2 - 3x + 2$

(c) Use your graph to solve the equation $x^2 - 3x + 2 = 0$

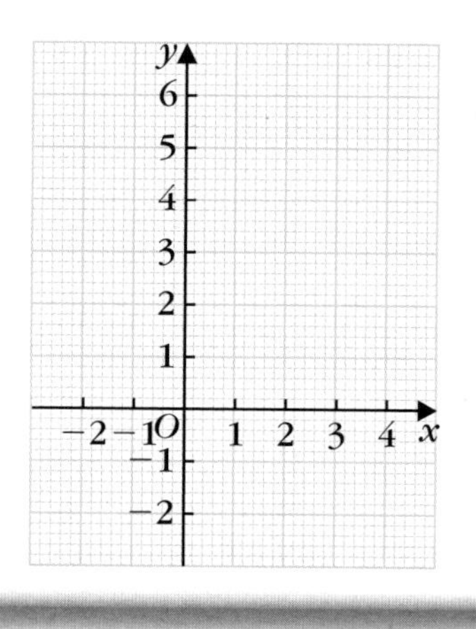

8 Make h the subject of the formula $A = 2\pi(r + h)$

9 Solve the simultaneous equations $2x + y = 7$, $2x - y = 3$

10 Solve $\frac{2y}{5} - \frac{3y}{4} = \frac{3}{10}$

11 A line passes through the points (2, 8) and (−1, 4).
Find the equation of this line.

12 A line has a gradient of $-\frac{1}{3}$ and passes through the point (3, −1).
What is the equation of the line?

13 A line L has an equation $y = 3x - 4$.

(a) Find the equation of the line that is parallel to L, and passes through the point (−4, −7).

(b) Find the equation of the line that is perpendicular to L, and passes through the point (4, 2).

14 Solve $3x^2 - 4x - 5 = 0$, giving your answers correct to 2 decimal places.

15 x is inversely proportional to the square of y. When $x = 10$, $y = 4$.

(a) Write down the rule connecting x and y.

(b) Find y when $x = 8$. Give your answer correct to 3 significant figures.

16 This is the sketch graph of $y = f(x)$.
Point A is the turning point of the graph at (−1, 5)
Write down the turning point of the graph for

(a) $y = f(x - 2)$

(b) $y = f(2x)$

Check your answers on page 176. For full worked solutions see the CD.
Tick the questions you got right.

Question	1	2	3	4	5	6	7	8	9	10	11	12	13a	13b	14	15	16
Grade	D	D	C	C	C	C	B	B	B	B	B	B	B	A	A	A	A*
Revise this on page	108	109	101	114	120	117	102	110	118	114	98–99	98–99	98–99	98–99	116	125	104

Mark the grade you are working at on your revision planner on page ix.
Go to the pages shown to revise for the ones you got wrong.

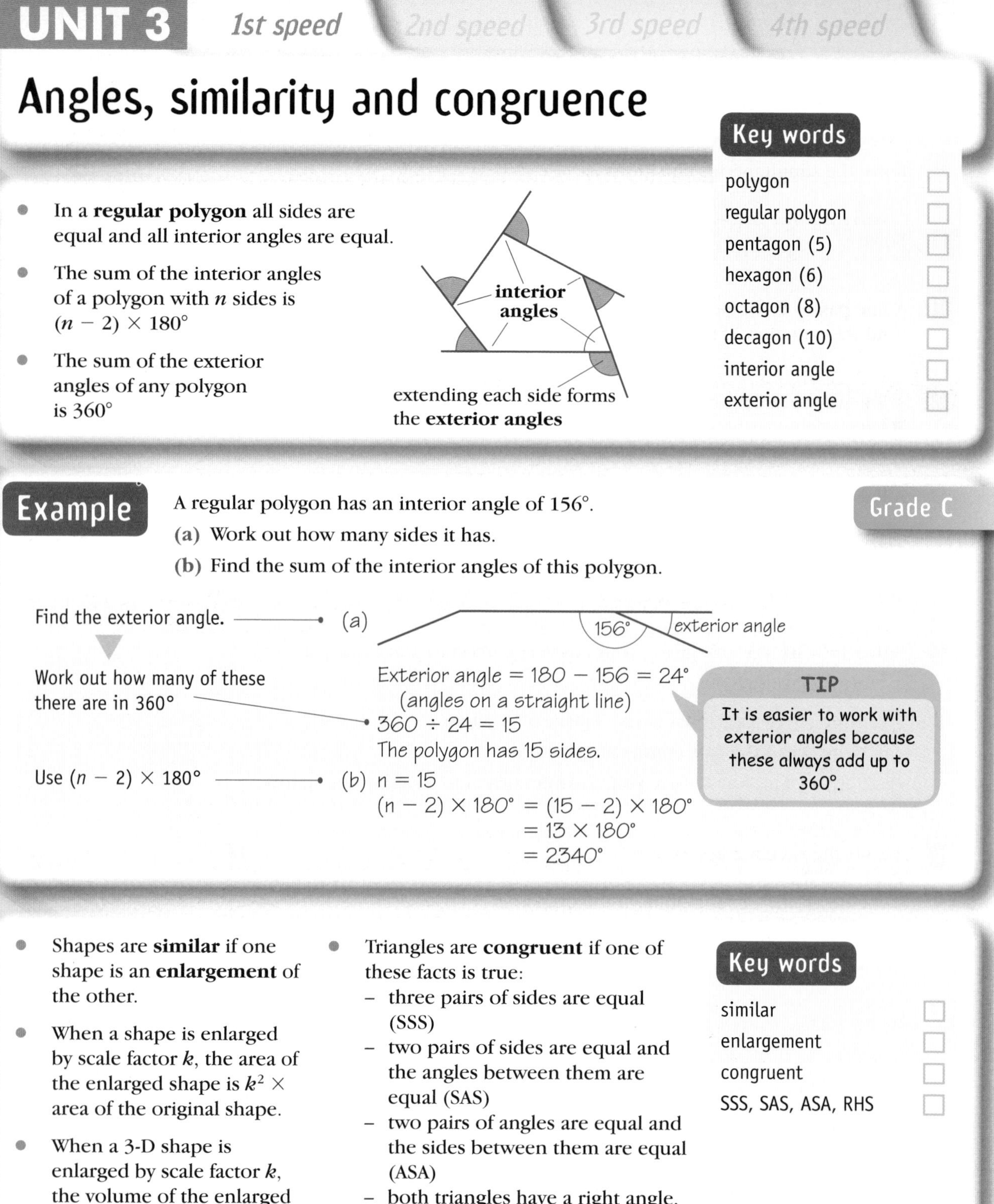

Angles, similarity and congruence

- In a **regular polygon** all sides are equal and all interior angles are equal.
- The sum of the interior angles of a polygon with n sides is $(n - 2) \times 180°$
- The sum of the exterior angles of any polygon is 360°

Key words

- polygon
- regular polygon
- pentagon (5)
- hexagon (6)
- octagon (8)
- decagon (10)
- interior angle
- exterior angle

Example

Grade C

A regular polygon has an interior angle of 156°.

(a) Work out how many sides it has.

(b) Find the sum of the interior angles of this polygon.

Find the exterior angle. → (a)

Work out how many of these there are in 360° →

Exterior angle = 180 − 156 = 24°
(angles on a straight line)
360 ÷ 24 = 15
The polygon has 15 sides.

Use $(n - 2) \times 180°$ → (b) n = 15

$$(n - 2) \times 180° = (15 - 2) \times 180° = 13 \times 180° = 2340°$$

TIP
It is easier to work with exterior angles because these always add up to 360°.

- Shapes are **similar** if one shape is an **enlargement** of the other.
- When a shape is enlarged by scale factor k, the area of the enlarged shape is $k^2 \times$ area of the original shape.
- When a 3-D shape is enlarged by scale factor k, the volume of the enlarged shape is $k^3 \times$ volume of the original shape.
- Triangles are **congruent** if one of these facts is true:
  - three pairs of sides are equal (SSS)
  - two pairs of sides are equal and the angles between them are equal (SAS)
  - two pairs of angles are equal and the sides between them are equal (ASA)
  - both triangles have a right angle, the hypotenuses are equal and one pair of corresponding sides are equal (RHS).

Key words

- similar
- enlargement
- congruent
- SSS, SAS, ASA, RHS

Example

Grade A

Two shapes are mathematically similar.
The surface area of the smaller shape is 4225 cm^2.
The surface area of the larger shape is 6084 cm^2.
The volume of the smaller shape is 15 000 cm^3.
Find the volume of the larger shape.

Use the areas to calculate the scale factor for length, k.

Larger area $= k^2 \times$ smaller area

$$6084 = k^2 \times 4225$$

$$k^2 = \frac{6084}{4225}$$

$$k = \frac{\sqrt{6084}}{\sqrt{4225}} = \frac{78}{65} = \frac{6}{5}$$

Calculate the larger volume.

Larger volume $= k^3 \times$ smaller volume

$$= \frac{6^3}{5^3} \times 15\,000$$

$$= \frac{216}{125} \times 15\,000$$

$$= 25\,920 \text{ cm}^3$$

WATCH OUT!

The scale factors for length, area and volume are not the same.

Practice

Grade C

1 $ABCD$ is part of a regular octagon. DCE is part of a regular hexagon.

(a) Work out the size of angle BCE.

(b) Explain why angle CBE = angle CEB.

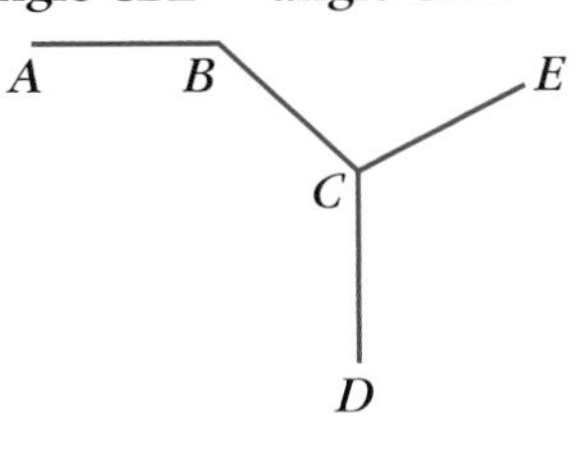

Grade C

2 $ABCD$ is part of a regular pentagon. Work out the size of angle BXC.

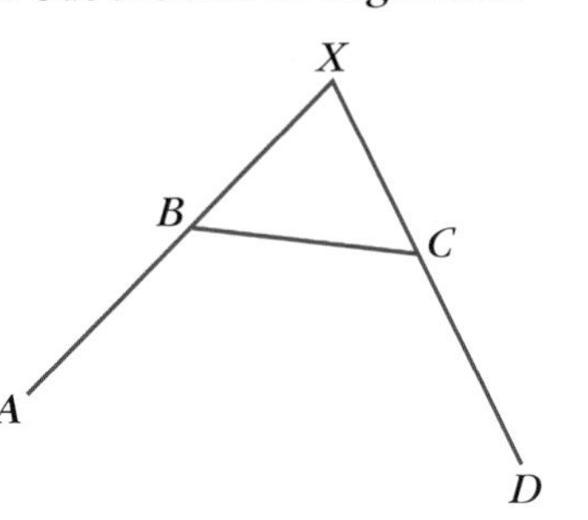

Grade C

3 A regular polygon has an exterior angle of 18°.

(a) Work out how many sides it has.

(b) Work out the total sum of the interior angles.

Grade A

4 $ABCDE$ is a regular pentagon.

(a) Prove that triangles DBC and CED are congruent.

(b) Hence show that triangles EBD and BEC are congruent.

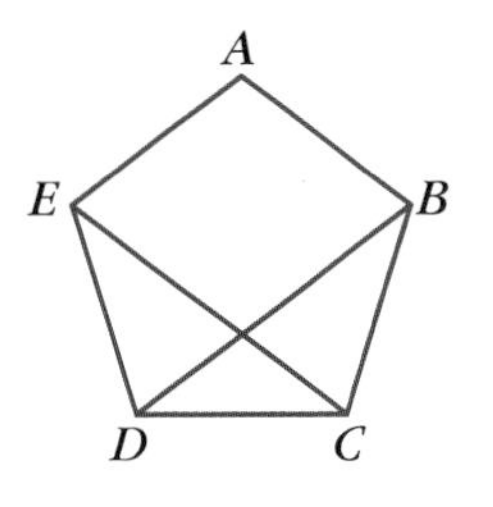

Grade A

5 Two solid cuboids are made from the same metal alloy.
The solids are mathematically similar.
The smaller cuboid weighs 729 g and the larger one weighs 1331 g.
The surface area of the larger cuboid is 242 cm^2.
Work out the surface area of the smaller cuboid.

Check your answers on page 176. For full worked solutions see the CD.
See the Student Book on the CD if you need more help.

Question	1	2	3	4	5
Grade	C	C	C	A	A
Student Book pages	U3 197–199	U3 197–199	U3 197–199	U3 207–208	U3 274–275

3-D shapes

- A **net** of a 3-D shape is a 2-D shape that can be folded to make the 3-D shape.
- A **prism** is a 3-D shape which has a uniform cross-section.
- Some 3-D shapes can be divided by a plane to produce two identical solid shapes. This plane is called a **plane of symmetry**.

Key words

net
prism
plane of symmetry

Example Grade D

Draw the net for this 3-D shape.

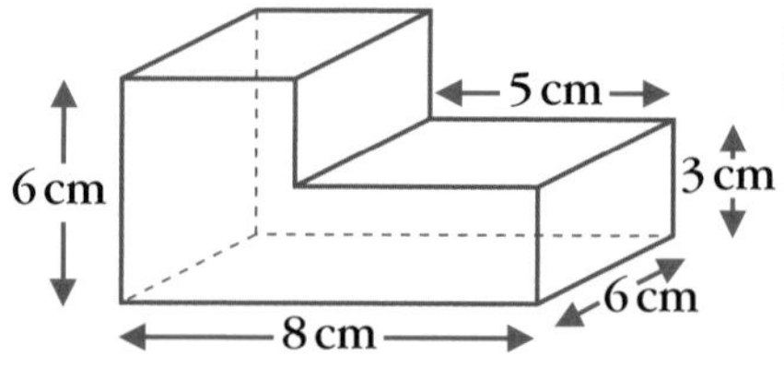

6 cm, 5 cm, 3 cm, 6 cm, 8 cm

① Draw the base.

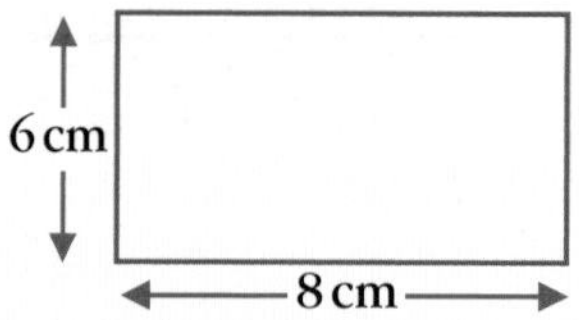

6 cm, 8 cm

② Draw the front and back faces.

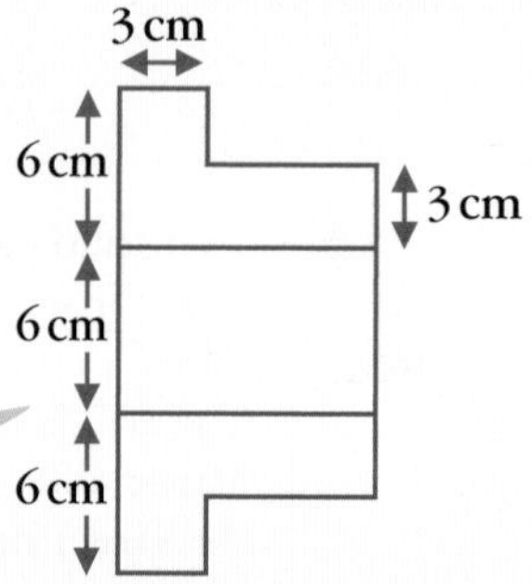

3 cm, 6 cm, 3 cm, 6 cm, 6 cm

TIP
Imagine folding down the faces one at a time.

③ Draw the remaining faces.

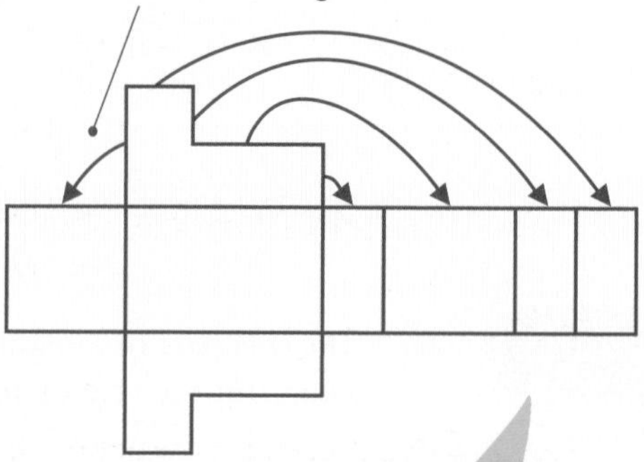

TIP
Remember that edges which will come together must be the same length.

Example Grade D

Draw the planes of symmetry for this prism.

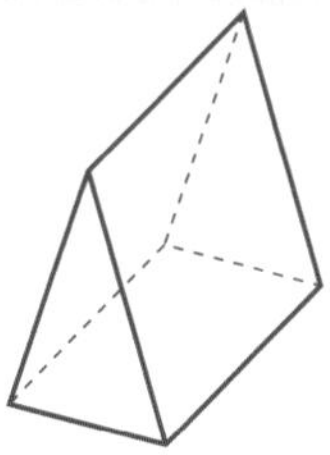

Draw a separate diagram for each plane.

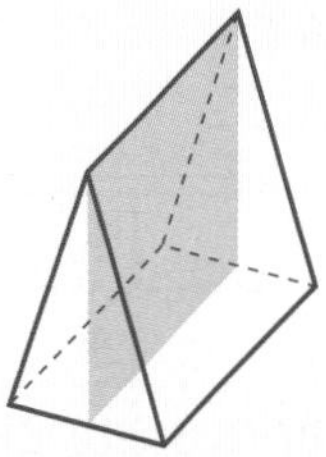

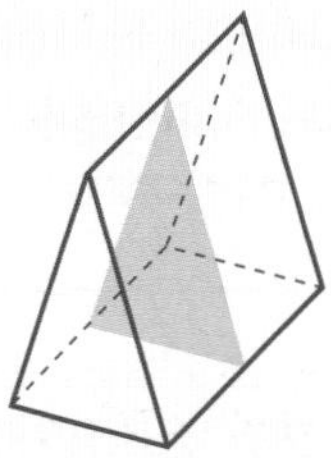

WATCH OUT!
Check that you have found *all* the planes of symmetry.

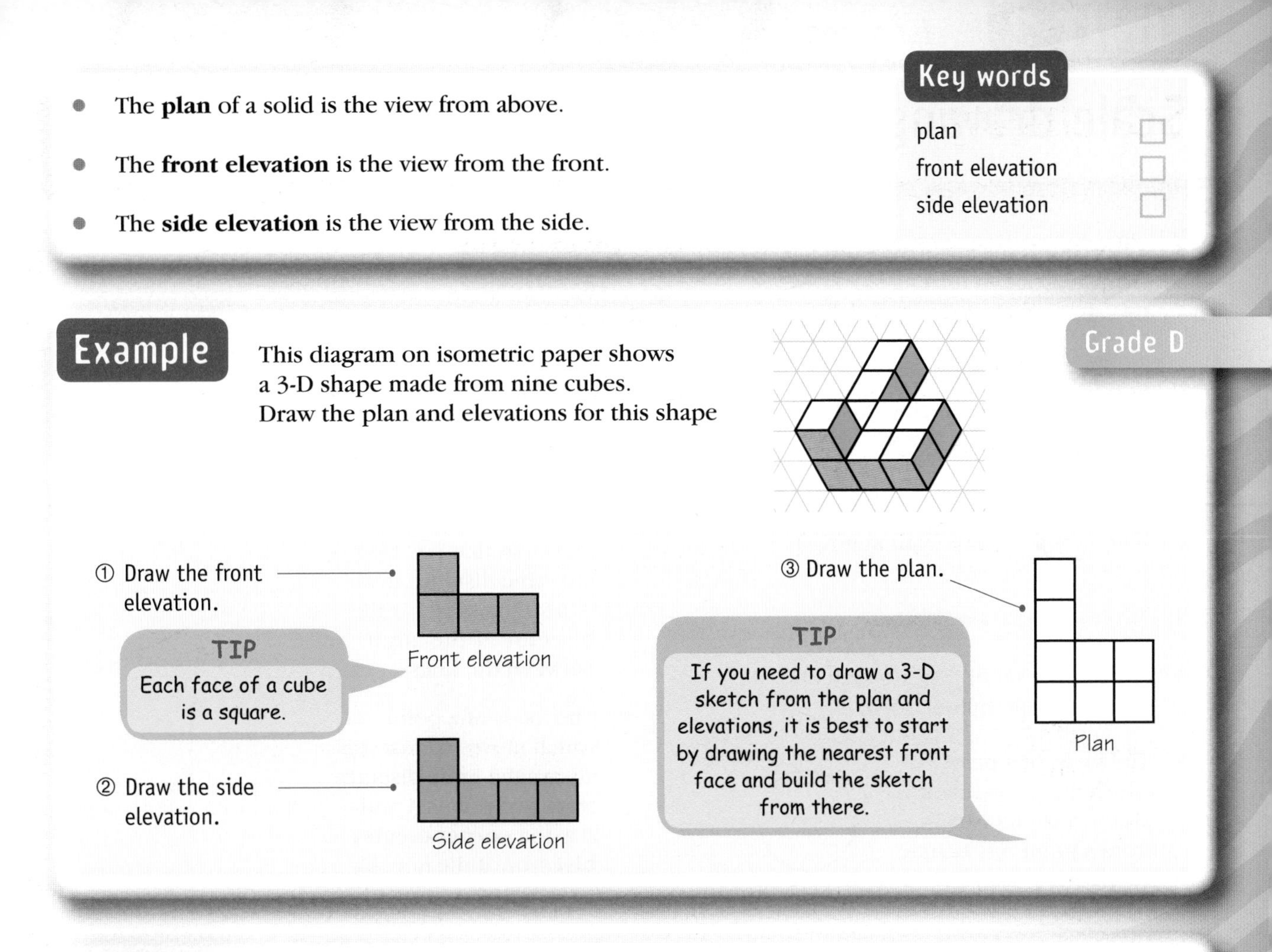

- The **plan** of a solid is the view from above.
- The **front elevation** is the view from the front.
- The **side elevation** is the view from the side.

Key words

plan ☐
front elevation ☐
side elevation ☐

Example

Grade D

This diagram on isometric paper shows a 3-D shape made from nine cubes.
Draw the plan and elevations for this shape

① Draw the front elevation.

Front elevation

TIP
Each face of a cube is a square.

② Draw the side elevation.

Side elevation

③ Draw the plan.

TIP
If you need to draw a 3-D sketch from the plan and elevations, it is best to start by drawing the nearest front face and build the sketch from there.

Plan

Practice

1 Here is a diagram of a 3-D shape.

Grade D

(a) Draw the plane of symmetry for the shape.
(b) Construct the net for the shape.
(c) Sketch the plan and elevations for the shape.

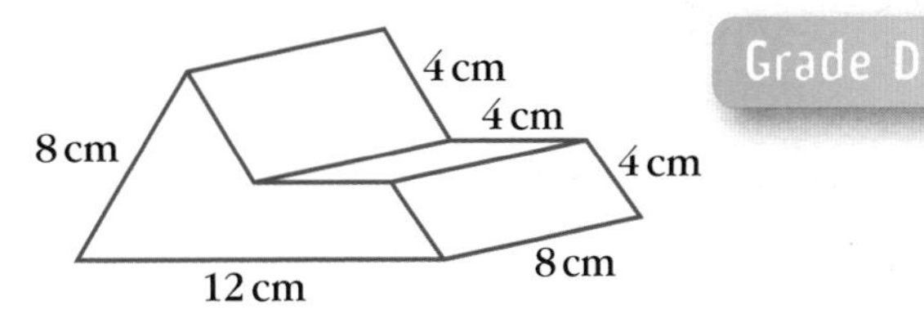

2 Draw a sketch of the 3-D prism with this plan and elevations.

Grade D

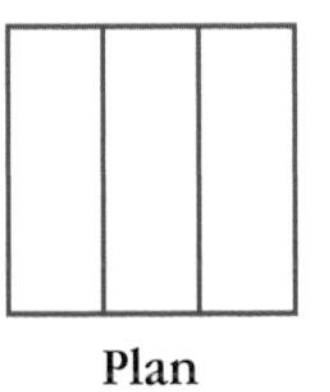
Plan

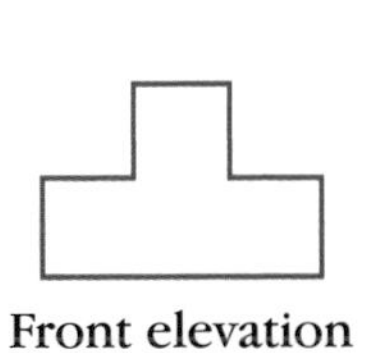
Front elevation

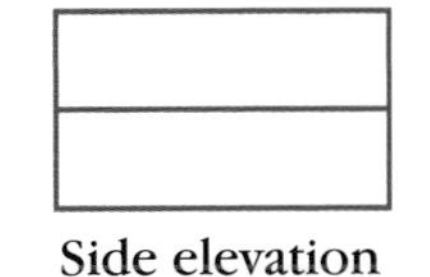
Side elevation

Check your answers on page 177. For full worked solutions see the CD.
See the Student Book on the CD if you need more help.

Question	1a	1bc	2
Grade	D	D	D
Student Book pages	U3 162–164	U3 189–193, 195–196	U3 195–196

Scale drawing, locus and bearings

- A **bearing** is the angle measured from facing North and turning clockwise. It is always a three-figure number.

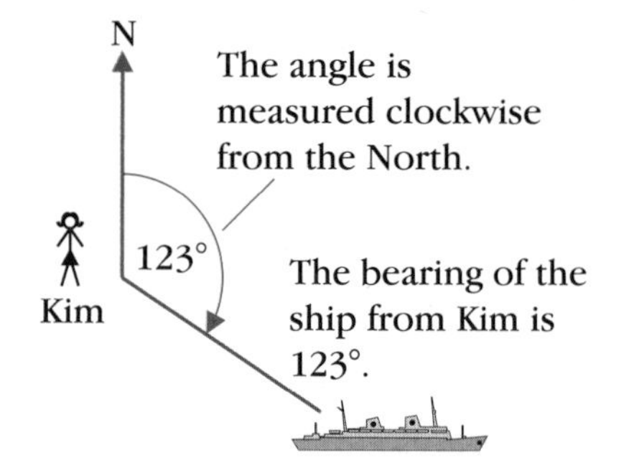

- A **locus** is a set of points which obey a particular rule (plural: loci).
- The locus of a point which moves so that it is always a fixed distance from a point A is a circle, centre A.

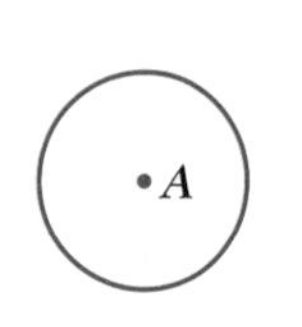

Key words

- bearing ☐
- locus ☐
- perpendicular bisector ☐
- angle bisector ☐

- The locus of a point that moves to that it is always the same distance from a straight line is two parallel lines with a semicircle at each end.
- The locus of a point that moves so that it is always the same distance from two straight lines is the **bisector of the angle** between the lines.
- The locus of a point which moves so that it is always the same distance from two points A and B is the **perpendicular bisector** of the line AB.

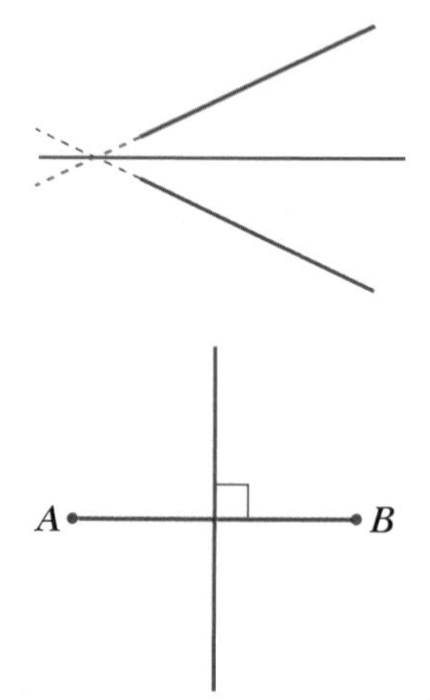

Example

Great Yarmouth is 30 kilometres from Norwich on a bearing of 095°.
Wells-next-the-Sea is 45 kilometres from Norwich on a bearing of 320°.

Grade D (a) Draw an accurate scale diagram using a scale of 1 cm for 5 kilometres.

Grade C (b) Radio X, based at Norwich, has a range of 20 kilometres. Shade the area that can receive broadcasts from Radio X.

Grade C (c) An aircraft flies on a path equidistant from Norwich and Wells-next-the-Sea. Construct this flight path on your diagram.

Use the scale to work out the lengths for the diagram.

(a) Scale 5 km is 1 cm
So ×6: 30 km is 6 cm
×9: 45 km is 9 cm

Mark North and your starting town.

Use a protractor to measure the bearing. Draw accurate lengths using a ruler.

Wells-next-the-Sea
N
9 cm
40°
95°
6 cm
Norwich
Great Yarmouth

TIP
A sketch helps you see roughly where to put the diagram.

TIP
320° clockwise is the same as 360° − 320° = 40° anticlockwise.

Locus of points 20 km from Norwich is a circle, radius 20 km. Using the scale, 20 km is 4 cm on the diagram.

The flight path is the perpendicular bisector of the line joining Norwich to Wells-next-the-Sea.

TIP

Draw the perpendicular bisector:

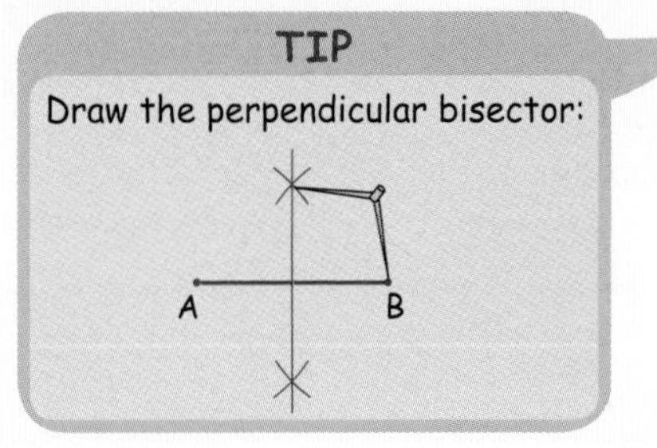

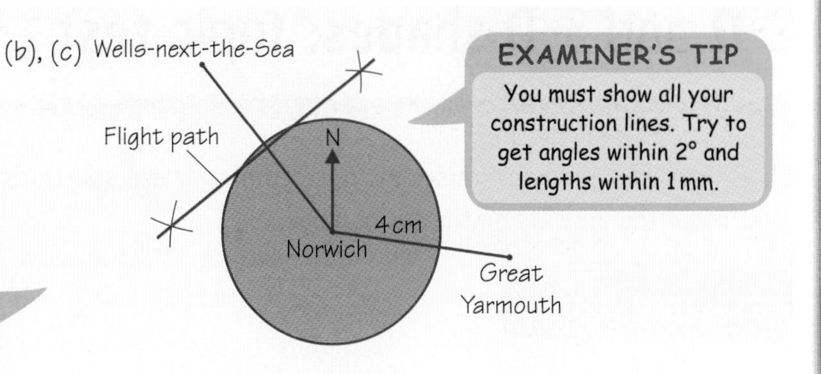

EXAMINER'S TIP

You must show all your construction lines. Try to get angles within 2° and lengths within 1 mm.

Practice

Grade D

1 (a) Use a ruler and protractor to make an accurate scale drawing of the diagram shown. Use a scale of 1 cm = 5 km.

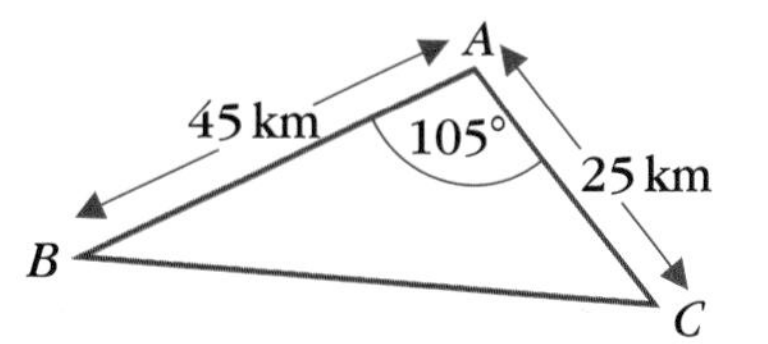

Grade C

(b) Shade all points that are 20 km or less from *C*.

Grade C

(c) Construct the locus of points that are the same distance from *AB* as from *AC*.

Grade C

2 Charlbury is 9 kilometres North of Witney. Woodstock is 11 kilometres from Witney on a bearing of 055°. Draw an accurate scale diagram using a scale of 1 cm to 2 km. Use your drawing to find the distance and bearing of Woodstock from Charlbury.

Grade C

3 Construct a triangle with sides 7 cm, 8 cm and 9 cm. Bisect the angle between the sides of length 8 cm and 9 cm. Construct the perpendicular bisector of the side of length 7 cm. Measure the obtuse angle where your bisector and perpendicular bisector intersect.

Check your answers on page 177. For full worked solutions see the CD. See the Student Book on the CD if you need more help.

Question	1a	1bc	2	3
Grade	D	C	C	C
Student Book pages	U3 182–186	U3 187–189	U3 179–181	U3 182–186

2-D and 3-D shapes: topic test

Check how well you know this topic by answering these questions.
First cover the answers on the facing page.

Test questions

1 Draw the plan and side elevation of this solid shape.

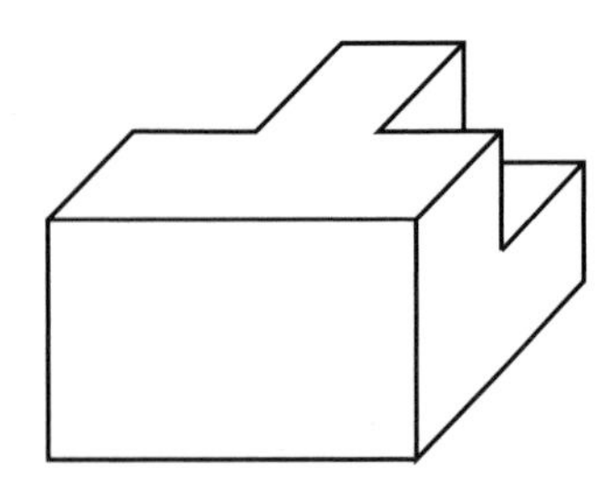

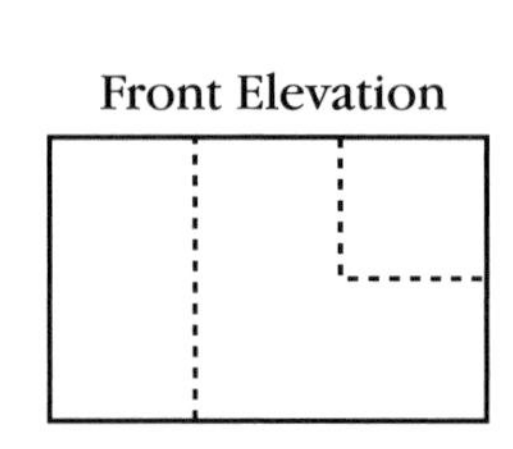

2 The scale of a map is 1 : 50 000.
The distance between two points on the map is 3.6 cm.
Work out the actual distance between the two places.

3 Mark two points, A and B, and construct the locus of the points that are equidistant from A and B.

4 Work out the sum of the interior angles of an irregular pentagon.

5 The sum of the interior angles of a polygon is 1260°.
Work out how many sides this polygon has.

6 Draw an angle of about 50°. Construct the bisector of this angle.

7 A, B and C are three towns.
The bearing of B from A is 070° and the distance is 50 km.
The bearing of C from B is 130° and the distance is 60 km.
Use a scale of 1 cm = 10 km to draw a scale diagram.
Use your diagram to find the distance and bearing of A from C.

8 The areas of two mathematically similar shapes are in the ratio 36 : 81.
Work out the ratio of the lengths and the ratio of the volumes.

9 $ABCD$ is a square.
APB and BQC are equilateral triangles.
Prove that triangles PAQ and PAD are congruent.

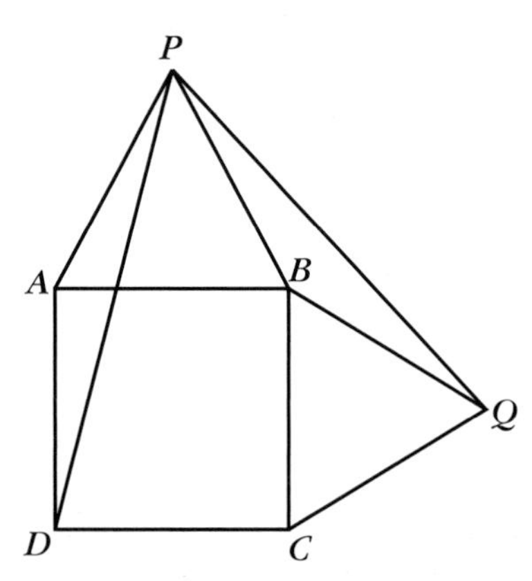

Now check your answers – see the facing page.

Cover this page while you answer the test questions opposite.

Worked answers

Revise this on...

D **1**

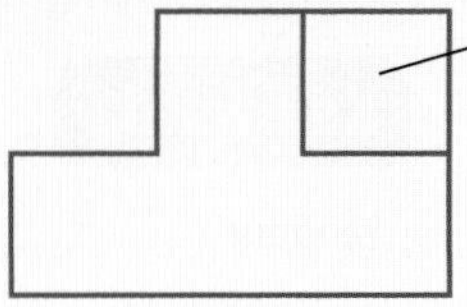

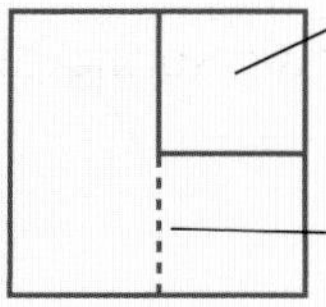

Plan — This face can be seen

Side Elevation — This face can be seen; This edge at the back cannot be seen, so it is dotted

D **2** Real distance = 50 000 × 3.6 cm = 180 000 cm = 1800 m = 1.8 km — pages 130–131

D **3** — pages 134–135

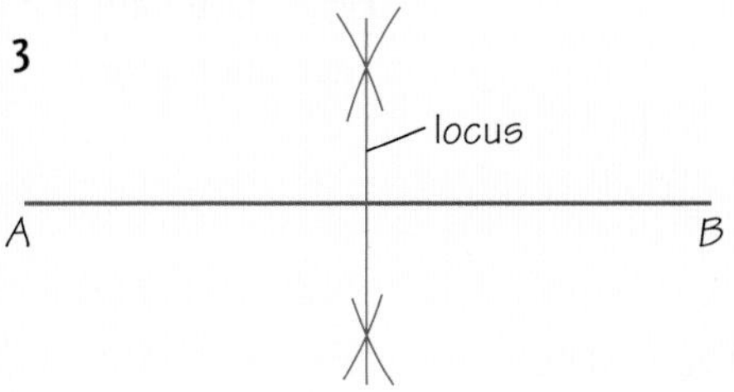

locus

A B

C **4** Angle sum = (n − 2) × 180° — page 130
n = 5 so angle sum = 3 × 180° = 540°

C **5** Angle sum = (n − 2) × 180° = 1260° → n − 2 = 7 → n = 9 — page 130
The polygon has 9 sides.

C **6**

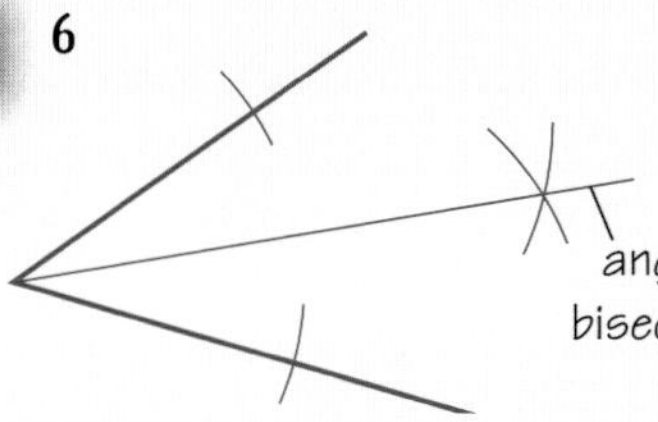

angle bisector

7 Diagram shown half size. — pages 134–135

The bearing of A from C is 283° and the distance is 95 km.

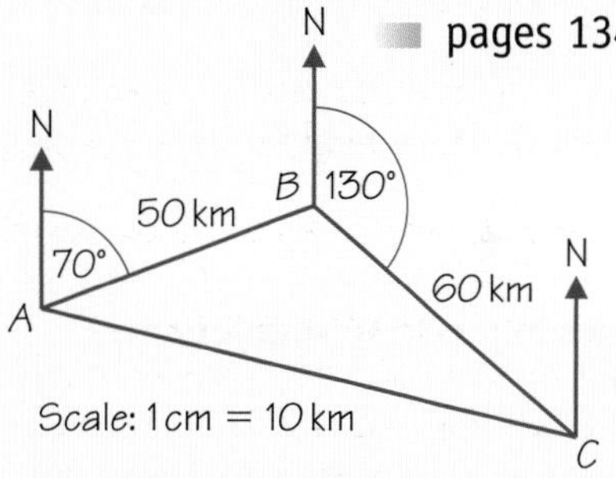

N, N, N, 50 km, B, 130°, 70°, A, 60 km, C

Scale: 1 cm = 10 km

B **8** Ratio of lengths $= \sqrt{36} : \sqrt{81} = 6:9 = 2:3$ — pages 130–131
Ratio of volumes $= 2^3 : 3^3 = 8:27$

A **9** PA = PB sides of an equilateral triangle — pages 130–131
AD = BQ both equal sies BC
opposite sides of a square; sides of an equilateral triangle
$P\hat{A}D = 90° + 60° = 150°$
$P\hat{A}Q = 360° - 90° - 90° - 60° = 150°$
So $P\hat{A}D = P\hat{B}Q$
So triangles PAD and PBQ are congruent.

Tick the questions you got right.

Question	1	2	3	4	5	6	7	8	9
Grade	D	D	D	C	C	C	C	B	A

Mark the grade you are working at on your revision planner on page x.

Circle theorems and proof

Key words

tangent
perpendicular
chord
subtend
segment
cyclic quadrilateral
alternate segment

- The angle between a **tangent** and a radius is 90°.
- The lengths of the two tangents from a point to a circle are equal.
- The **perpendicular** from the centre of a circle to a **chord** bisects the chord.
- The angle **subtended** at the centre of a circle is twice the angle at the circumference.

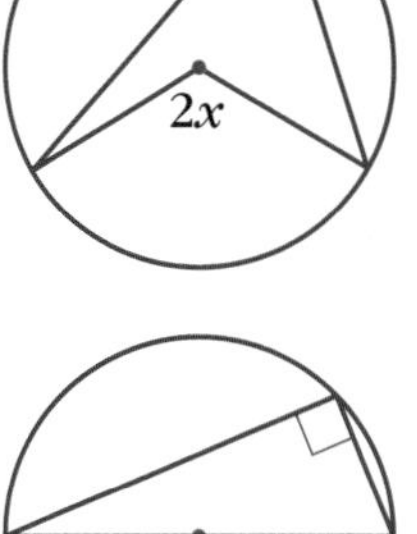

- The angle in a semicircle is a right angle.

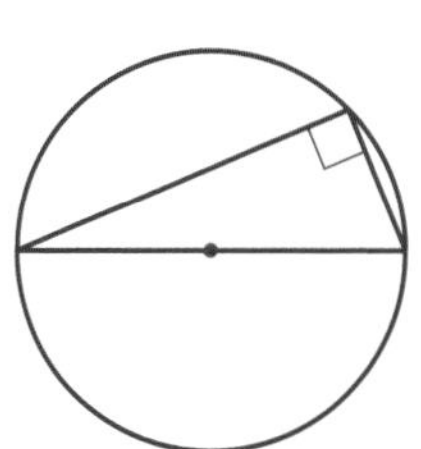

- Angles in the same **segment** are equal.
- Opposite angles of a **cyclic quadrilateral** add up to 180°.
  $a + c = b + d = 180°$

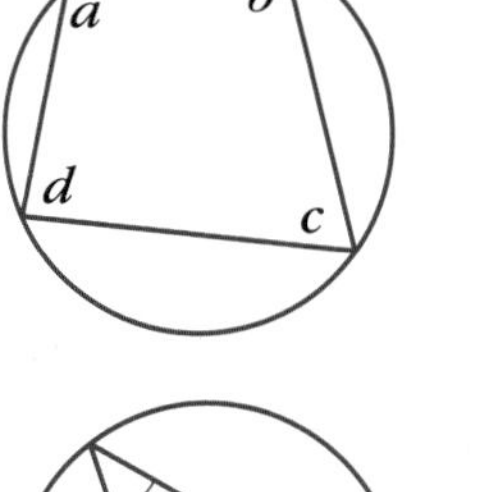

- The angle between a tangent and a chord is equal to the angle in the **alternate segment**.

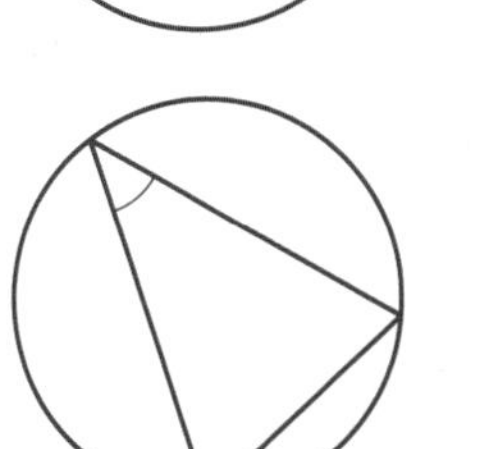

EXAMINER'S TIP

See Student Book pages 251–254 on the CD to remind yourself how to prove these theorems.
You may need to do this in your GCSE exam.

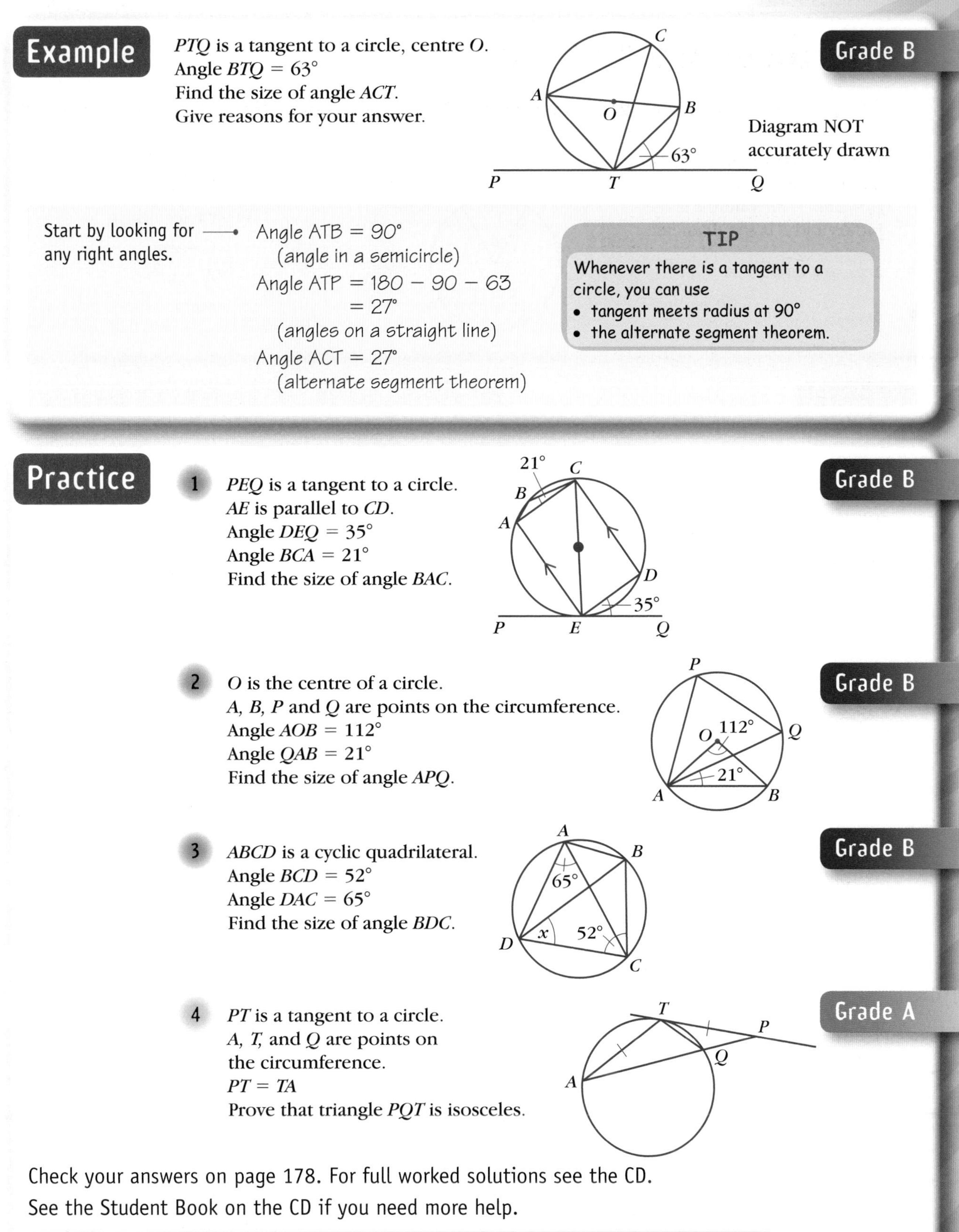

Example — Grade B

PTQ is a tangent to a circle, centre *O*.
Angle $BTQ = 63°$
Find the size of angle *ACT*.
Give reasons for your answer.

Diagram NOT accurately drawn

Start by looking for any right angles. → Angle ATB = 90°
(angle in a semicircle)
Angle ATP = 180 − 90 − 63
= 27°
(angles on a straight line)
Angle ACT = 27°
(alternate segment theorem)

TIP

Whenever there is a tangent to a circle, you can use
- tangent meets radius at 90°
- the alternate segment theorem.

Practice

1 (Grade B) *PEQ* is a tangent to a circle.
AE is parallel to *CD*.
Angle $DEQ = 35°$
Angle $BCA = 21°$
Find the size of angle *BAC*.

2 (Grade B) *O* is the centre of a circle.
A, *B*, *P* and *Q* are points on the circumference.
Angle $AOB = 112°$
Angle $QAB = 21°$
Find the size of angle *APQ*.

3 (Grade B) *ABCD* is a cyclic quadrilateral.
Angle $BCD = 52°$
Angle $DAC = 65°$
Find the size of angle *BDC*.

4 (Grade A) *PT* is a tangent to a circle.
A, *T*, and *Q* are points on the circumference.
$PT = TA$
Prove that triangle *PQT* is isosceles.

Check your answers on page 178. For full worked solutions see the CD.
See the Student Book on the CD if you need more help.

Question	1	2	3	4
Grade	B	B	B	A
Student Book pages	U3 248–249	U3 245–247	U3 248–249	U3 251–254

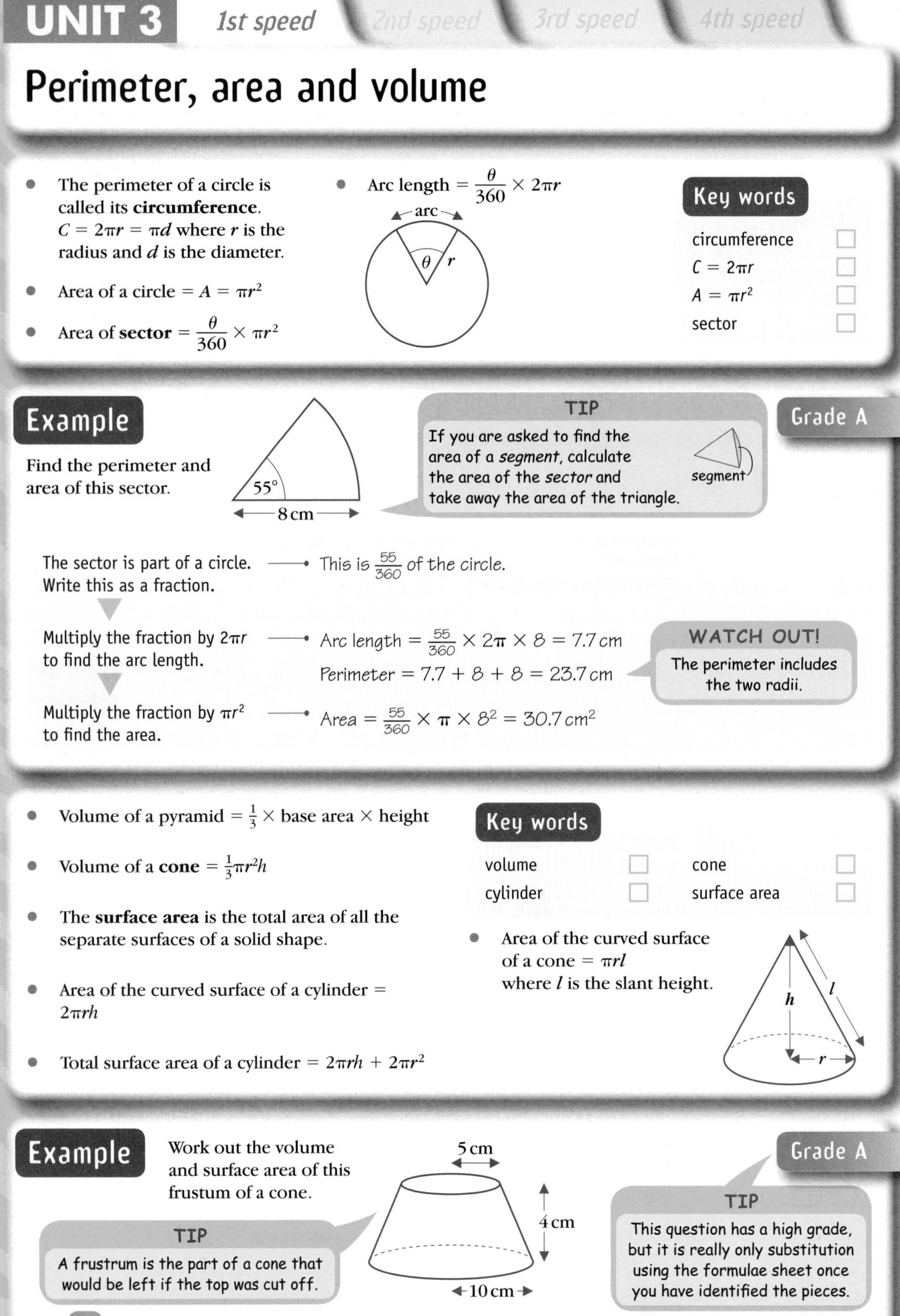

Perimeter, area and volume

- The perimeter of a circle is called its **circumference**. $C = 2\pi r = \pi d$ where r is the radius and d is the diameter.
- Area of a circle $= A = \pi r^2$
- Area of **sector** $= \frac{\theta}{360} \times \pi r^2$
- Arc length $= \frac{\theta}{360} \times 2\pi r$

Key words

- circumference
- $C = 2\pi r$
- $A = \pi r^2$
- sector

Example

Grade A

Find the perimeter and area of this sector.

TIP

If you are asked to find the area of a *segment*, calculate the area of the *sector* and take away the area of the triangle.

The sector is part of a circle. Write this as a fraction. → This is $\frac{55}{360}$ of the circle.

Multiply the fraction by $2\pi r$ to find the arc length. → Arc length $= \frac{55}{360} \times 2\pi \times 8 = 7.7$ cm

Perimeter $= 7.7 + 8 + 8 = 23.7$ cm

WATCH OUT!

The perimeter includes the two radii.

Multiply the fraction by πr^2 to find the area. → Area $= \frac{55}{360} \times \pi \times 8^2 = 30.7\ \text{cm}^2$

- Volume of a pyramid $= \frac{1}{3} \times$ base area $\times$ height
- Volume of a **cone** $= \frac{1}{3}\pi r^2 h$
- The **surface area** is the total area of all the separate surfaces of a solid shape.
- Area of the curved surface of a cylinder $= 2\pi rh$
- Total surface area of a cylinder $= 2\pi rh + 2\pi r^2$

Key words

- volume
- cylinder
- cone
- surface area

- Area of the curved surface of a cone $= \pi rl$ where l is the slant height.

Example

Grade A

Work out the volume and surface area of this frustum of a cone.

TIP

A frustrum is the part of a cone that would be left if the top was cut off.

TIP

This question has a high grade, but it is really only substitution using the formulae sheet once you have identified the pieces.

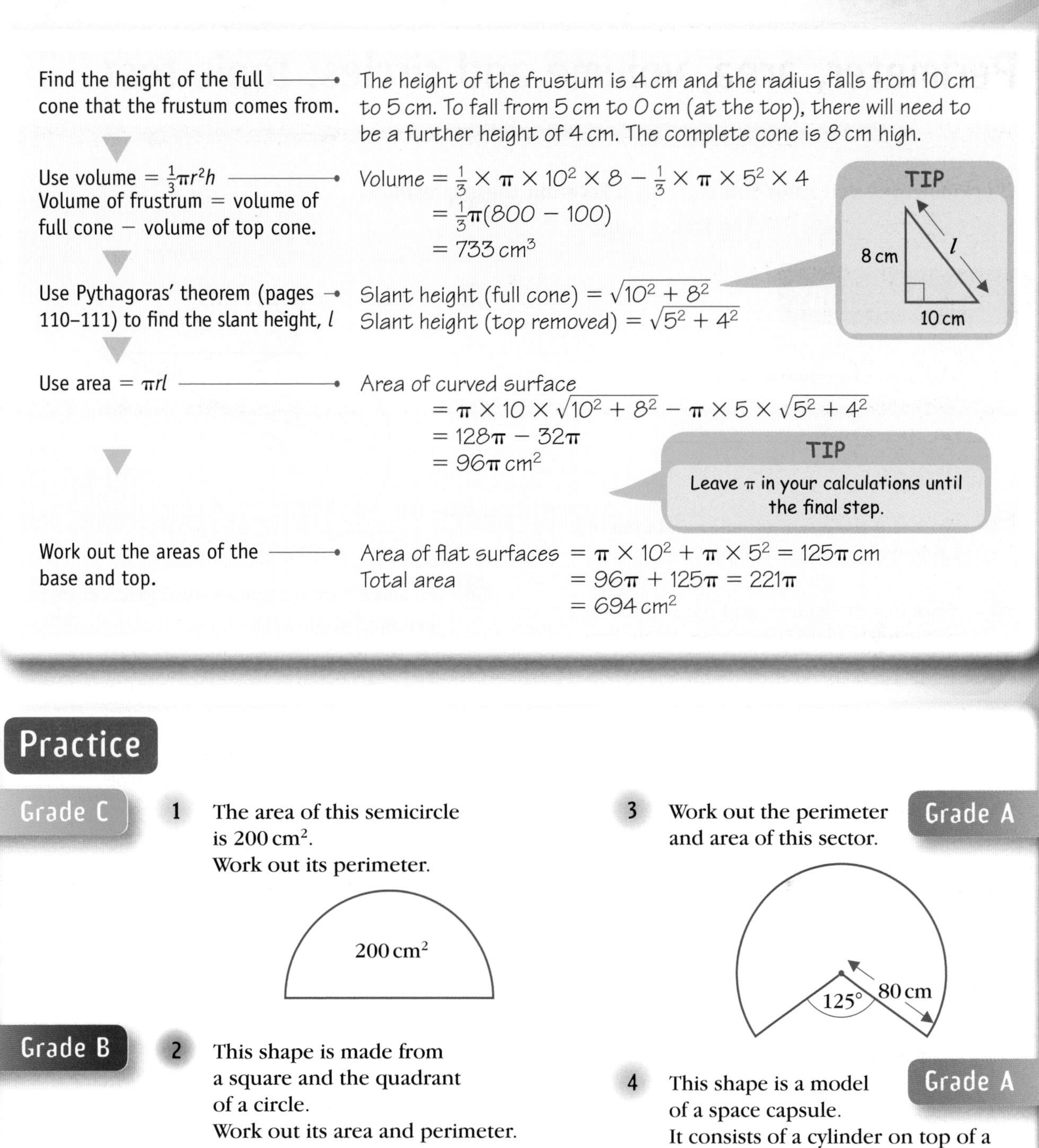

Find the height of the full cone that the frustum comes from.	The height of the frustum is 4 cm and the radius falls from 10 cm to 5 cm. To fall from 5 cm to 0 cm (at the top), there will need to be a further height of 4 cm. The complete cone is 8 cm high.
Use volume $= \frac{1}{3}\pi r^2 h$ Volume of frustrum = volume of full cone − volume of top cone.	Volume $= \frac{1}{3} \times \pi \times 10^2 \times 8 - \frac{1}{3} \times \pi \times 5^2 \times 4$ $= \frac{1}{3}\pi(800 - 100)$ $= 733\text{ cm}^3$
Use Pythagoras' theorem (pages 110–111) to find the slant height, l	Slant height (full cone) $= \sqrt{10^2 + 8^2}$ Slant height (top removed) $= \sqrt{5^2 + 4^2}$
Use area $= \pi r l$	Area of curved surface $= \pi \times 10 \times \sqrt{10^2 + 8^2} - \pi \times 5 \times \sqrt{5^2 + 4^2}$ $= 128\pi - 32\pi$ $= 96\pi\text{ cm}^2$
Work out the areas of the base and top.	Area of flat surfaces $= \pi \times 10^2 + \pi \times 5^2 = 125\pi$ cm Total area $= 96\pi + 125\pi = 221\pi$ $= 694\text{ cm}^2$

TIP

TIP

Leave π in your calculations until the final step.

Practice

Grade C

1 The area of this semicircle is 200 cm^2.
Work out its perimeter.

Grade B

2 This shape is made from a square and the quadrant of a circle.
Work out its area and perimeter.

Grade A

3 Work out the perimeter and area of this sector.

Grade A

4 This shape is a model of a space capsule.
It consists of a cylinder on top of a frustum.
Work out its volume and surface area.

Check your answers on page 178.
For full worked solutions see the CD.
See the Student Book on the CD if you need more help.

Question	1	2	3	4
Grade	C	B	A	A
Student Book pages	U3 259–262	U3 259–262	U3 268–269	U3 275–277

Perimeter, area, volume and circles: topic test

Check how well you know this topic by answering these questions.
First cover the answers on the facing page.

Test questions

1 The area of a circle is 30.6 cm^2.
Work out
(a) the radius
(b) the circumference.

2 Work out the circumference and area of a circle with diameter 15 cm.

3 Work out the volume and the surface area of a cylinder with base radius 12 cm and height 20 cm.

4 Work out the perimeter of a quadrant of a circle with radius 20 cm.

5 Work out the area of the shaded segment.

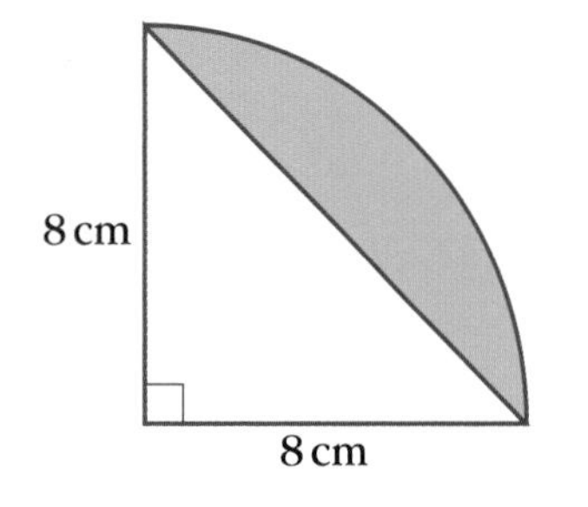

6 Work out the value of y.

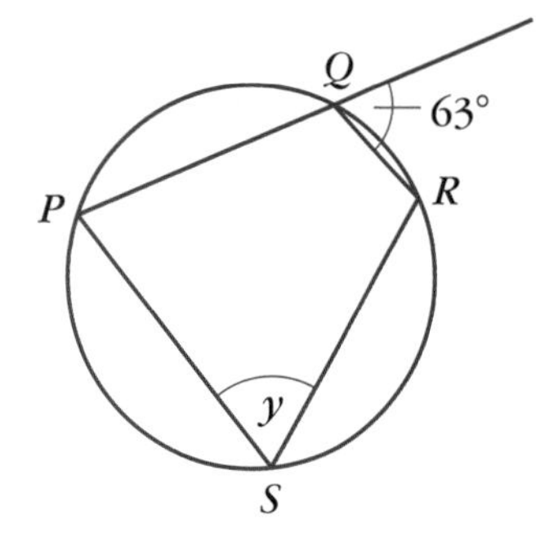

7 Work out the value of x.

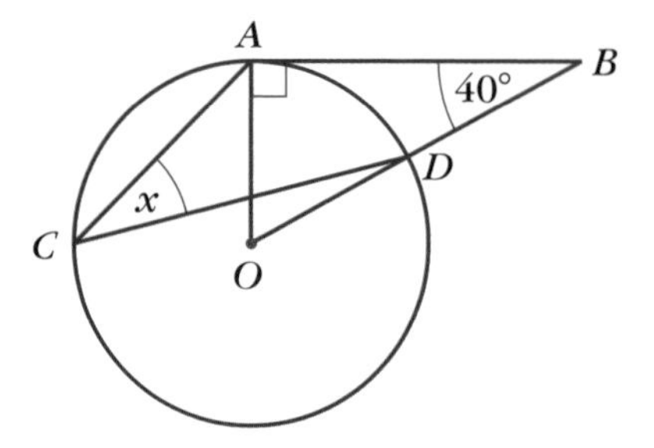

8 PS and PT are tangents to the circle, centre O.
(a) Find angle STO.
(b) Work out the size of angle SQT.
Give reasons for your answers.

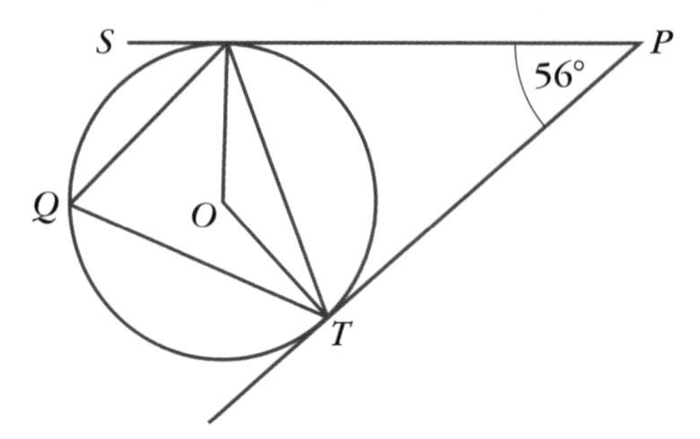

9 Work out the perimeter of this shape, which is made from a square and two sectors. The lengths of the sides of the square are 32 cm and the angles of the sectors are 35°.

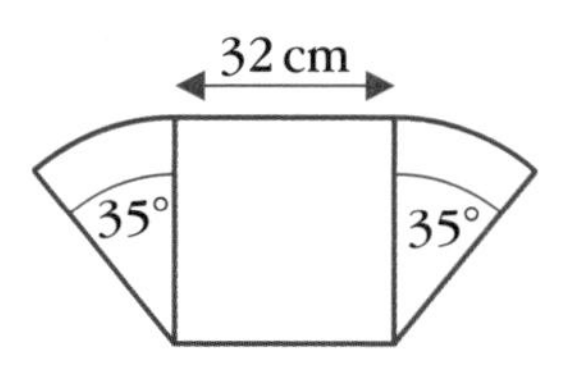

10 A cone rests on a cylinder. The cone and the cylinder both have radius 5 cm and height 12 cm. The slant height of the cone is 13 cm.
Work out the volume of the solid shape. Leave your answer in terms of π.

Now check your answers – see the facing page.

Cover this page while you answer the test questions opposite.

Worked answers

Revise this on...

C **1** (a) Area $= \pi r^2 = 30.6 \rightarrow r^2 = 30.6 \div \pi = 9.74 \rightarrow r = 3.12$ cm — page 140

(b) Circumference $= 2\pi r = 2\pi \times 3.12 = 19.6$ cm

C **2** Circumference $= \pi d = 15\pi = 47.1$ cm — page 140

Area $= \pi r^2 = \pi \times 7.5^2 = 176.7$ cm^2

C **3** Volume $= \pi r^2 h = \pi \times 12^2 \times 20 = 2880\pi = 9048$ cm^3 — page 140

Surface area $= 2\pi rh + 2\pi r^2$ (curved surface and two ends)

$= 2\pi(12 \times 20 + 12^2) = 2413$ cm^2

B **4** Perimeter = radius + radius + arc — page 140

$= 20 + 20 + \frac{1}{4} \times 2\pi \times 20 = 40 + 10\pi = 71.4$ cm

B **5** Area of segment = Area of quadrant − Area of triangle — page 140

$= \frac{1}{4}\pi(8)^2 - \frac{1}{2} \times 8 \times 8$

$= 50.3 - 32$

$= 17.7$ cm^2

B **6** $P\hat{Q}R = 180° - 63° = 127°$ (angles on a straight line) — pages 138–139

$P\hat{S}R = 180 - 127$

$y = 63°$ (opposite angles of a cyclic quadrilateral)

B **7** $O\hat{A}B = 90°$ (radius perpendicular to tangent) — pages 138–139

$A\hat{O}B = 180° - 90° - 40° = 50°$ (angles in a triangle)

$A\hat{C}D = \frac{1}{2} \times 50° = 25°$ (angle at the circumference)

B **8** (a) Angle PSO = angle PTO = 9° — page 140

(angle between radius and tangent)

Angle SOT $= 360° - 90° - 90° - 56° = 124°$

(angles of a quadrilateral = 360°)

Angle STO $= \frac{1}{2}(180° - 124°) = 28°$

(base angle of an isosceles triangle)

(b) Angle SQT $= \frac{1}{2} \times 124 = 62°$

(angle at circumference $= \frac{1}{2}$ angle at centre)

A **9** The two sectors have radius equal to the side of the square = 32 cm — page 140

Each sector is $\frac{35}{360}$ of a circle.

Length of each arc $= \frac{35}{360} \times 2\pi \times 32 = 19.55$ cm

Perimeter = two arcs + two radii + two sides of the square

$= 39.1 + 64 + 64 = 167$ cm

A **10** Volume $= \pi r^2 h + \frac{1}{3}\pi r^2 h = \pi \times 25 \times 12 + \frac{1}{3} \times \pi \times 25 \times 12 = 400\pi$ cm^3 — page 140

Tick the questions you got right.

Question	1	2	3	4	5	6	7	8	9	10
Grade	C	C	B	B	B	B	B	B	A	A

Mark the grade you are working at on your revision planner on page x.

Reflections, rotations and enlargements

- A **reflection** in a line produces a mirror image. The mirror line is a **line of symmetry**.
- To describe a reflection fully you need to give the equation of the mirror line.
- Images formed by turning are called **rotations**.
- To describe a rotation fully you need to give the **centre of rotation**, the amount of turn (the angle), and whether it is clockwise or anticlockwise.

Key words

- reflection ☐
- line of symmetry ☐
- rotation ☐
- centre of rotation ☐

Example

Grade C

Describe fully the single transformation which maps shape **P** on to **Q**.

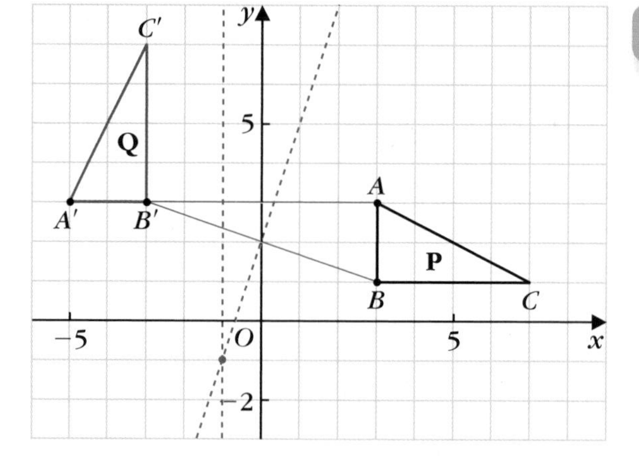

Has the shape been flipped over? (Reflection) → No.

Has it been turned? (Rotation) → Yes. It is a rotation.

Compare corresponding sides on the object and image. → A′B′ is perpendicular to AB. It has turned through 90° clockwise.

Find the centre of rotation. → The centre is at (−1, −1)

TIP

Method 1: Test different coordinate points using tracing paper.

Method 2: Draw the perpendicular bisectors of lines joining corresponding points.

- An **enlargement** changes the size but not the shape of an object. The **scale factor** of the enlargement is the value that the lengths of the original object are multiplied by.
- To describe an enlargement fully you need to give the **centre of enlargement** and the scale factor.
- An enlargement by a scale factor smaller than 1 gives a reduced image.
- An enlargement by a negative scale factor gives an image on the opposite side of the centre of enlargement.

Key words

- enlargement ☐
- scale factor ☐
- centre of enlargement ☐

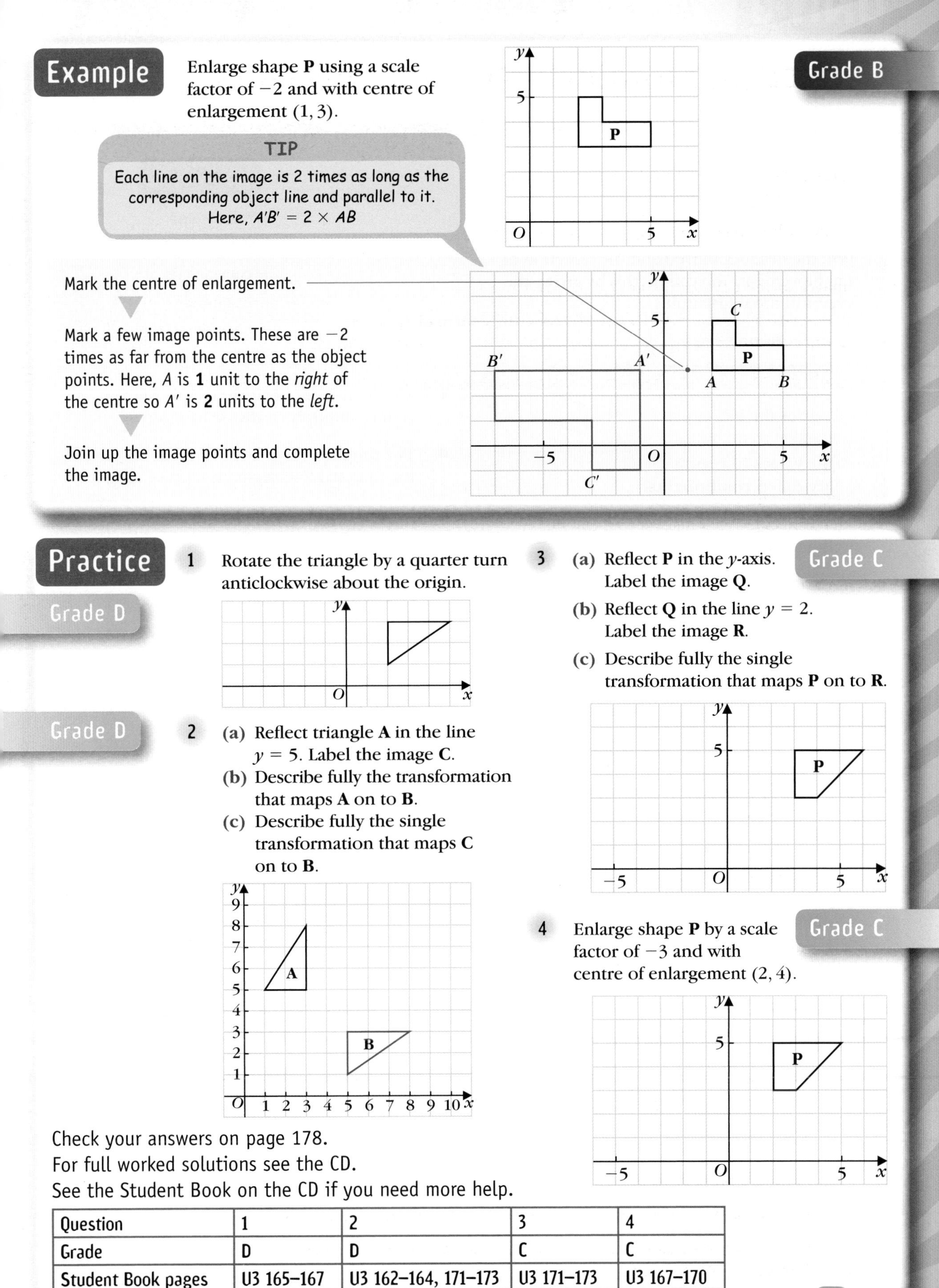

Example

Grade B

Enlarge shape **P** using a scale factor of -2 and with centre of enlargement $(1, 3)$.

TIP
Each line on the image is 2 times as long as the corresponding object line and parallel to it. Here, $A'B' = 2 \times AB$

Mark the centre of enlargement.

Mark a few image points. These are -2 times as far from the centre as the object points. Here, A is **1** unit to the *right* of the centre so A' is **2** units to the *left*.

Join up the image points and complete the image.

Practice

Grade D

1 Rotate the triangle by a quarter turn anticlockwise about the origin.

Grade D

2 (a) Reflect triangle **A** in the line $y = 5$. Label the image **C**.
(b) Describe fully the transformation that maps **A** on to **B**.
(c) Describe fully the single transformation that maps **C** on to **B**.

Grade C

3 (a) Reflect **P** in the y-axis. Label the image **Q**.
(b) Reflect **Q** in the line $y = 2$. Label the image **R**.
(c) Describe fully the single transformation that maps **P** on to **R**.

Grade C

4 Enlarge shape **P** by a scale factor of -3 and with centre of enlargement $(2, 4)$.

Check your answers on page 178.
For full worked solutions see the CD.
See the Student Book on the CD if you need more help.

Question	1	2	3	4
Grade	D	D	C	C
Student Book pages	U3 165–167	U3 162–164, 171–173	U3 171–173	U3 167–170

Vectors and translations

Key words

- translation
- column vector
- components
- magnitude
- position vector

- A **translation** moves every point on a shape the same distance and direction.
- To describe a translation fully you need to give the distance moved and the direction of the movement. You can do this by giving the vector of the translation.
- In the **column vector** $\begin{pmatrix} x \\ y \end{pmatrix}$
  - x gives the movement parallel to the x-axis
  - y gives the movement parallel to the y-axis
  - the values of x and y are called **components**.
- The translation that takes A to B can be written as a translation vector $\overrightarrow{AB}$.
- If $\overrightarrow{AD} = \overrightarrow{BC}$, then AD and BC are parallel and have equal length.
- A vector described as $\mathbf{a}$ has a unique length and direction. The vector with the same length but opposite direction is $-\mathbf{a}$.
- For any two vectors $\mathbf{a}$ and $\mathbf{b}$, $\mathbf{a} + \mathbf{b} = \mathbf{b} + \mathbf{a}$ and $\mathbf{a} - \mathbf{b} = \mathbf{a} + (-\mathbf{b})$
- If a vector $\mathbf{a}$ is multiplied by a scalar k then the vector $k\mathbf{a}$ is parallel to $\mathbf{a}$ and is equal to k times $\mathbf{a}$.
- The basic rules of algebra apply but you cannot multiply or divide by a vector.
- The **magnitude** of a vector is the length of the directed line segment representing it.
- In general, the magnitude of a vector $\begin{pmatrix} x \\ y \end{pmatrix}$ is $\sqrt{x^2 + y^2}$
- The **position vector** of a point P is $\overrightarrow{OP}$, where O is usually the origin.
- If A and B have position vectors $\mathbf{a}$ and $\mathbf{b}$ respectively, then the position vector of the mid-point, M, of the line joining A to B is $\overrightarrow{OM} = \mathbf{m} = \frac{1}{2}(\mathbf{a} + \mathbf{b})$

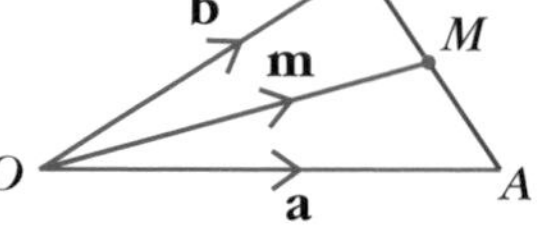

Example

Grade A

Find the values of a and b in

$$a\begin{pmatrix} 2 \\ 3 \end{pmatrix} + b\begin{pmatrix} 7 \\ -3 \end{pmatrix} = \begin{pmatrix} -8 \\ 15 \end{pmatrix}$$

Write down the x and y components separately to give two equations.

$$2a + 7b = -8 \quad (1)$$
$$3a - 3b = 15 \quad (2)$$

TIP Each component defines one equation.

Solve these as simultaneous equations.

$$(2) \times \tfrac{2}{3}: 2a - 2b = 10 \quad (3)$$
$$(1) - (3): 9b = -18$$
$$b = -2$$

Substitute $b = -2$ into (1):

$$2a - 14 = -8$$
$$2a = 6$$
$$a = 3$$

TIP For more on simultaneous equations see page 118.

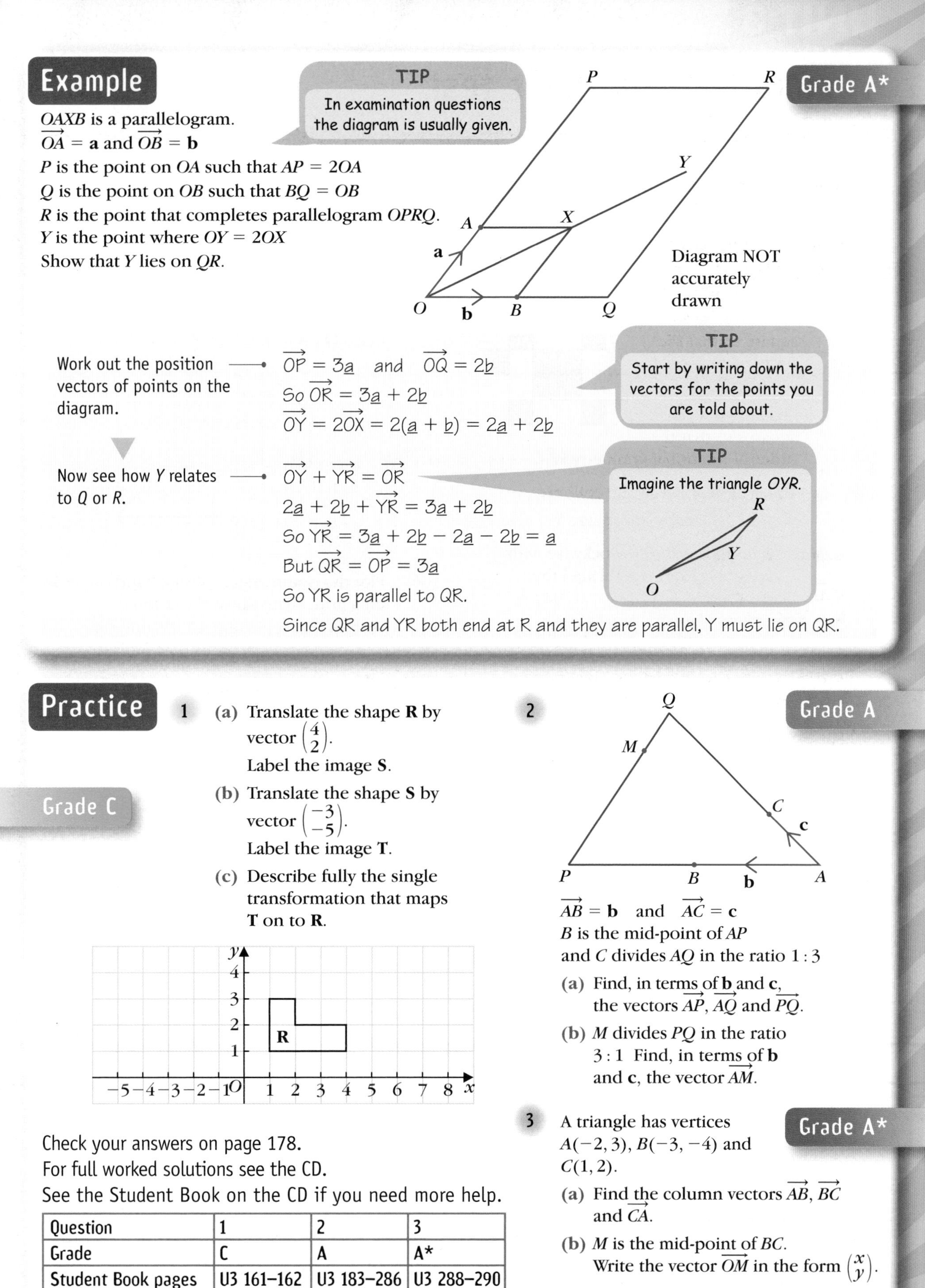

Example

Grade A*

$OAXB$ is a parallelogram.
$\overrightarrow{OA} = \mathbf{a}$ and $\overrightarrow{OB} = \mathbf{b}$
P is the point on OA such that $AP = 2OA$
Q is the point on OB such that $BQ = OB$
R is the point that completes parallelogram $OPRQ$.
Y is the point where $OY = 2OX$
Show that Y lies on QR.

TIP
In examination questions the diagram is usually given.

Work out the position vectors of points on the diagram. → $\overrightarrow{OP} = 3\underline{a}$ and $\overrightarrow{OQ} = 2\underline{b}$

So $\overrightarrow{OR} = 3\underline{a} + 2\underline{b}$

$\overrightarrow{OY} = 2\overrightarrow{OX} = 2(\underline{a} + \underline{b}) = 2\underline{a} + 2\underline{b}$

TIP
Start by writing down the vectors for the points you are told about.

Now see how Y relates to Q or R. → $\overrightarrow{OY} + \overrightarrow{YR} = \overrightarrow{OR}$

$2\underline{a} + 2\underline{b} + \overrightarrow{YR} = 3\underline{a} + 2\underline{b}$

So $\overrightarrow{YR} = 3\underline{a} + 2\underline{b} - 2\underline{a} - 2\underline{b} = \underline{a}$

But $\overrightarrow{QR} = \overrightarrow{OP} = 3\underline{a}$

So YR is parallel to QR.

Since QR and YR both end at R and they are parallel, Y must lie on QR.

TIP
Imagine the triangle OYR.

Practice

1 **Grade C**

(a) Translate the shape **R** by vector $\begin{pmatrix} 4 \\ 2 \end{pmatrix}$.
Label the image **S**.

(b) Translate the shape **S** by vector $\begin{pmatrix} -3 \\ -5 \end{pmatrix}$.
Label the image **T**.

(c) Describe fully the single transformation that maps **T** on to **R**.

2 **Grade A**

$\overrightarrow{AB} = \mathbf{b}$ and $\overrightarrow{AC} = \mathbf{c}$
B is the mid-point of AP
and C divides AQ in the ratio $1:3$

(a) Find, in terms of **b** and **c**, the vectors $\overrightarrow{AP}$, $\overrightarrow{AQ}$ and $\overrightarrow{PQ}$.

(b) M divides PQ in the ratio $3:1$ Find, in terms of **b** and **c**, the vector $\overrightarrow{AM}$.

3 **Grade A***

A triangle has vertices $A(-2, 3)$, $B(-3, -4)$ and $C(1, 2)$.

(a) Find the column vectors $\overrightarrow{AB}$, $\overrightarrow{BC}$ and $\overrightarrow{CA}$.

(b) M is the mid-point of BC.
Write the vector $\overrightarrow{OM}$ in the form $\begin{pmatrix} x \\ y \end{pmatrix}$.

Check your answers on page 178.
For full worked solutions see the CD.
See the Student Book on the CD if you need more help.

Question	1	2	3
Grade	C	A	A*
Student Book pages	U3 161–162	U3 183–286	U3 288–290

Transformations: topic test

Check how well you know this topic by answering these questions.
First cover the answers on the facing page.

Test questions

1 **(a)** Shade **one** more square so that the order of rotational symmetry is 2.

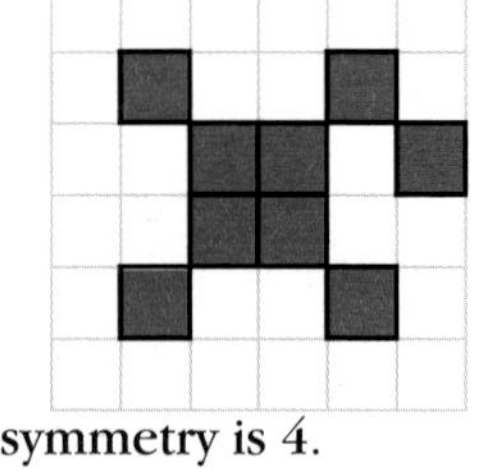

(b) Shade **two** further squares so that the order of rotational symmetry is 4.

2 **(a)** Translate shape **A** by the column vector $\begin{pmatrix} 2 \\ -3 \end{pmatrix}$. Label the image **Y**.

(b) Rotate shape **A** by 90° clockwise with centre of rotation (1, 3). Label the image **Z**.

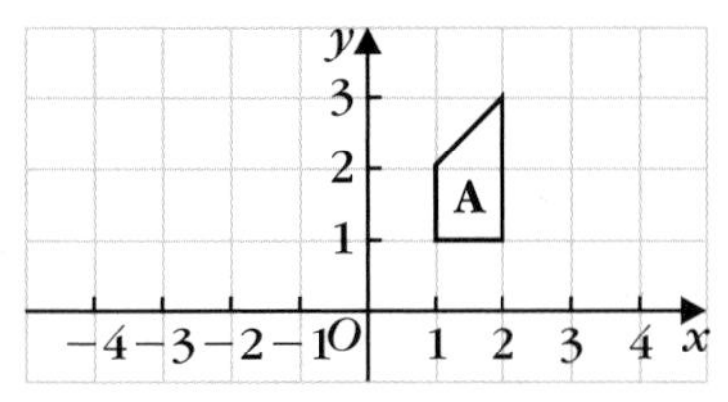

3 Enlarge shape **B** by a scale factor of −1.5 using (1, 4) as the centre.

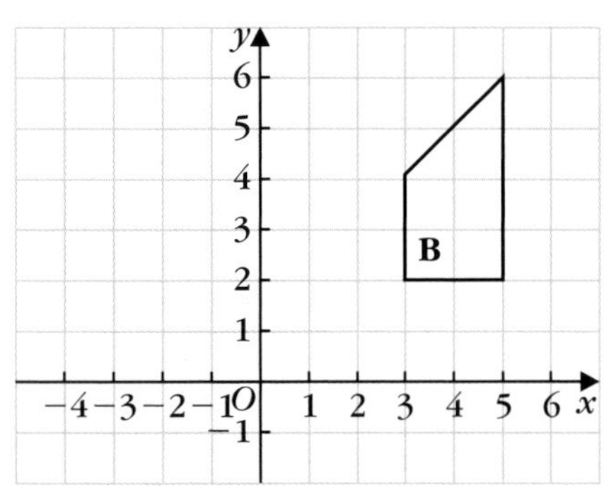

4 Reflect shape **C** in the line $x = y$.

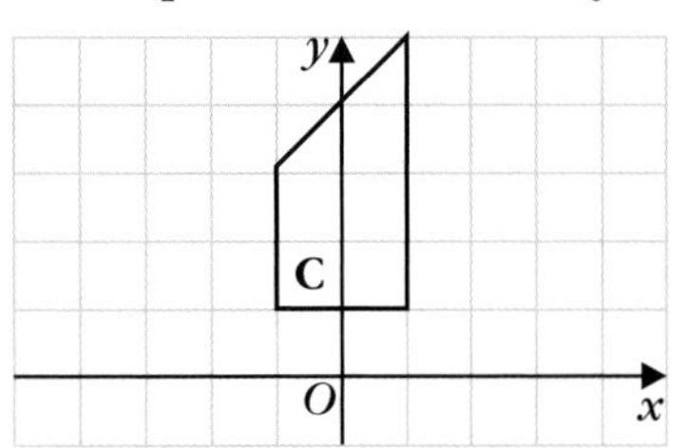

5 Describe fully the transformation which maps **D** on to **E**.

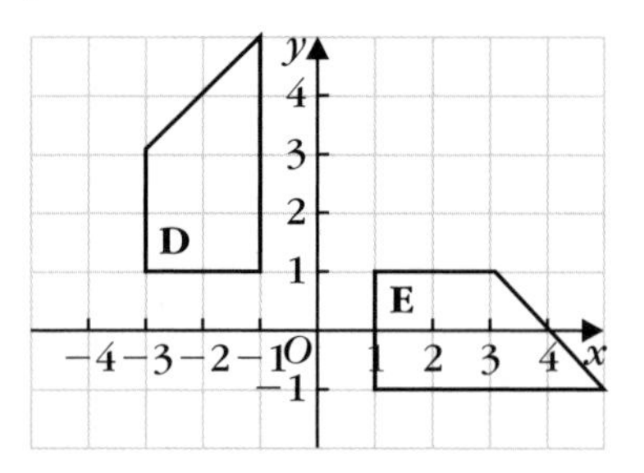

6 Plot the points $A(1, 2)$, $B(4, 4)$ and $C(6, -1)$ on a grid. Write down the column vectors that represent $\overrightarrow{AB}$, $\overrightarrow{BC}$ and $\overrightarrow{CA}$.

7

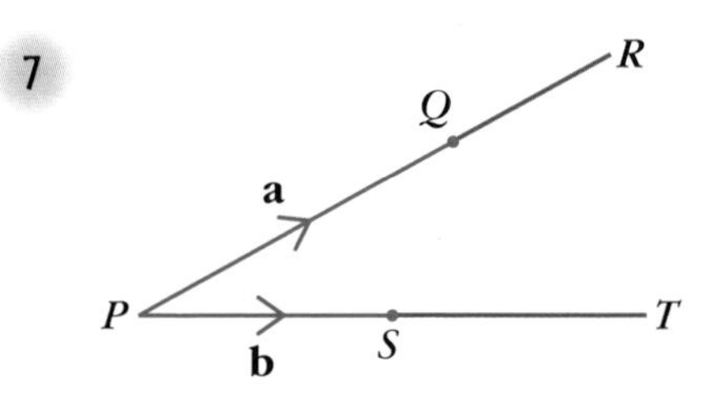

$\overrightarrow{PQ} = \mathbf{a}$ and $\overrightarrow{PS} = \mathbf{b}$

$ST = PS$ and $QR = \frac{1}{2}PQ$

Find, in terms of **a** and/or **b**, the vectors representing $\overrightarrow{PR}$, $\overrightarrow{PT}$ and $\overrightarrow{RT}$.

8 $\mathbf{a} = \begin{pmatrix} 2 \\ 5 \end{pmatrix}$ and $\mathbf{b} = \begin{pmatrix} 6 \\ 8 \end{pmatrix}$

Find the magnitude of

(a) $\mathbf{b}$

(b) $\mathbf{b} - \mathbf{a}$

(c) $2\mathbf{a} + \mathbf{b}$

(d) $3\mathbf{a} - \mathbf{b}$

9 The position vector of B is $\begin{pmatrix} 1 \\ 5 \end{pmatrix}$ and the position vector of C is $\begin{pmatrix} 5 \\ 3 \end{pmatrix}$. Work out the position vector for M, the mid-point of BC.

Now check your answers – see the facing page.

Cover this page while you answer the test questions opposite.

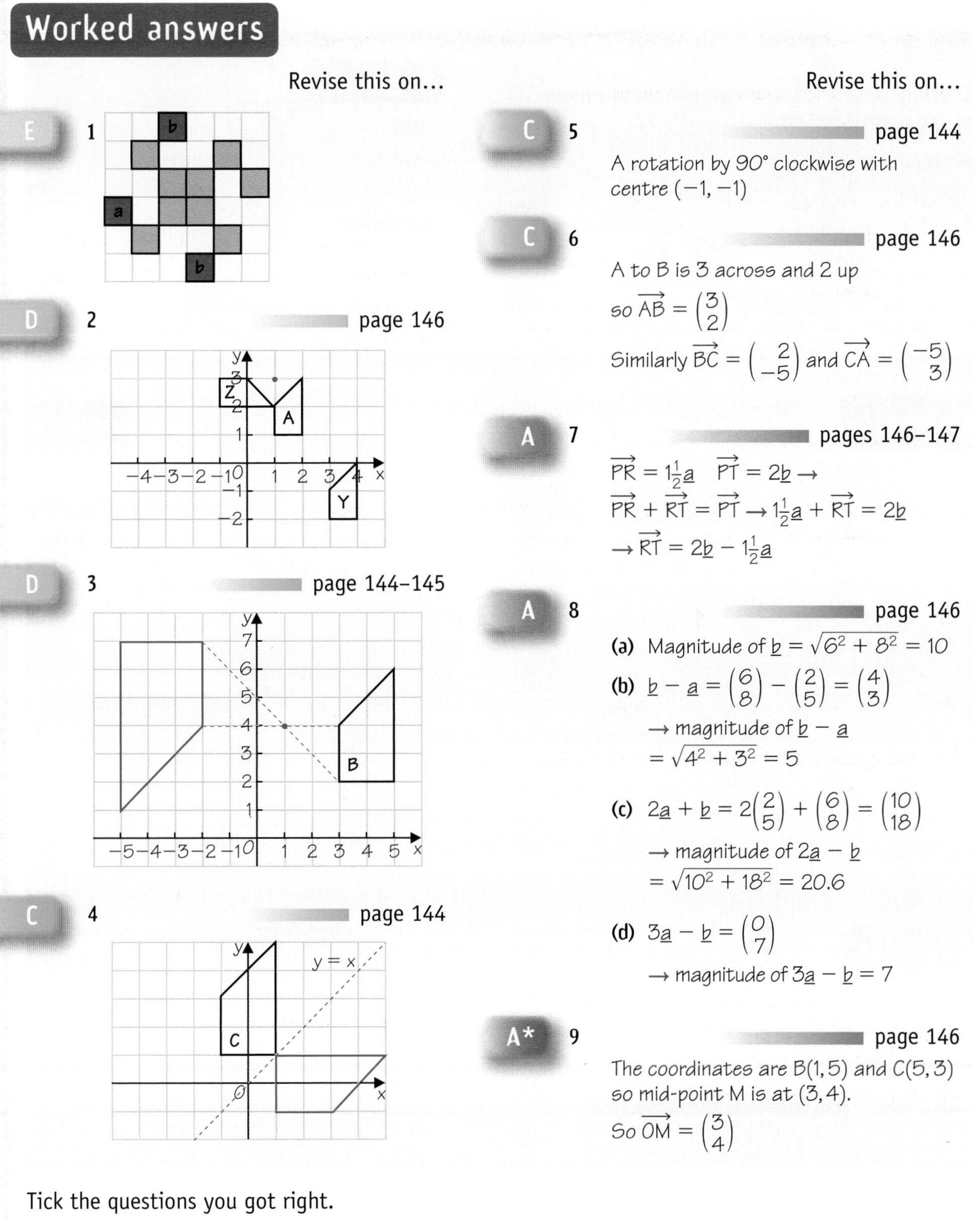

Worked answers

Revise this on...

E 1

D 2 page 146

D 3 page 144–145

C 4 page 144

C 5 page 144

A rotation by 90° clockwise with centre $(-1, -1)$

C 6 page 146

A to B is 3 across and 2 up

so $\overrightarrow{AB} = \begin{pmatrix} 3 \\ 2 \end{pmatrix}$

Similarly $\overrightarrow{BC} = \begin{pmatrix} 2 \\ -5 \end{pmatrix}$ and $\overrightarrow{CA} = \begin{pmatrix} -5 \\ 3 \end{pmatrix}$

A 7 pages 146–147

$\overrightarrow{PR} = 1\frac{1}{2}\underline{a} \quad \overrightarrow{PT} = 2\underline{b} \rightarrow$

$\overrightarrow{PR} + \overrightarrow{RT} = \overrightarrow{PT} \rightarrow 1\frac{1}{2}\underline{a} + \overrightarrow{RT} = 2\underline{b}$

$\rightarrow \overrightarrow{RT} = 2\underline{b} - 1\frac{1}{2}\underline{a}$

A 8 page 146

(a) Magnitude of $\underline{b} = \sqrt{6^2 + 8^2} = 10$

(b) $\underline{b} - \underline{a} = \begin{pmatrix} 6 \\ 8 \end{pmatrix} - \begin{pmatrix} 2 \\ 5 \end{pmatrix} = \begin{pmatrix} 4 \\ 3 \end{pmatrix}$

$\rightarrow$ magnitude of $\underline{b} - \underline{a}$

$= \sqrt{4^2 + 3^2} = 5$

(c) $2\underline{a} + \underline{b} = 2\begin{pmatrix} 2 \\ 5 \end{pmatrix} + \begin{pmatrix} 6 \\ 8 \end{pmatrix} = \begin{pmatrix} 10 \\ 18 \end{pmatrix}$

$\rightarrow$ magnitude of $2\underline{a} - \underline{b}$

$= \sqrt{10^2 + 18^2} = 20.6$

(d) $3\underline{a} - \underline{b} = \begin{pmatrix} 0 \\ 7 \end{pmatrix}$

$\rightarrow$ magnitude of $3\underline{a} - \underline{b} = 7$

A* 9 page 146

The coordinates are B(1, 5) and C(5, 3) so mid-point M is at (3, 4).

So $\overrightarrow{OM} = \begin{pmatrix} 3 \\ 4 \end{pmatrix}$

Tick the questions you got right.

Question	1	2	3	4	5	6	7	8	9
Grade	E	D	D	C	C	C	A	A	A*

Mark the grade you are working at on your revision planner on page x.

Basic trigonometry

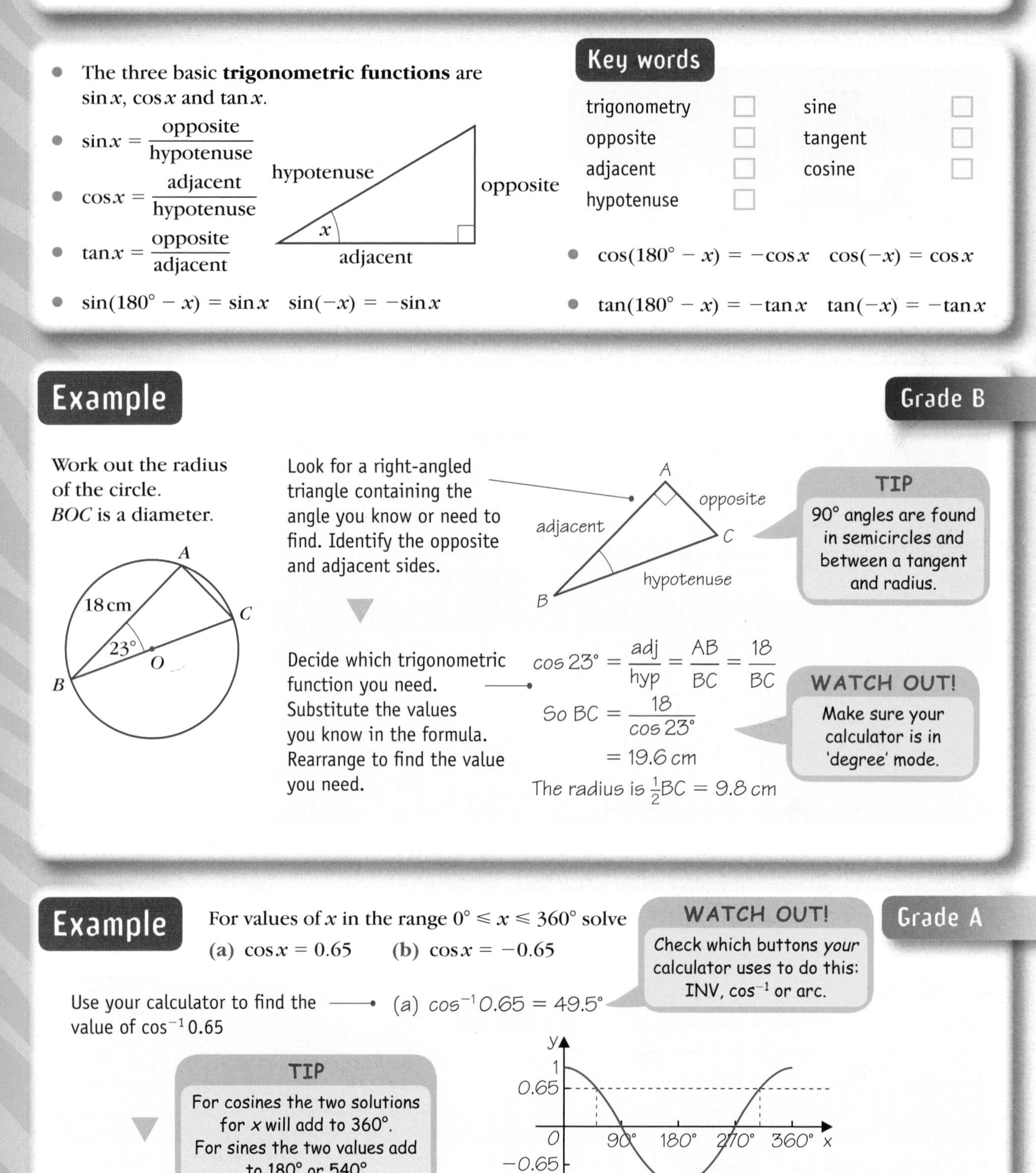

- The three basic **trigonometric functions** are $\sin x$, $\cos x$ and $\tan x$.
- $\sin x = \dfrac{\text{opposite}}{\text{hypotenuse}}$
- $\cos x = \dfrac{\text{adjacent}}{\text{hypotenuse}}$
- $\tan x = \dfrac{\text{opposite}}{\text{adjacent}}$
- $\sin(180° - x) = \sin x \quad \sin(-x) = -\sin x$
- $\cos(180° - x) = -\cos x \quad \cos(-x) = \cos x$
- $\tan(180° - x) = -\tan x \quad \tan(-x) = -\tan x$

Key words

trigonometry ☐
opposite ☐
adjacent ☐
hypotenuse ☐
sine ☐
tangent ☐
cosine ☐

Example

Grade B

Work out the radius of the circle. *BOC* is a diameter.

Look for a right-angled triangle containing the angle you know or need to find. Identify the opposite and adjacent sides.

TIP
90° angles are found in semicircles and between a tangent and radius.

Decide which trigonometric function you need. Substitute the values you know in the formula. Rearrange to find the value you need.

$\cos 23° = \dfrac{\text{adj}}{\text{hyp}} = \dfrac{AB}{BC} = \dfrac{18}{BC}$

So $BC = \dfrac{18}{\cos 23°}$

$= 19.6$ cm

The radius is $\frac{1}{2}BC = 9.8$ cm

WATCH OUT!
Make sure your calculator is in 'degree' mode.

Example

Grade A

For values of x in the range $0° \leqslant x \leqslant 360°$ solve

(a) $\cos x = 0.65$ **(b)** $\cos x = -0.65$

WATCH OUT!
Check which buttons *your* calculator uses to do this: INV, $\cos^{-1}$ or arc.

Use your calculator to find the value of $\cos^{-1} 0.65$ → (a) $\cos^{-1} 0.65 = 49.5°$

TIP
For cosines the two solutions for x will add to 360°.
For sines the two values add to 180° or 540°

Use the symmetry of the graph of $\cos x$ to find the other value in the range. → The other solution is $360 - 49.5 = 310.5°$
So $x = 49.5°$ or $x = 310.5°$

Use the symmetry of the graph and your answers from (a).

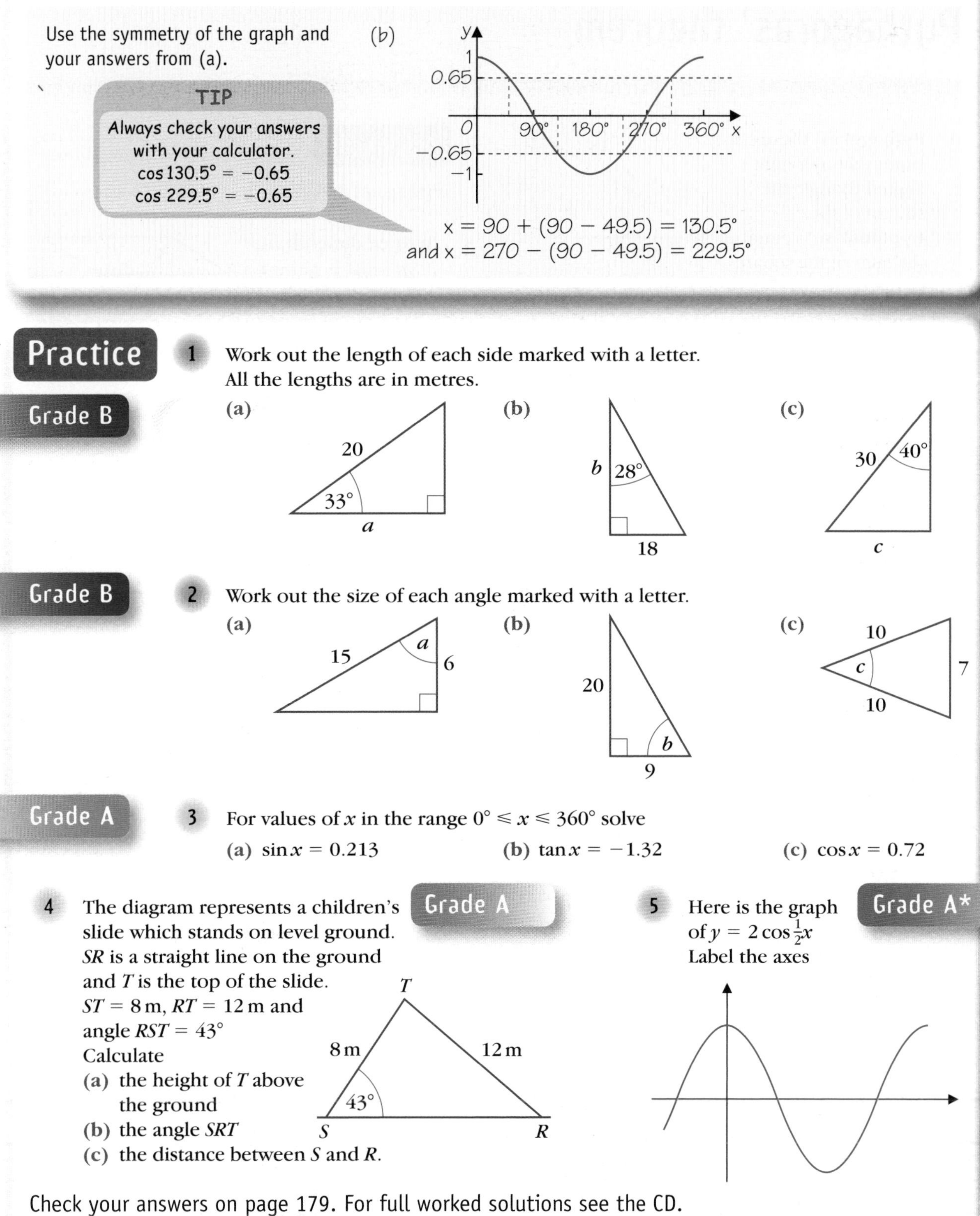

$x = 90 + (90 - 49.5) = 130.5°$
and $x = 270 - (90 - 49.5) = 229.5°$

Practice

Grade B

1 Work out the length of each side marked with a letter.
All the lengths are in metres.

(a) **(b)** **(c)**

Grade B

2 Work out the size of each angle marked with a letter.

(a) **(b)** **(c)**

Grade A

3 For values of x in the range $0° \leqslant x \leqslant 360°$ solve

(a) $\sin x = 0.213$ **(b)** $\tan x = -1.32$ **(c)** $\cos x = 0.72$

4 **Grade A** The diagram represents a children's slide which stands on level ground.
SR is a straight line on the ground and T is the top of the slide.
$ST = 8$ m, $RT = 12$ m and angle $RST = 43°$
Calculate

(a) the height of T above the ground

(b) the angle SRT

(c) the distance between S and R.

5 **Grade A*** Here is the graph of $y = 2\cos\frac{1}{2}x$
Label the axes

Check your answers on page 179. For full worked solutions see the CD.
See the Student Book on the CD if you need more help.

Question	1	2	3	4	5
Grade	B	B	A	A	A*
Student Book pages	U3 220–227	U3 220–227	U3 238	U3 220–227	U3 236–237

Pythagoras' theorem

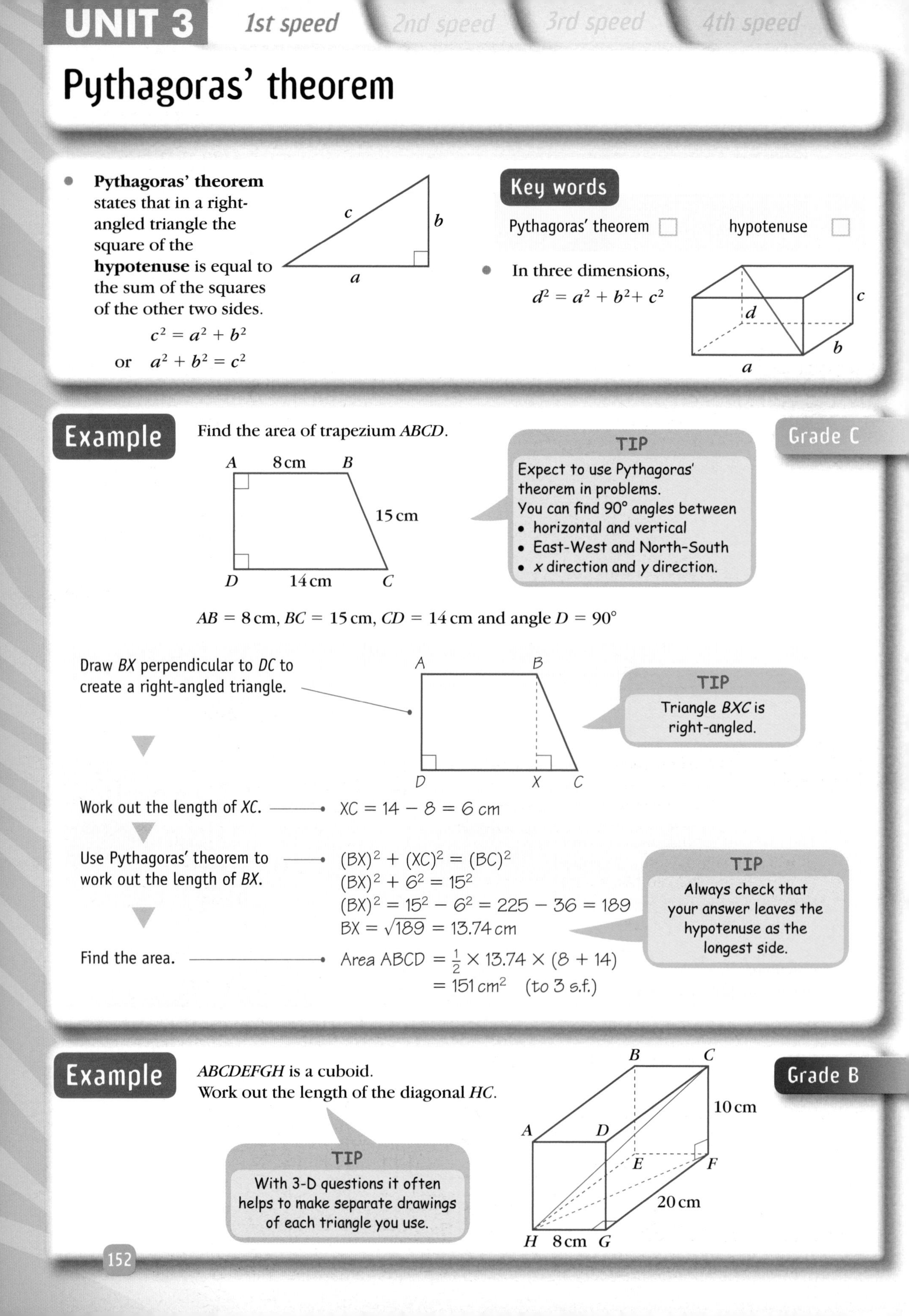

- **Pythagoras' theorem** states that in a right-angled triangle the square of the **hypotenuse** is equal to the sum of the squares of the other two sides.

$$c^2 = a^2 + b^2$$

or $$a^2 + b^2 = c^2$$

Key words

Pythagoras' theorem ☐ hypotenuse ☐

- In three dimensions, $d^2 = a^2 + b^2 + c^2$

Example

Grade C

Find the area of trapezium *ABCD*.

TIP
Expect to use Pythagoras' theorem in problems.
You can find 90° angles between
- horizontal and vertical
- East-West and North-South
- *x* direction and *y* direction.

$AB = 8\text{ cm}$, $BC = 15\text{ cm}$, $CD = 14\text{ cm}$ and angle $D = 90°$

Draw *BX* perpendicular to *DC* to create a right-angled triangle.

TIP
Triangle *BXC* is right-angled.

Work out the length of *XC*. → $XC = 14 - 8 = 6\text{ cm}$

Use Pythagoras' theorem to work out the length of *BX*. →
$(BX)^2 + (XC)^2 = (BC)^2$
$(BX)^2 + 6^2 = 15^2$
$(BX)^2 = 15^2 - 6^2 = 225 - 36 = 189$
$BX = \sqrt{189} = 13.74\text{ cm}$

TIP
Always check that your answer leaves the hypotenuse as the longest side.

Find the area. → Area ABCD $= \frac{1}{2} \times 13.74 \times (8 + 14)$
$= 151\text{ cm}^2$ (to 3 s.f.)

Example

Grade B

ABCDEFGH is a cuboid.
Work out the length of the diagonal *HC*.

TIP
With 3-D questions it often helps to make separate drawings of each triangle you use.

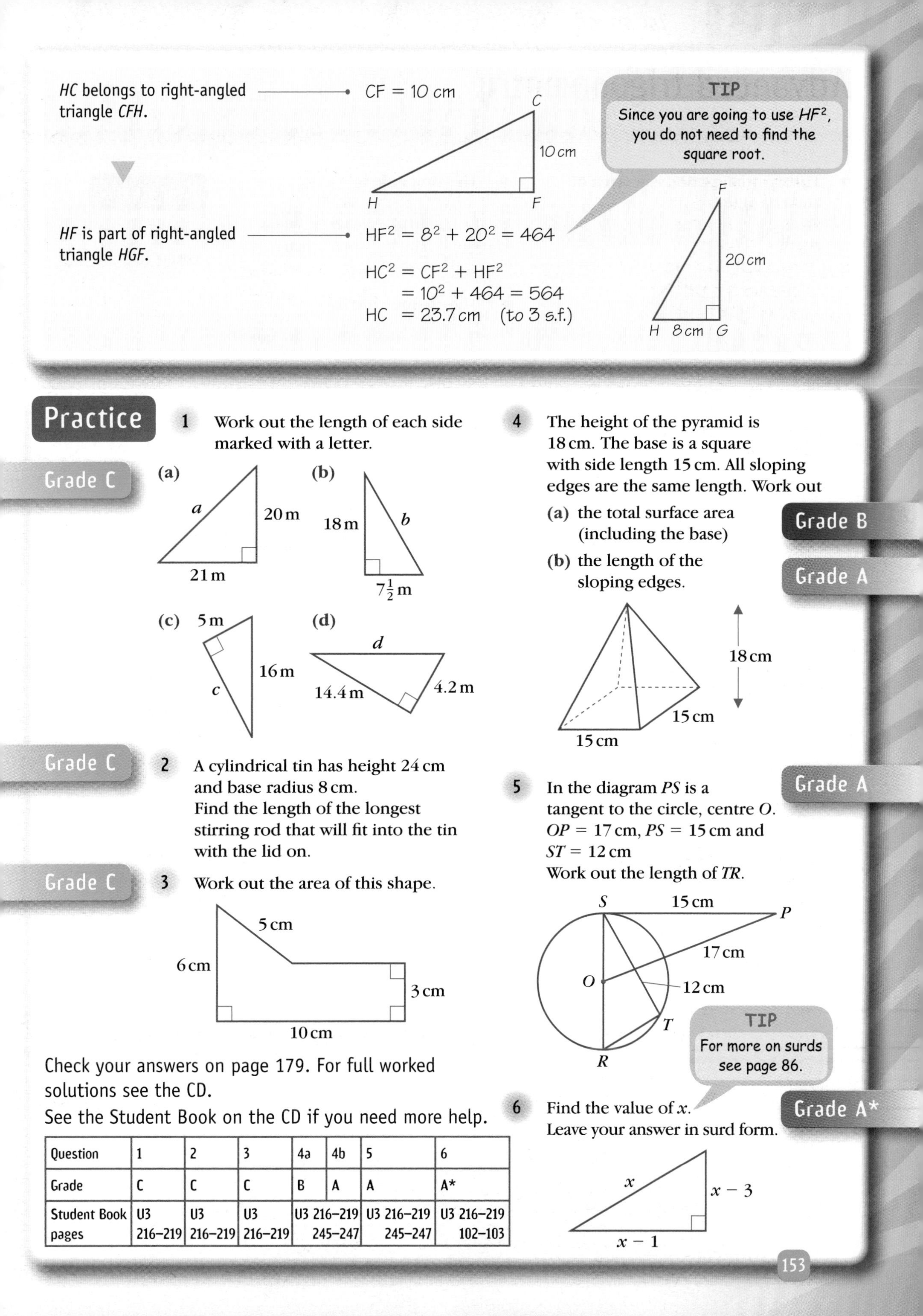

HC belongs to right-angled triangle *CFH*. → $CF = 10$ cm

HF is part of right-angled triangle *HGF*. → $HF^2 = 8^2 + 20^2 = 464$

$HC^2 = CF^2 + HF^2$

$= 10^2 + 464 = 564$

$HC = 23.7$ cm (to 3 s.f.)

TIP

Since you are going to use HF^2, you do not need to find the square root.

Practice

Grade C

1 Work out the length of each side marked with a letter.

(a)

(b)

(c)

(d)

Grade C

2 A cylindrical tin has height 24 cm and base radius 8 cm. Find the length of the longest stirring rod that will fit into the tin with the lid on.

Grade C

3 Work out the area of this shape.

4 The height of the pyramid is 18 cm. The base is a square with side length 15 cm. All sloping edges are the same length. Work out

(a) the total surface area (including the base) **Grade B**

(b) the length of the sloping edges. **Grade A**

Grade A

5 In the diagram *PS* is a tangent to the circle, centre *O*. $OP = 17$ cm, $PS = 15$ cm and $ST = 12$ cm

Work out the length of *TR*.

TIP

For more on surds see page 86.

Grade A*

6 Find the value of x. Leave your answer in surd form.

Check your answers on page 179. For full worked solutions see the CD.

See the Student Book on the CD if you need more help.

Question	1	2	3	4a	4b	5	6
Grade	C	C	C	B	A	A	A*
Student Book pages	U3 216–219	U3 216–219	U3 216–219	U3 216–219	245–247	U3 216–219 245–247	U3 216–219 102–103

Advanced trigonometry

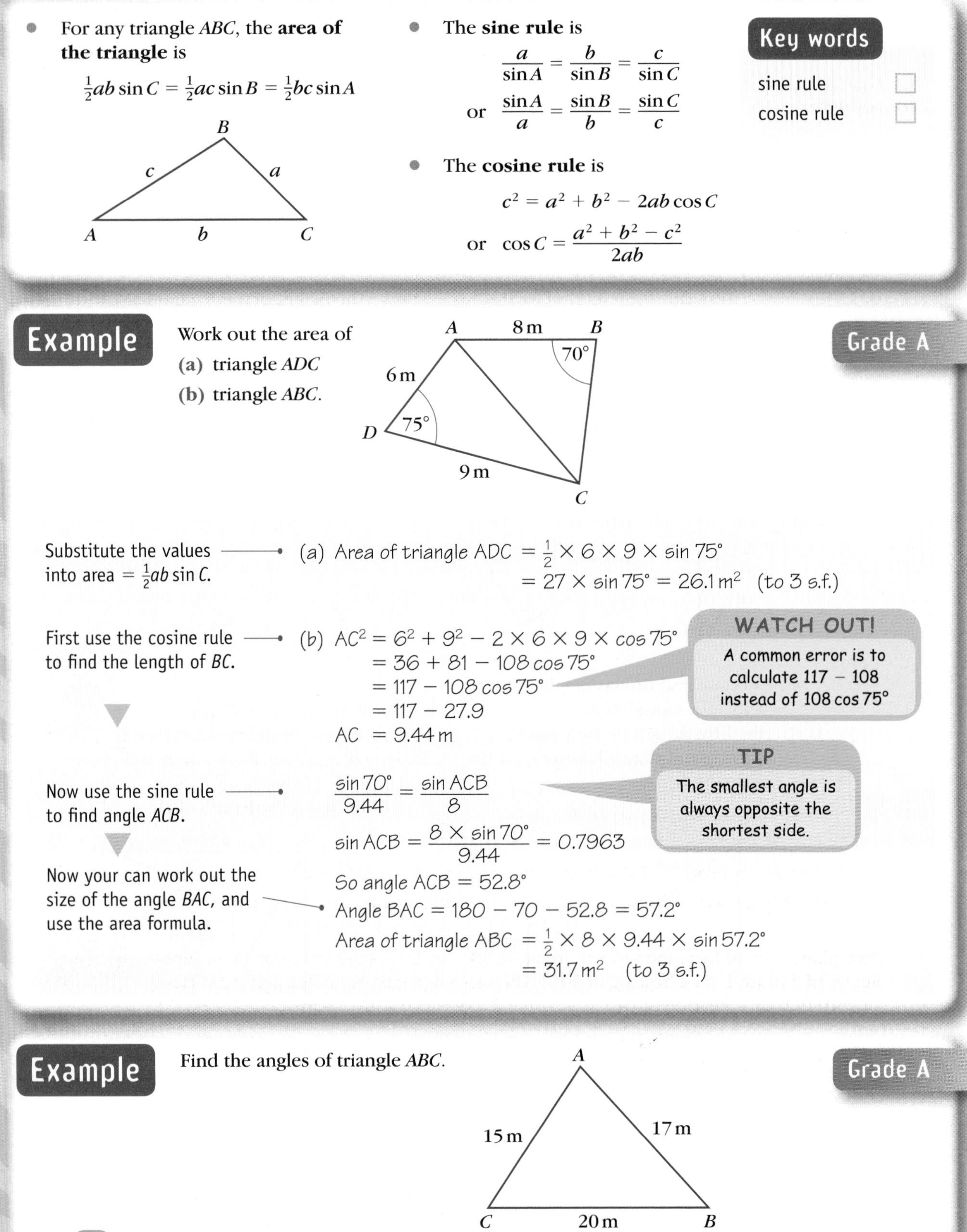

- For any triangle ABC, the **area of the triangle** is

  $\frac{1}{2}ab\sin C = \frac{1}{2}ac\sin B = \frac{1}{2}bc\sin A$

- The **sine rule** is

  $$\frac{a}{\sin A} = \frac{b}{\sin B} = \frac{c}{\sin C}$$

  or $$\frac{\sin A}{a} = \frac{\sin B}{b} = \frac{\sin C}{c}$$

- The **cosine rule** is

  $$c^2 = a^2 + b^2 - 2ab\cos C$$

  or $$\cos C = \frac{a^2 + b^2 - c^2}{2ab}$$

Key words

sine rule

cosine rule

Example

Grade A

Work out the area of

(a) triangle ADC

(b) triangle ABC.

Substitute the values into area $= \frac{1}{2}ab\sin C$.

(a) Area of triangle ADC $= \frac{1}{2} \times 6 \times 9 \times \sin 75°$

$= 27 \times \sin 75° = 26.1\,\text{m}^2$ (to 3 s.f.)

First use the cosine rule to find the length of BC.

(b) $AC^2 = 6^2 + 9^2 - 2 \times 6 \times 9 \times \cos 75°$

$= 36 + 81 - 108\cos 75°$

$= 117 - 108\cos 75°$

$= 117 - 27.9$

$AC = 9.44\,\text{m}$

WATCH OUT!

A common error is to calculate 117 − 108 instead of 108 cos 75°

Now use the sine rule to find angle ACB.

$$\frac{\sin 70°}{9.44} = \frac{\sin ACB}{8}$$

$$\sin ACB = \frac{8 \times \sin 70°}{9.44} = 0.7963$$

TIP

The smallest angle is always opposite the shortest side.

Now your can work out the size of the angle BAC, and use the area formula.

So angle ACB $= 52.8°$

Angle BAC $= 180 - 70 - 52.8 = 57.2°$

Area of triangle ABC $= \frac{1}{2} \times 8 \times 9.44 \times \sin 57.2°$

$= 31.7\,\text{m}^2$ (to 3 s.f.)

Example

Grade A

Find the angles of triangle ABC.

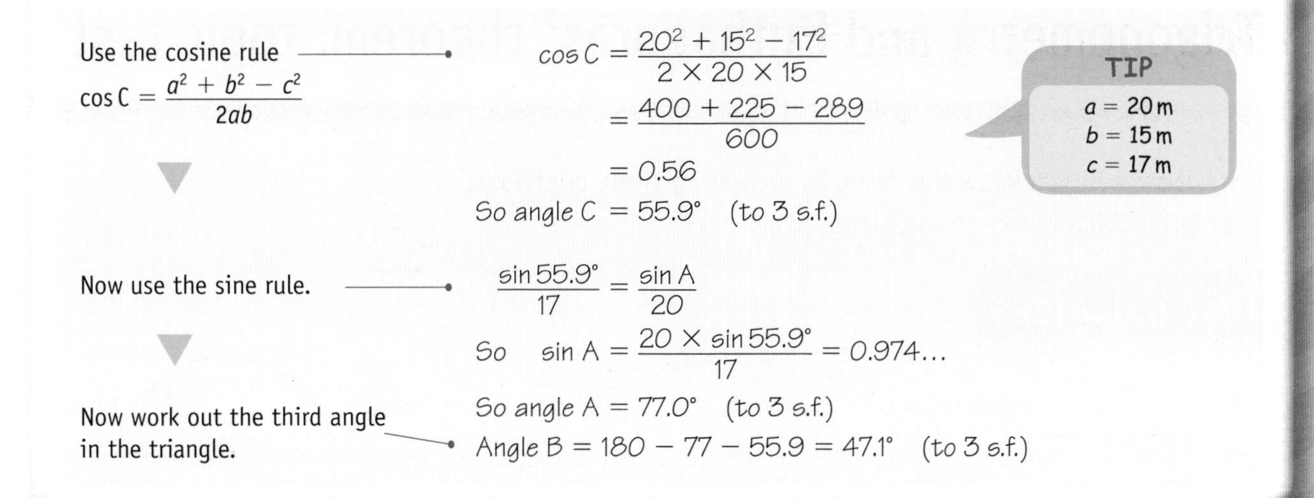
Use the cosine rule $\cos C = \dfrac{a^2 + b^2 - c^2}{2ab}$

$$\cos C = \frac{20^2 + 15^2 - 17^2}{2 \times 20 \times 15} = \frac{400 + 225 - 289}{600} = 0.56$$

So angle $C = 55.9°$ (to 3 s.f.)

TIP

$a = 20\,\text{m}$
$b = 15\,\text{m}$
$c = 17\,\text{m}$

Now use the sine rule.

$$\frac{\sin 55.9°}{17} = \frac{\sin A}{20}$$

$$\text{So} \quad \sin A = \frac{20 \times \sin 55.9°}{17} = 0.974\ldots$$

So angle $A = 77.0°$ (to 3 s.f.)

Now work out the third angle in the triangle.

Angle $B = 180 - 77 - 55.9 = 47.1°$ (to 3 s.f.)

Practice

1 For each triangle, calculate the area and the length of the side marked x and the size of the angle marked y. All lengths are in cm. **Grade A**

(a)

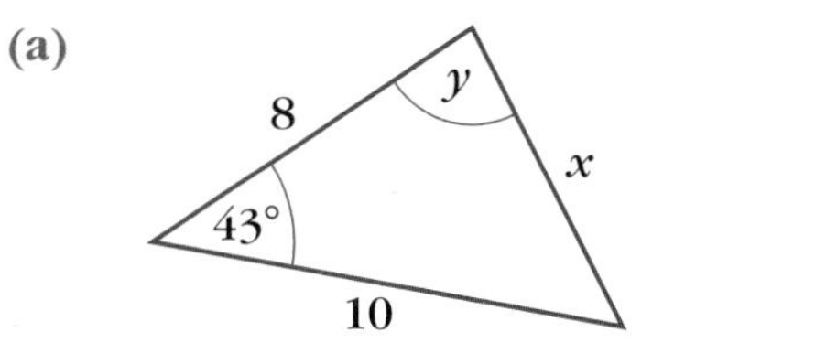

(b)

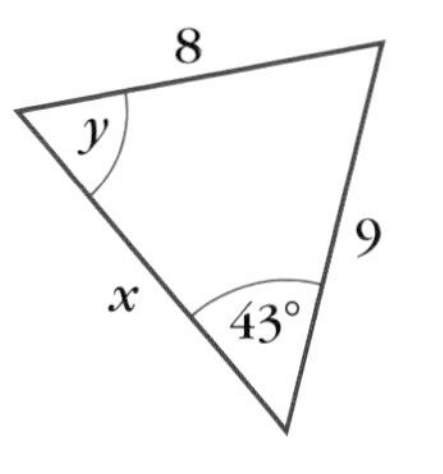

2 The lengths of the sides of a triangle are 12 cm, 15 cm and 16 cm. Calculate the area of the triangle. **Grade A**

3 A speedboat races 600 metres due East to a marker buoy for the first leg of a race. It then continues to the second marker buoy which is 800 metres away on a bearing of 127° for the second leg. It then returns to its starting position for the third leg, which completes one circuit. **Grade A**

(a) Calculate the total distance for one circuit.

(b) Calculate the bearing of the second marker buoy from the start.

(c) Calculate the boat's shortest distance from the first marker buoy during the third leg of the race.

4 Two planes set off from the same point at 12:00. The first plane travels at a constant speed of 650 km/h on a bearing of 043°. The second plane travels at a constant speed of 740 km/h on a bearing of 290°. Calculate the distance between the planes at 14:00. **Grade A**

Check your answers on page 179. For full worked solutions see the CD.
See the Student Book on the CD if you need more help.

Question	1	2	3	4
Grade	A	A	A	A
Student Book pages	U3 232–234	U3 231–232	U3 232–234	U3 232–234

Trigonometry and Pythagoras' theorem: topic test

Check how well you know this topic by answering these questions.
First cover the answers on the facing page.

Test questions

1 Work out the length of each side marked with a letter.

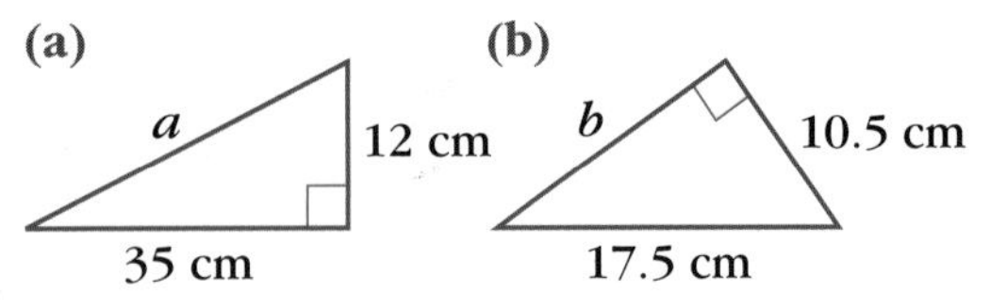

2 Work out the length of each side marked with a letter.

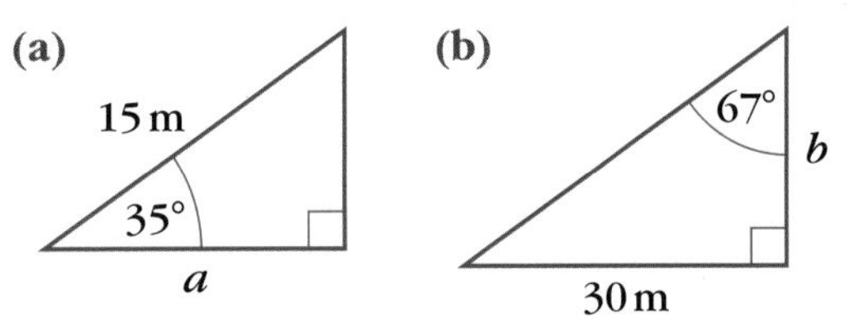

3 Work out the size of each angle marked with a letter.

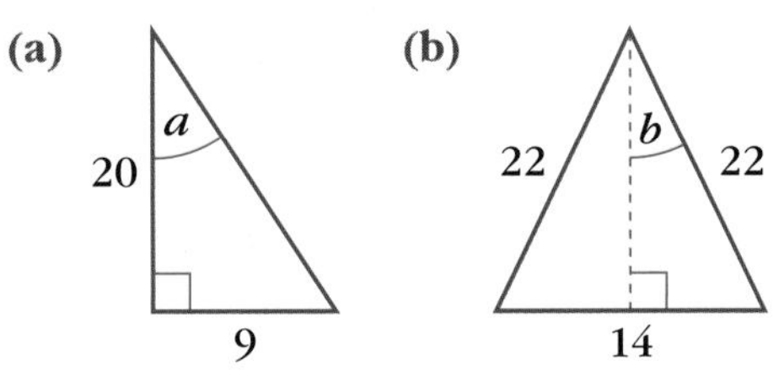

4 Find all values of x in the range $0° \leqslant x \leqslant 360°$ that are solutions to

(a) $\sin x = 0.2$ **(b)** $\cos x = -0.3$

(c) $\tan x = 2.7$

5 For values of x in the given range, find all solutions to the equation

(a) $7 \sin x = 2 \quad (0° \leqslant x \leqslant 180°)$

(b) $3 \tan 2x = 5 \quad (0° \leqslant x \leqslant 360°)$

(c) $3 \cos x + 4 = 1 \quad (0° \leqslant x \leqslant 360°)$

6 The shaded area is $140\,\text{cm}^2$. Find the radius of the circle.

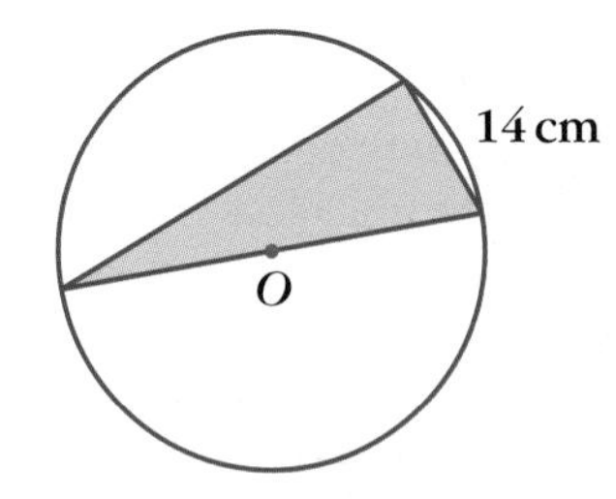

7 The height of this pyramid is 12 cm. Work out the length of the sloping edges.

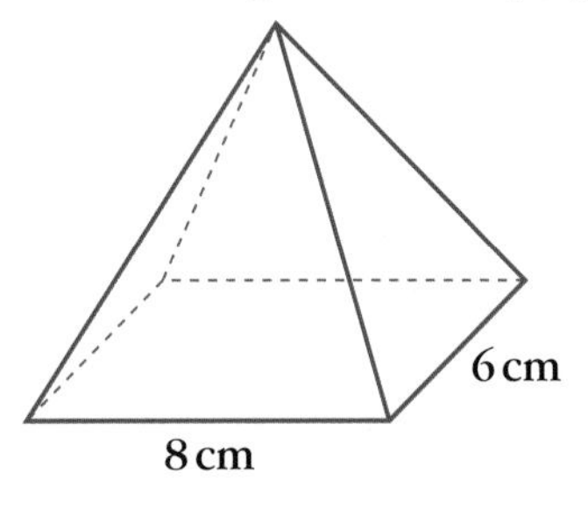

8 **(a)** Find the area of this triangle.

(b) Use the cosine rule to find the length of the third side.

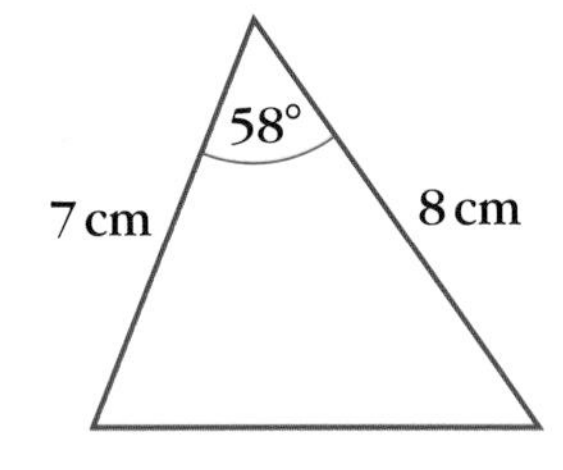

9 Calculate the unlabelled lengths and angles in triangle ABC.

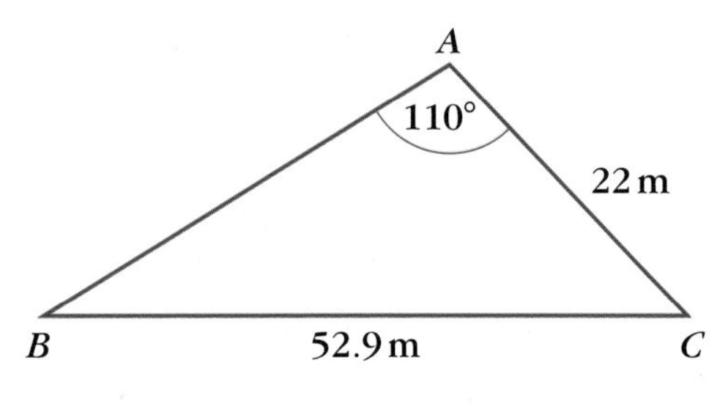

Now check your answers – see the facing page.

Cover this page while you answer the test questions opposite.

Worked answers

Revise this on...

C **1** (a) $a^2 = 12^2 + 35^2 = 144 + 1225 = 1369 \rightarrow a = 37\text{ cm}$ — page 150

(b) $b^2 + 10.5^2 = 17.5^2 \rightarrow b^2 = 17.5^2 - 10.5^2 = (17.5 + 10.5)(17.5 - 10.5)$
$= 28 \times 7 \rightarrow b = 14\text{ cm}$

B **2** (a) $a = 15\cos 35° = 12.3\text{ m}$ (b) $b\tan 67° = 30 \rightarrow b = 12.7\text{ m}$ — page 150

B **3** (a) $\tan a = \frac{9}{20} \rightarrow a = 24.2°$ (b) $\sin b = \frac{7}{22} \rightarrow b = 18.6°$ — page 150

A **4** (a) $x = 11.5°, 168.5°$ (b) $x = 107.5°, 252.5°$ (c) $x = 69.7°, 249.7°$ — pages 150–151

A **5** (a) $\sin x = \frac{2}{7} \rightarrow x = 16.6°, 163.4°$ — pages 150–151

(b) $\tan 2x = \frac{5}{3} \rightarrow 2x = 59°, 239°, 419°, 599° \rightarrow x = 29.5°, 119.5°, 209.5°, 299.5°$

(c) $3\cos x = 1 - 4 = -3 \rightarrow \cos x = -1 \rightarrow x = 180°$

A **6** Area $= \frac{1}{2} \times$ base $\times$ height $\rightarrow 140 = \frac{1}{2} \times 14 \times a = 7a \rightarrow a = 20\text{ cm}$ — page 152

Diameter$^2 = 14^2 + 20^2 = 196 + 400 = 596 \rightarrow$ diameter $= 24.4$
$\rightarrow$ radius $= 12.2\text{ cm}$

A **7** (Diagonal of base)$^2 = 6^2 + 8^2 = 100 \rightarrow$ base diagonal $= 10\text{ cm}$ — pages 152–153

Now use the right-angled triangle formed by height, half of base diagonal and the sloping edge:

(Sloping edge)$^2 = 5^2 + 12^2 \rightarrow$ sloping edge $= 13\text{ cm}$

A **8** (a) Area $= \frac{1}{2}ab\sin C = \frac{1}{2} \times 7 \times 8 \times \sin 58° = 23.75\text{ cm}^2$ — pages 154–155

(b) $c^2 = a^2 + b^2 - 2bc\cos C = 7^2 + 8^2 - 2 \times 7 \times 8 \times \cos 58°$
$= 49 + 64 - 112\cos 58°$
$c = 7.32\text{ cm}$

A **9** $\frac{\sin 110°}{52.9} = \frac{\sin B}{22} \rightarrow$ angle $B = 23°$ — pages 154–155

Angle $C = 180 - 110 - 23 = 47°$

$\frac{52.9}{\sin 110°} = \frac{AB}{\sin 47°} \rightarrow AB = 41.2\text{ m}$ (to 3 s.f.)

Tick the questions you got right.

Question	1	2	3	4	5	6	7	8	9
Grade	C	B	B	A	A	A	A	A	A

Mark the grade you are working at on your revision planner on page x.

Shape, space and measure: subject test

Exam practice questions

1 A is 8 km due North of B. A ship leaves A and travels on a bearing of 120°. Another ship leaves B and travels on a bearing of 068°.
Using a scale of 1 cm to represent 1 km draw a scale drawing and use it to find how far from A the ship's paths cross.

2 Work out the sum of the interior angles of a heptagon (7 sides).

3 Draw a sketch of a net for this shape.

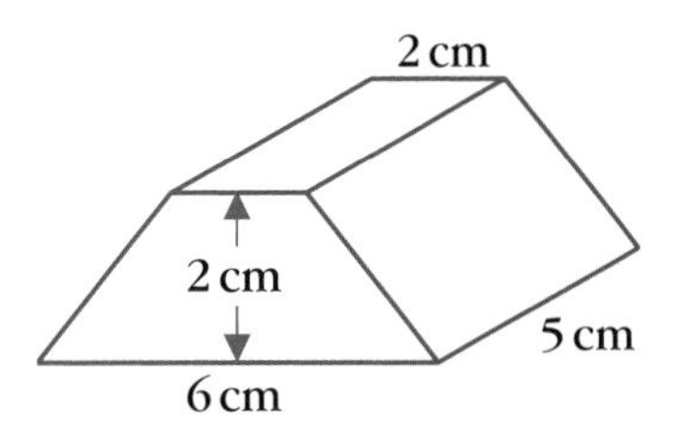

4 Describe fully the single transformation that maps A onto B.

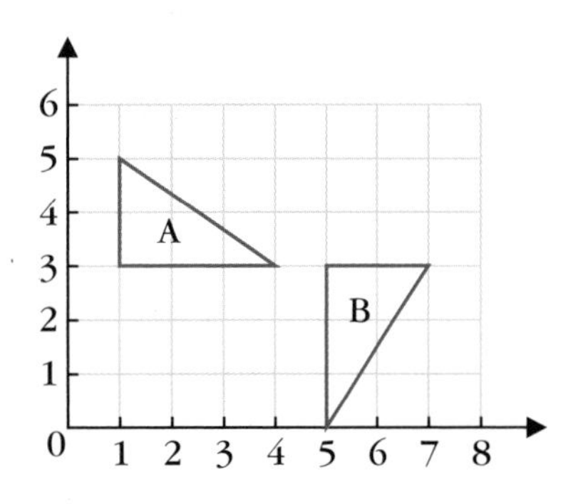

5 Mark two points P and Q, 5 cm apart. Shade in the locus of the points that are less than 3.5 cm from P and nearer to Q than they are to P.

6 Enlarge the shape P. Using a scale factor of $-1\frac{1}{2}$ and centre (3, 1).

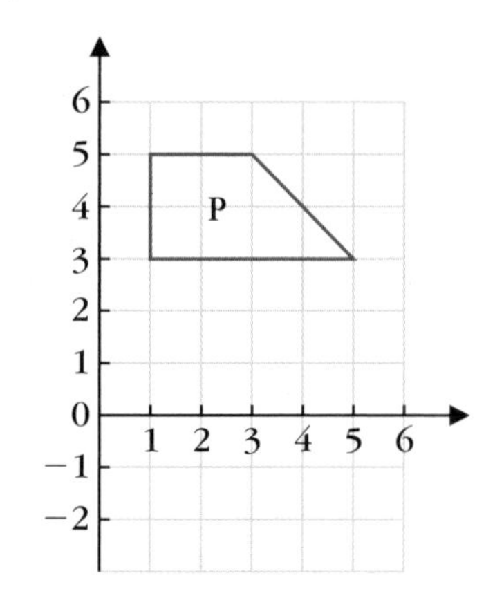

7 Angle $ABC = 90°$
$AB = 3.7$ cm
$BC = 6.9$ cm

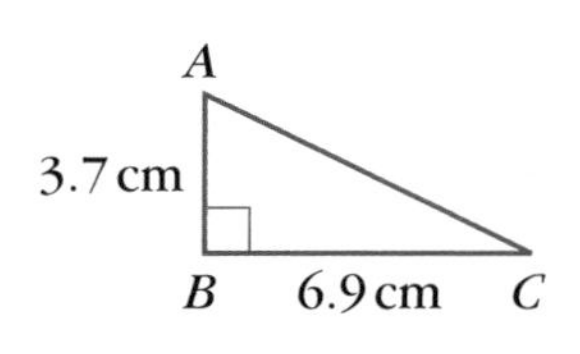

(a) Work out the length of AC.

(b) Find the size of angle ACB.

8 $ABCD$ is a cyclic quadrilateral.
SAT is a tangent to the circle at A.
$AB = BC$ and $BAC = 62°$

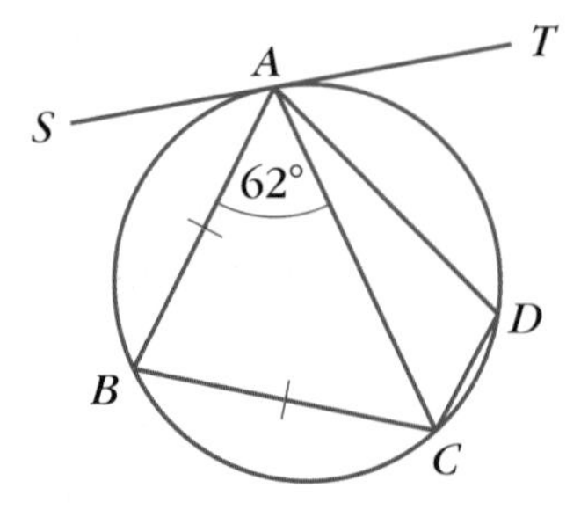

Find, showing all your reasons,

(a) angle BCA

(b) angle SAB

(c) angle ADC.

9 *ABDE* is a square.
$BC = CD$
Prove that triangles *ABC* and *EDC* are congruent.
Hence show that *ACE* is an isosceles triangle.

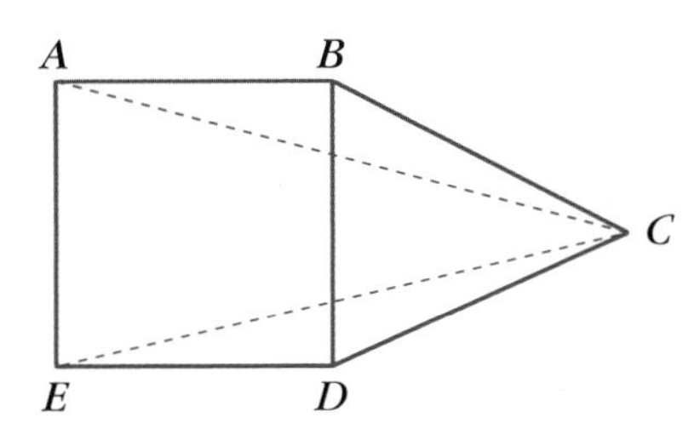

10 **(a)** **(i)** Find the shaded area.
(ii) Find the arc length AB.

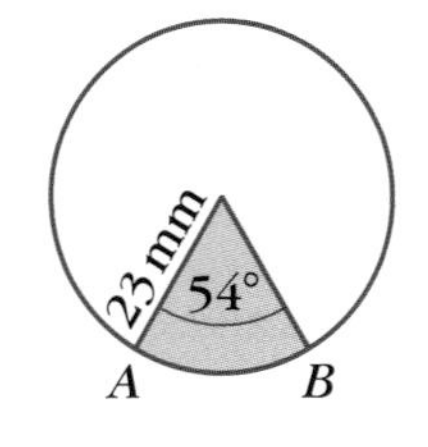

(b) Find the shaded area.

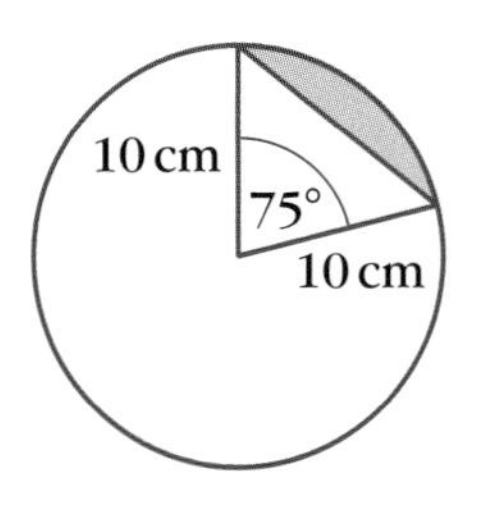

11 *PQRSTU* is a regular octahedron.
$\overrightarrow{PQ}$ is vector **a**,
$\overrightarrow{QR}$ is vector **b**, and
$\overrightarrow{RT}$ is vector **c**.

(a) Write down in terms of **a**, **b** and **c** the vectors
(i) $\overrightarrow{PT}$ **(ii)** $\overrightarrow{TQ}$
(iii) $\overrightarrow{UP}$ **(iv)** $\overrightarrow{SU}$

(b) Write down two lines that have the vectors
(i) $\mathbf{b} + \mathbf{c}$ **(ii)** $\mathbf{a} + \mathbf{c}$ **(iii)** $\mathbf{a} + \mathbf{b} + \mathbf{c}$

12 *VABCD* is a pyramid.
The horizontal base *ABCD* is a rhombus of side 8 cm.
The angle *ABC* is 135°.

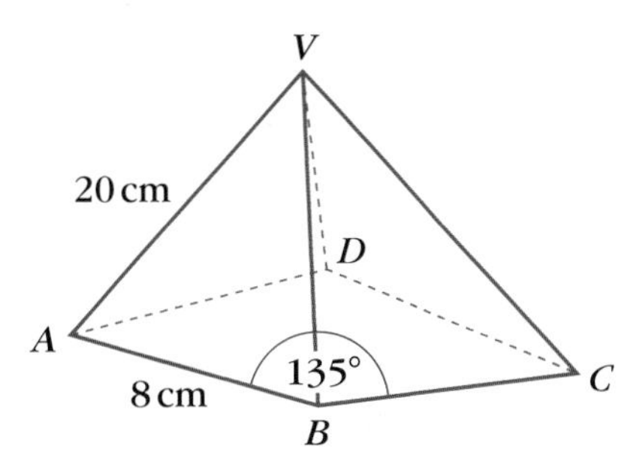

(a) Calculate the length of each diagonal of the base of the pyramid.

The vertex *V* is vertically above the centre of the base. $VA = 20$ cm

(b) Calculate the height of *V* above the base.

Check your answers on pages 179–180.
For full worked solutions see the CD.
Tick the questions you got right.

Question	1	2	3	4	5	6	7	8	9	10	11	12
Grade	D	D	D	C	C	C	B	B	A	A	A	A*
Revise this on page	134–135	130–131	132	146–147	134–135	144–145	150, 152	138–139	130	140	146–147	150–153

Mark the grade you are working at on your revision planner on page x.
Go to the pages shown to revise for the ones you got wrong.

Number

Powers, surds, bounds and percentages

- A number is in **standard form** when

$$7.2 \times 10^6 = 7\,200\,000$$
$$7.2 \times 10^{-6} = 0.000\,007\,2$$

This part is written as a number between 1 and 10

This part is written as a **power of 10**

- A number written exactly using square roots is called a **surd**: $\sqrt{5}$ and $2 - \sqrt{3}$. Simplified surds should never have a square root in the denominator.

- In addition and multiplication calculations:
  - Use the two lower bounds to find the **lower bound** of the result.
  - Use the two upper bounds to find the **upper bound** of the result.

- In subtraction and division calculations you need to use one upper bound and one lower bound to find each bound of the result.

- To **increase** or **decrease** a number by a percentage, you find the percentage of that amount and then add it to or subtract it from the starting amount.

- **Simple interest** is when interest is paid just on the original amount.

- **Compound interest** is when interest is paid on the original amount *and* on the interest already earned.

- A **price index** shows how the price of something changes over time. The index always starts at 100. An index greater than 100 shows a price rise, while an index less than 100 shows a price fall.

Ratio and proportion

- Two quantities are in **direct proportion** if their **ratio** stays the same as the quantities increase or decrease.

- In the **unitary method**, you find the value of one item first.

- To share an amount in a given ratio find the total of all the ratio numbers, then split the amount into fractions with a denominator that is the total.

Algebra

Graphs

- **Gradient** $= \dfrac{\text{change in } y\text{-direction}}{\text{change in } x\text{-direction}}$
- The **straight line** with equation $y = mx + c$ has a gradient of m and its **intercept** on the y-axis is $(0, c)$.
- **Simultaneous equations** can be solved graphically by drawing the graphs for the two equations and finding **coordinates** of their **point of intersection**.
- The solutions of a **quadratic equation** are the values of x where the graph cuts the x-axis.
- For any **function** f the graph of $y = f(x + a) + b$ is the graph of $y = f(x)$ **translated** a units horizontally (in the negative x-direction if $a > 0$, in the positive x-direction if $a < 0$) followed by a translation of b units vertically (upwards if $b > 0$, downwards if $b < 0$).

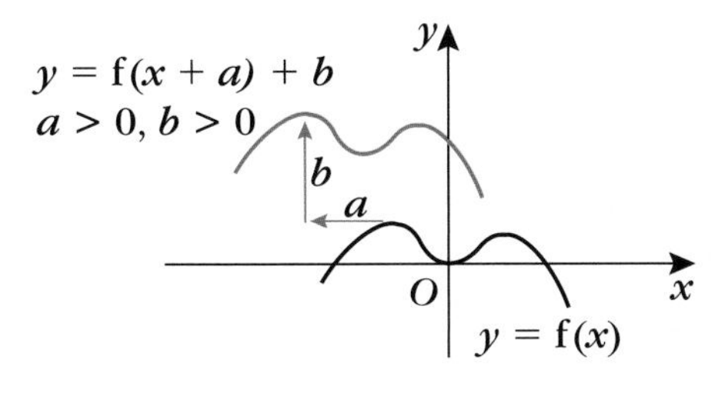

- For any function f, the graph of $y = -f(x)$ is obtained by

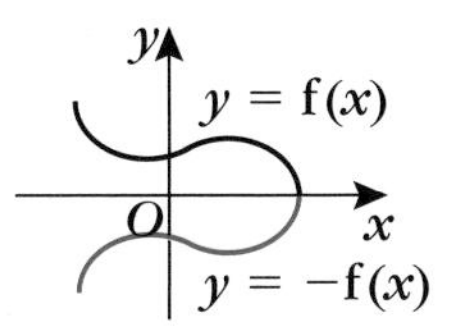

Formulae

- x is the **subject** of a formula when it appears on its own on one side of the formula and does not appear on the other side.
- To keep an **equation** balanced you must do the same to each side.

$$a + 4 = 7 \rightarrow a + 4 - 4 = 7 - 4 \rightarrow a = 3$$
$$a - 3 = 1 \rightarrow a - 3 + 3 = 1 + 3 \rightarrow a = 4$$
$$5a = 30 \rightarrow 5a \div 5 = 30 \div 5 \rightarrow a = 6$$
$$\frac{a}{2} = 7 \rightarrow \frac{a}{2} \times 2 = 7 \times 2 \rightarrow a = 14$$

Solving equations and inequalities

- The roots of the **quadratic equation** $ax^2 + bx + c = 0$, where $a \neq 0$, are given by the formula

$$x = \frac{-b \pm \sqrt{b^2 - 4ac}}{2a}$$

- You can find approximate solutions of more complicated equations by **trial and improvement**.

- **Simultaneous equations** can be solved algebraically by multiplying one or both of the equations by a number, if necessary, and then adding or subtracting before dividing by the coefficient in front of the unknown.

- You can use a **number line** to show an **inequality** such as $-2 \leqslant x < 1$:

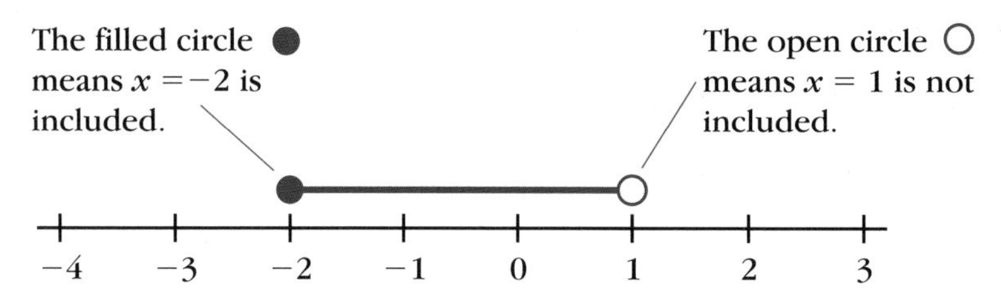

- To solve an inequality you **can**:
  - add the same quantity to both sides
  - subtract the same quantity from both sides
  - multiply both sides by the same *positive* quantity
  - divide both sides by the same *positive* quantity.

  But you **must not**:
  - multiply both sides by a *negative* quantity
  - divide both sides by a *negative* quantity

- **Regions** on a graph can be used to represent inequalities. For example, the shaded region represents the inequality $5x + 3y > 15$. The line $5x + 3y = 15$ is broken to show it is not included in the region.

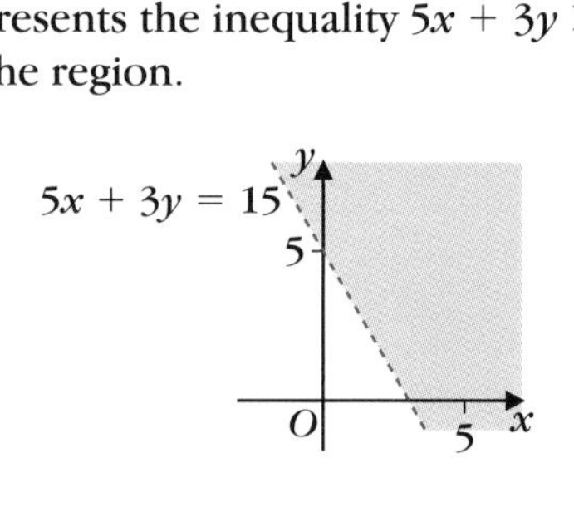

Proportion

- When y is **directly proportional** to x, you can write a proportionality statement and a formula connecting y and x:
  - the proportionality statement is $y \propto x$
  - the proportionality formula is $y = kx$, where k is the **constant of proportionality**.

- $y \propto \frac{1}{x}$ means 'y is **inversely proportional** to x'.

- When y is inversely proportional to x:
  - the proportionality statement is $y \propto \frac{1}{x}$
  - the proportionality formula can be written as $y = k \times \frac{1}{x}$ or $y = \frac{k}{x}$, where k is the constant of inverse proportionality.

Shape, space and measure

2-D and 3-D shapes

- The sum of the **exterior angles** of a polygon always add up to 360°.
- The sum of the **interior angles** of a polygon with n sides is $(n - 2) \times 180°$.

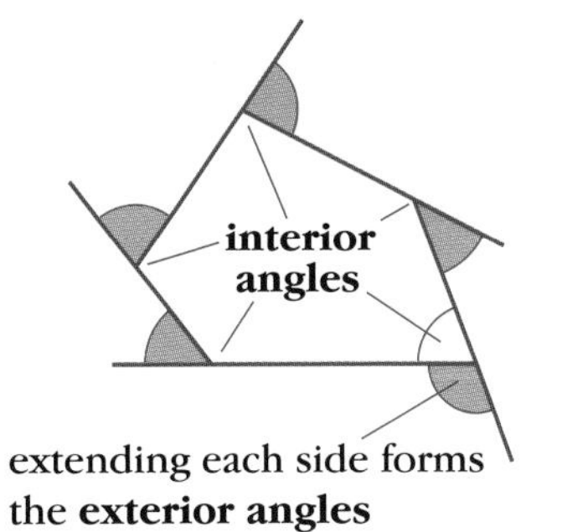

Perimeter, area and volume

- **Circumference** of a circle $= 2\pi r$
- Area of a **circle** $= \pi r^2$
- **Arc length** $= \dfrac{\theta}{360} \times 2\pi r$
- Area of **sector** $= \dfrac{\theta}{360} \times \pi r^2$

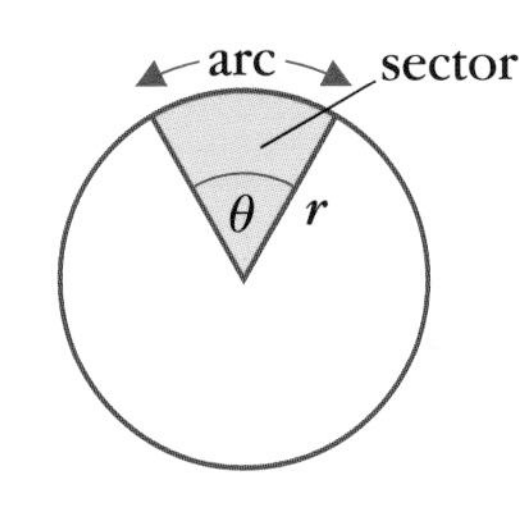

Transformations

- You should be able to:
  - perform a **rotation** with a given centre and angle of rotation.
  - perform a **reflection** in a given mirror line.
  - describe rotations and reflections fully.
- You should be able to:
  - perform an **enlargement** with a given centre and scale factor.
  - perform a **translation** described by a column vector.
  - describe translations and enlargements fully.

Trigonometry and Pythagoras' theorem

- **Pythagoras' theorem** states that $a^2 + b^2 = c^2$
  - Take care to make sure that your answer is sensible. Remember, the side opposite the right angle must be the longest side.

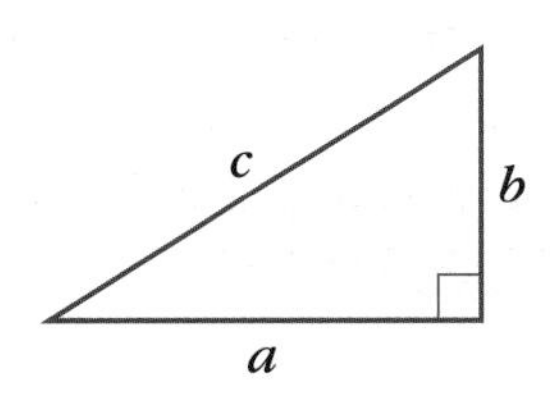

- The formulae of the Sine rule and Cosine rule are on the formulae sheet.

Unit 3 Examination practice paper

A formula sheet can be found on page 167.

Section A (calculator)

1 The cost of having a car serviced is £56.40 before VAT at $17\frac{1}{2}\%$ is added.
Find the total cost after VAT is added. **(3 marks)**

2 The diameter of a wheel is 70 centimetres.
Work out how many revolutions the wheel makes when travelling 1 kilometre. **(4* marks)**

3 Simplify

(a) **(i)** $a^5 \times a^2$ **(ii)** $15x^2y \div 3xy^2$

(b) Find the value of n in
$a^n \times a^3 \div a^7 = a^2$ **(5 marks)**

4 $x^3 + 3x - 16 = 0$

Use trial and improvement to find the positive solution of this equation.
Give your answer correct to 1 decimal place. **(4 marks)**

5 The shape shown is a triangular prism.
The total surface area, including the base, is $2100\,\text{cm}^2$.

Work out the width, x cm, of the prism.

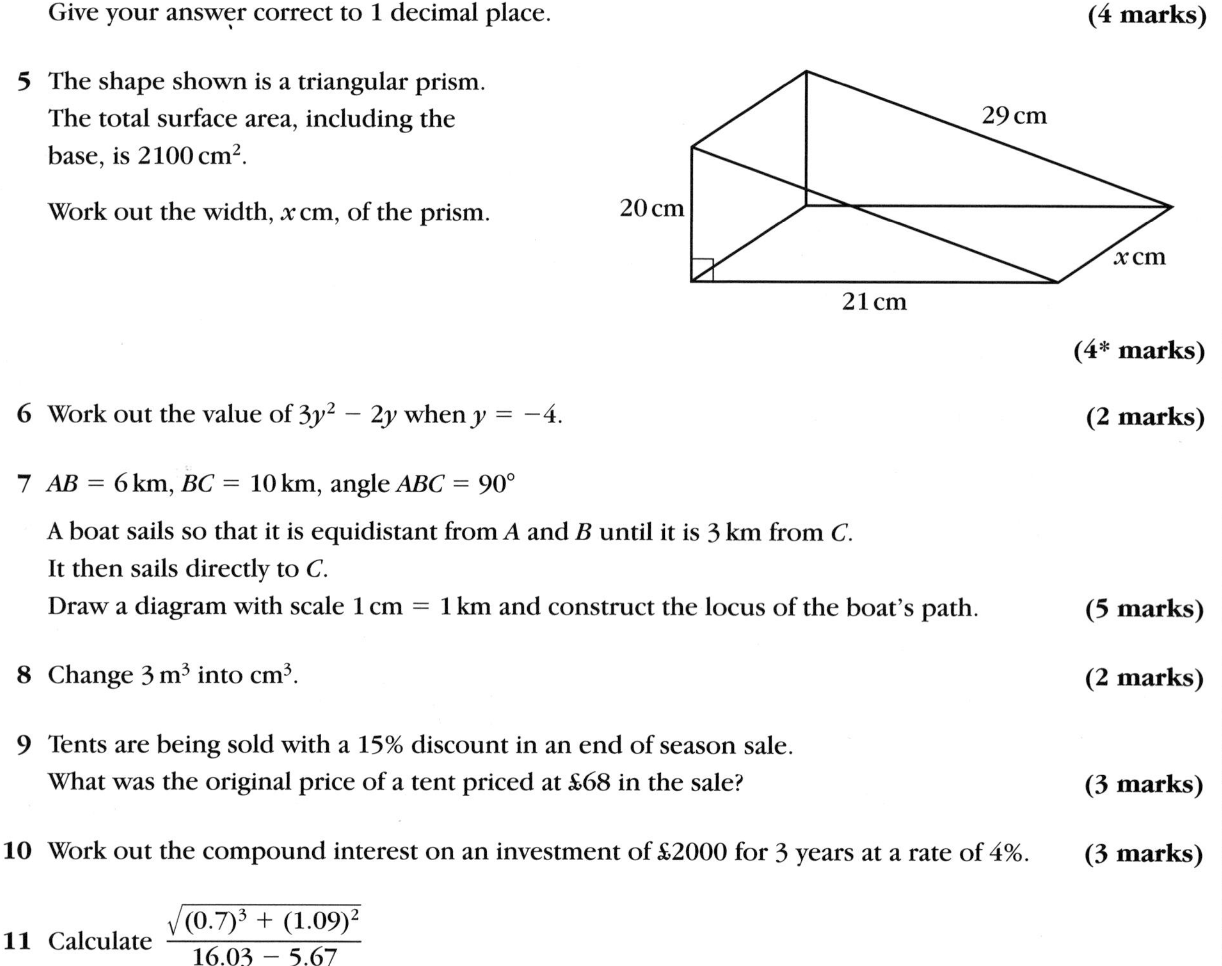

(4* marks)

6 Work out the value of $3y^2 - 2y$ when $y = -4$. **(2 marks)**

7 $AB = 6\,\text{km}$, $BC = 10\,\text{km}$, angle $ABC = 90°$

A boat sails so that it is equidistant from A and B until it is 3 km from C.
It then sails directly to C.
Draw a diagram with scale 1 cm = 1 km and construct the locus of the boat's path. **(5 marks)**

8 Change $3\,\text{m}^3$ into cm^3. **(2 marks)**

9 Tents are being sold with a 15% discount in an end of season sale.
What was the original price of a tent priced at £68 in the sale? **(3 marks)**

10 Work out the compound interest on an investment of £2000 for 3 years at a rate of 4%. **(3 marks)**

11 Calculate $\dfrac{\sqrt{(0.7)^3 + (1.09)^2}}{16.03 - 5.67}$

Write down all the figures on your calculator display. **(3 marks)**

12 The fast train travels the 26.4 kilometres from Morpeth to Newcastle in 23 minutes.
Work out the average speed of the train. **(3 marks)**

13 Work out the length of BC. **(3 marks)**

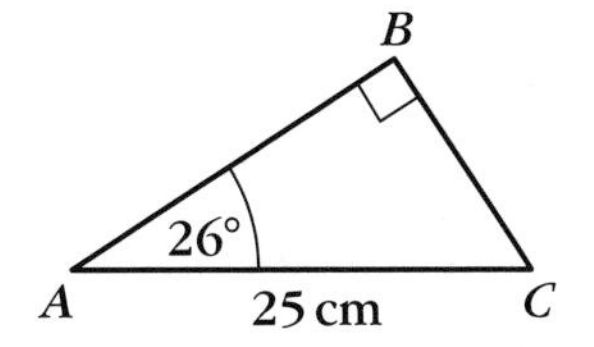

14 $ABCD$ and $GDEF$ are squares, where $GD = DA$.
Prove triangles AFG and GBA are congruent.

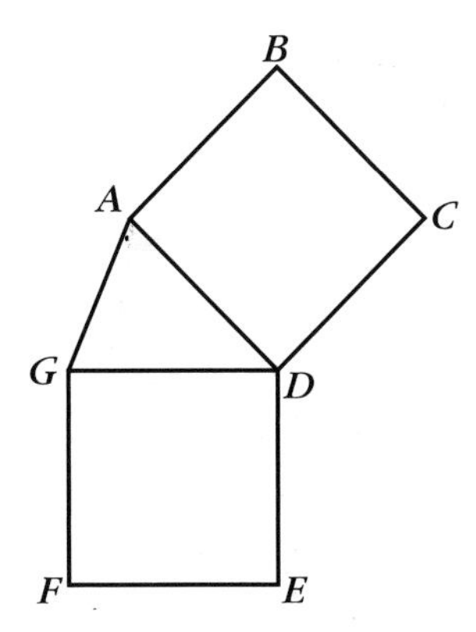

(4* marks)

15 Find the area of the sector of a circle with radius 16 cm and angle 125°. **(3 marks)**

16 A stone is dropped from a cliff.
After t seconds the stone has fallen a distance s metres.
s is directly proportional to t^2.

When $t = 2$ seconds, $s = 20$ metres.

(a) Find an equation connecting s and t.

The cliff is 80 metres high.

(b) Find how long it takes the stone to fall to the foot of the cliff. **(2 marks)**

17 $ABCD$ is a rectangle.
DEC is an equilateral triangle.
DEA and BEC are right angles.
$BC = 2x$

Work out the area of $ABCD$ in terms of x. **(4 marks)**

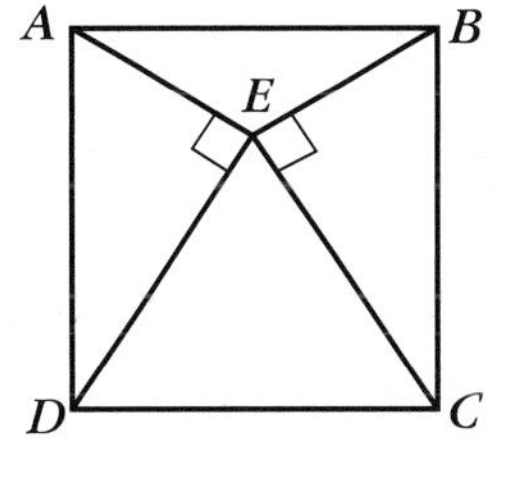

Check your answers on page 180. For full worked solutions see the CD.

Section B (non-calculator)

1 A packet of readimix cement contains weights of sand and cement in the ratio 4 : 1.
Find the weight of sand in a 10 kilogram bag of readimix. **(2 marks)**

2 (a) Solve $3x + 1 = 16$

(b) Solve $\frac{2y}{5} = 3$

(c) Solve $3(4 - 3x) = x + 7$ **(3 marks)**

3 $-3 < x \leq 1$ where x is an integer.
List all the possible values of x. **(2 marks)**

4 (a) Find the mid-point of the line segment joining the points $A(-1, -2)$ to $B(1, 4)$.

(b) Find the equation of the line which passes through these points.

(c) Find the equation of the line perpendicular to AB which passes through $(1, 4)$. **(7 marks)**

5 (a) Simplify $\frac{3}{4} + \frac{4}{5}$ (b) Simplify $1\frac{3}{5} \times 2\frac{1}{4}$

Leave your answers as fractions in their simplest form. **(5 marks)**

6 The equilateral triangle is surrounded by regular polygons.
Work out the number of sides each regular polygon has. **(3* marks)**

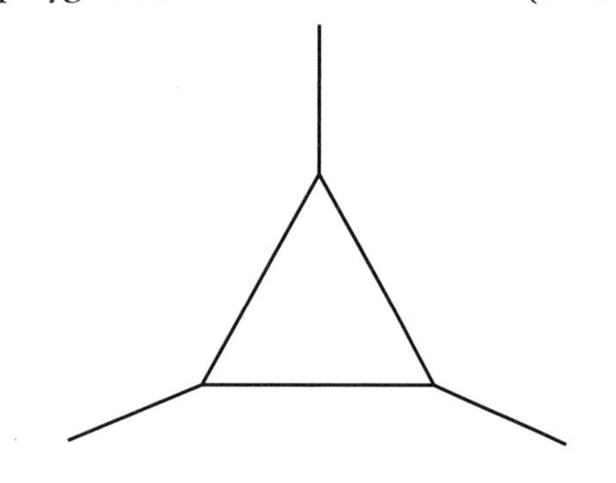

7 Sketch the shape shown by the plan and elevations. **(2* marks)**

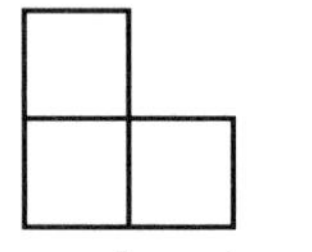

Front elevation

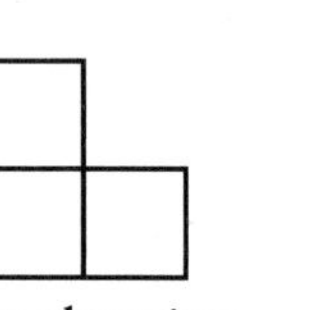

Side elevation

Plan

8 Solve the simultaneous equations

$$2x + 3y = -6$$
$$6x + y = 6$$

(4 marks)

9 Make r the subject of $A = \pi r^2$ **(2 marks)**

10 $(3 + 2\sqrt{5})^2 = a + b\sqrt{5}$
Find the value of a and b. **(2 marks)**

11

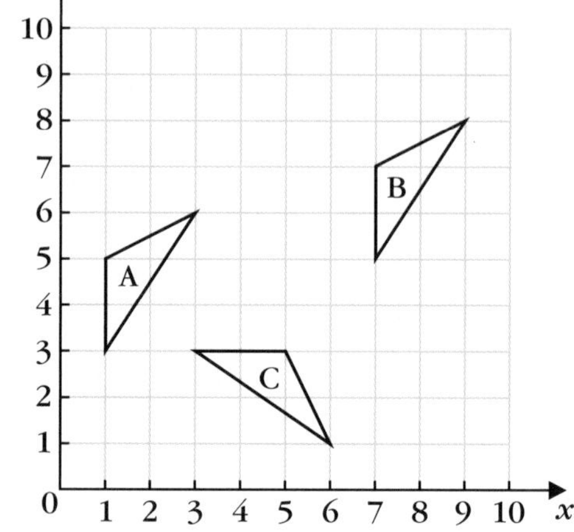

(a) Describe fully the transformation that maps A onto B.

(b) Describe fully the transformation that maps A onto C. **(5* marks)**

12 $\overrightarrow{OA} = \underline{a}$
and $\overrightarrow{OB} = \underline{b}$

$AP = 3OA$

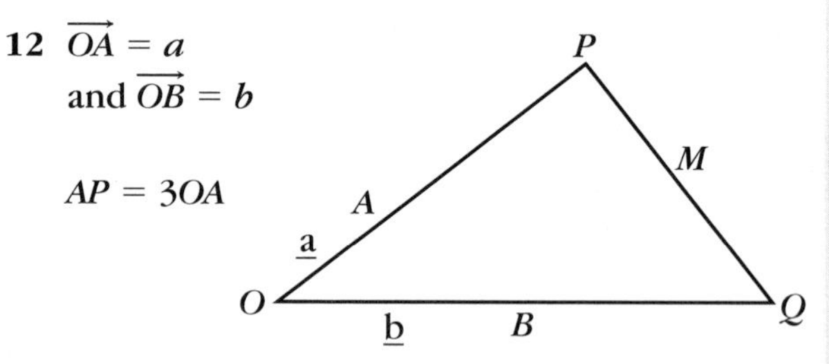

B is the mid-point of OQ. M is the mid-point of PQ.
Express, in terms of $\underline{a}$ and/or $\underline{b}$, the vectors

(a) $\overrightarrow{OP}$ (b) $\overrightarrow{PQ}$ (c) $\overrightarrow{OM}$ (d) $\overrightarrow{BM}$ **(4 marks)**

13 $AB = AC$
Angle $BAD = 90°$

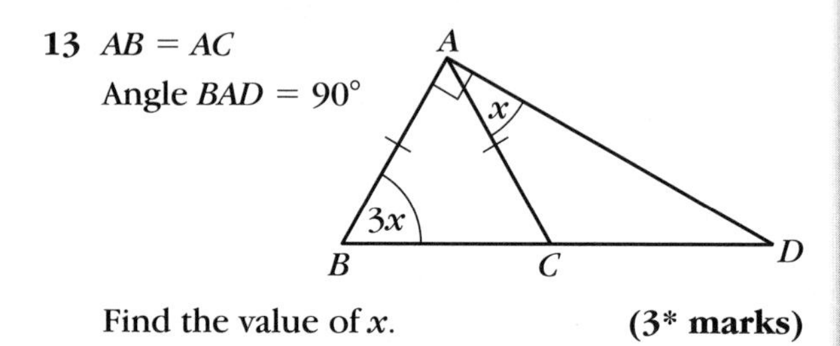

Find the value of x. **(3* marks)**

14 **(a)** Write 2.6×10^{-3} as an ordinary number.

(b) Simplify $2.6 \times 10^{-3} + 1.2 \times 10^{-4}$
Give your answer in standard form.

(3 marks)

15 Solve the simultaneous equations

$$y = 3x + 2$$
$$x^2 + y^2 = 2$$

(6 marks)

16 *ABC* and *PQR* are similar triangles.

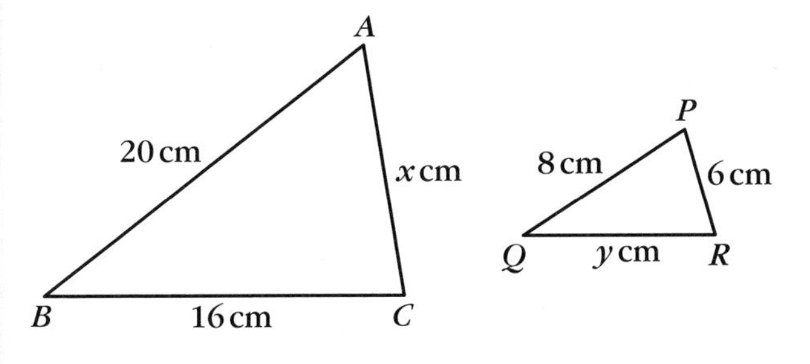

Diagrams **NOT** accurately drawn

Find the value of **(a)** x **(b)** y **(4 marks)**

17

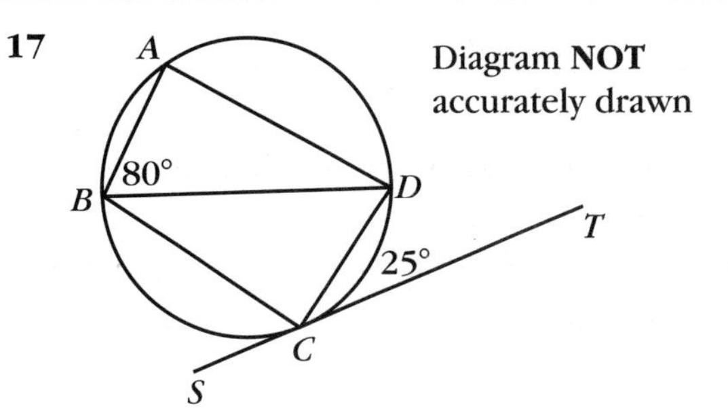

Diagram **NOT** accurately drawn

ABCD are points on the circumference of a circle.
SCT is a tangent to the circle.
Angle $DCT = 25°$ Angle $ABD = 80°$

(a) Find angle *DBC*.

(b) Find angle *ADC*.

Give a reason for your answer.

(2 + 1* mark)

Check your answers on page 180. For full worked solutions see the CD.

Formulae

Volume of prism = area of cross section × length

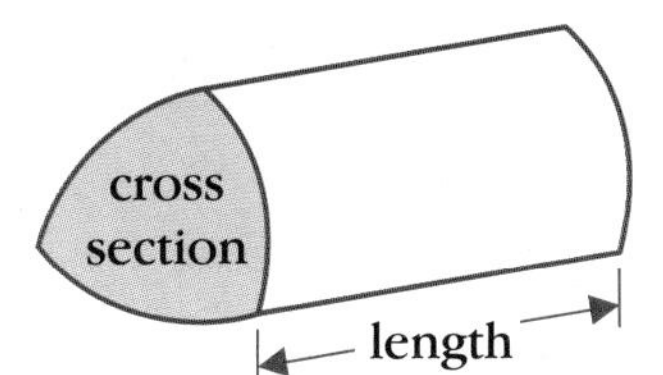

Volume of sphere $= \frac{4}{3}\pi r^3$

Surface of a sphere $= 4\pi r^2$

Volume of cone $= \frac{1}{3}\pi r^2 h$

Curved surface area of cone $= \pi r l$

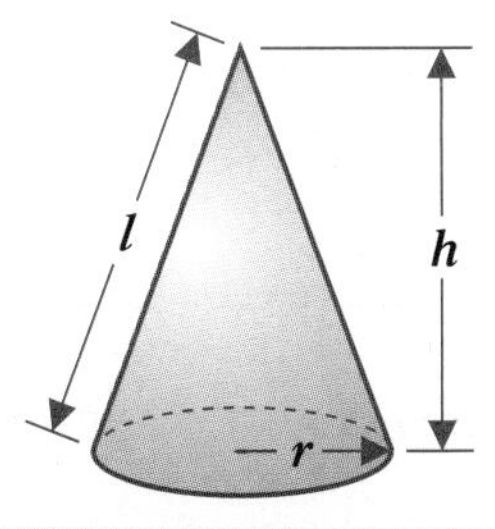

In any triangle *ABC*

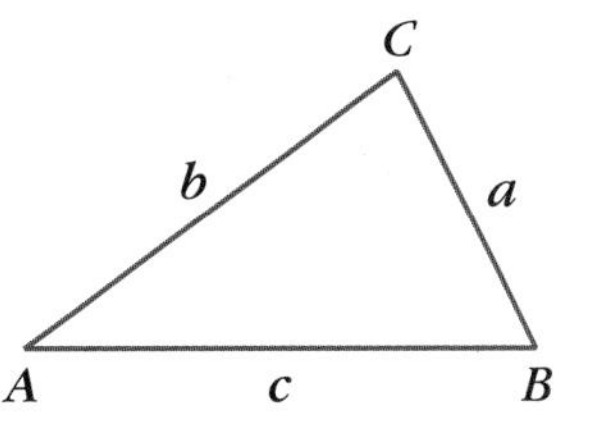

Sine rule $\dfrac{a}{\sin A} = \dfrac{b}{\sin B} = \dfrac{c}{\sin C}$

Cosine rule $a^2 = b^2 + c^2 - 2bc \cos A$

Area of a triangle $= \frac{1}{2}ab \sin C$

The quadratic equation
The solutions of $ax^2 + bx + c = 0$ where $a \neq 0$, are given by

$$x = \frac{-b \pm \sqrt{(b^2 - 4ac)}}{2a}$$

Answers

Unit 1

Collecting and organising data

1 (a) 64 miles
(b) Hull or Sheffield
(c) 155 miles

2 (a) Options are too vague; no time period specified.
(b) How much money do you spend in the café in a week?
£0 ☐ £0.01–£4.99 ☐ £5–£9.99 ☐
£10–£20 ☐ more than £20 ☐

3

Type of pet	Tally	Frequency
Dog		
Cat		
Hamster		
Rabbit		
Goldfish		

4 (a) 18 boys (b) 21 girls

Averages

1 (a) 34 (b) 2 (c) 2 (d) 2.12
(e) The most tries the team scored in any match was 4, so the average cannot be higher than 4.

2 (a) $20 \leq t < 25$ (b) $20 \leq t < 25$
(c) 23.5 minutes

3 30, 29, 25.25

4 (a) 12, 9
(b)

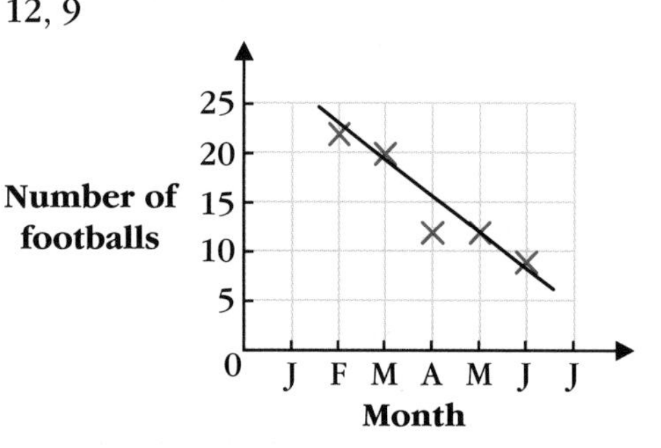

(c) The trend line shows a steep decrease in the number of footballs sold each month over this period.

Probability

1 0.35

2 (a) (B1, R1), (B1, R2), (B1, R3), (B1, R4), (B1, R5), (B1, R6)
(B2, R1), (B2, R2), (B2, R3), (B2, R4), (B2, R5), (B2, R6)
(B3, R1), (B3, R2), (B3, R3), (B3, R4), (B3, R5), (B3, R6)
(B4, R1), (B4, R2), (B4, R3), (B4, R4), (B4, R5), (B4, R6)
(B5, R1), (B5, R2), (B5, R3), (B5, R4), (B5, R5), (B5, R6)
(B6, R1), (B6, R2), (B6, R3), (B6, R4), (B6, R5), (B6, R6)
(b) $\frac{6}{36} = \frac{1}{6}$

3 $\frac{3}{6} = \frac{1}{2}$

4 (a)

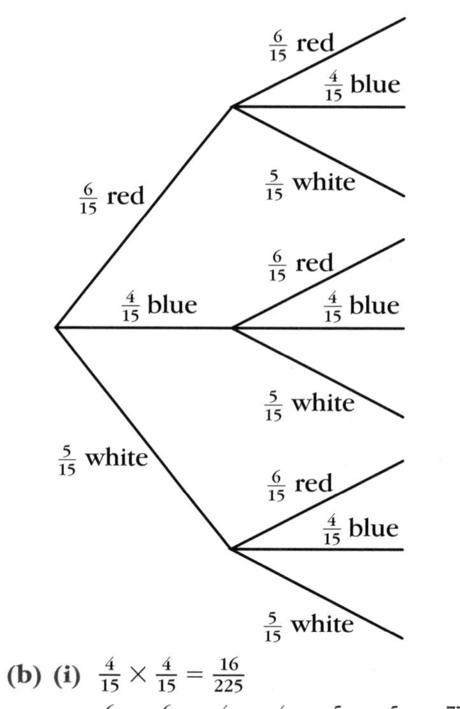

(b) (i) $\frac{4}{15} \times \frac{4}{15} = \frac{16}{225}$
(ii) $\frac{6}{15} \times \frac{6}{15} + \frac{4}{15} \times \frac{4}{15} + \frac{5}{15} \times \frac{5}{15} = \frac{77}{225}$

5 (a)

1st attempt 2nd attempt 3rd attempt
pass 0.7
fail 0.3
pass 0.8
fail 0.2
pass 0.8

(b) $0.7 + 0.3 \times 0.8 + 0.3 \times 0.2 \times 0.8 = 0.988$

6 (a) $\frac{13}{51}$

(b) $\frac{3}{663}$

Frequency charts

1 (a)

0	0, 2, 3, 3, 6, 7, 8, 9
1	1, 2, 3, 5, 9
2	2, 4, 9
3	1, 1, 8
4	7

Key: 2 | 4 means 24

(b) 12.5

(c) Lower quartile 6.25, upper quartile 27.75, interquartile range 21.5

2

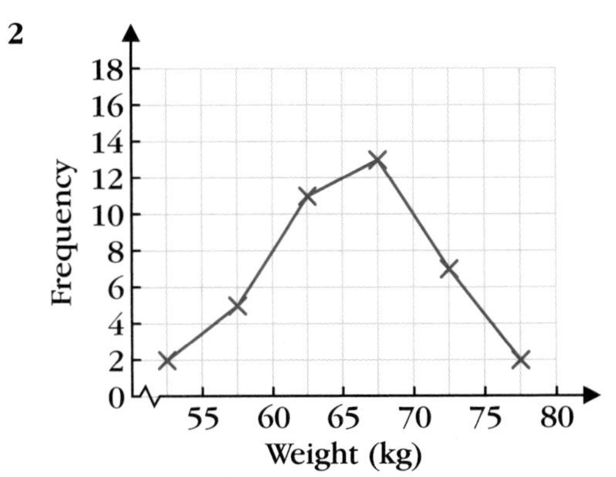

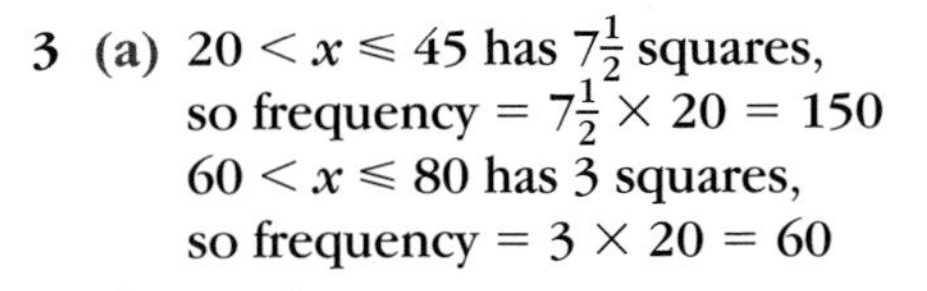

3 (a) $20 < x \leqslant 45$ has $7\frac{1}{2}$ squares, so frequency $= 7\frac{1}{2} \times 20 = 150$

$60 < x \leqslant 80$ has 3 squares, so frequency $= 3 \times 20 = 60$

(b)

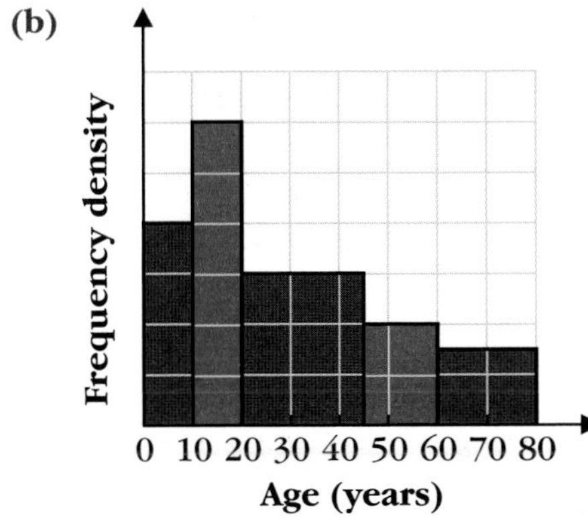

Scatter graphs, correlation and the RPI

1 (a) (c)

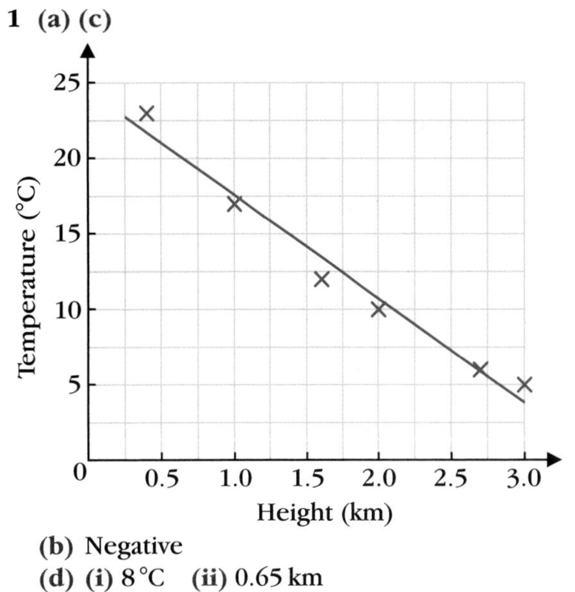

(b) Negative

(d) (i) 8 °C (ii) 0.65 km

2 Generally positive during their working lives

3 £167.62

Cumulative frequency and box plots

1 (a)

Weekly rainfall, d (mm)	Cumulative frequency
$0 \leqslant d < 10$	20
$0 \leqslant d < 20$	38
$0 \leqslant d < 30$	44
$0 \leqslant d < 40$	48
$0 \leqslant d < 50$	50
$0 \leqslant d < 60$	52

(b)

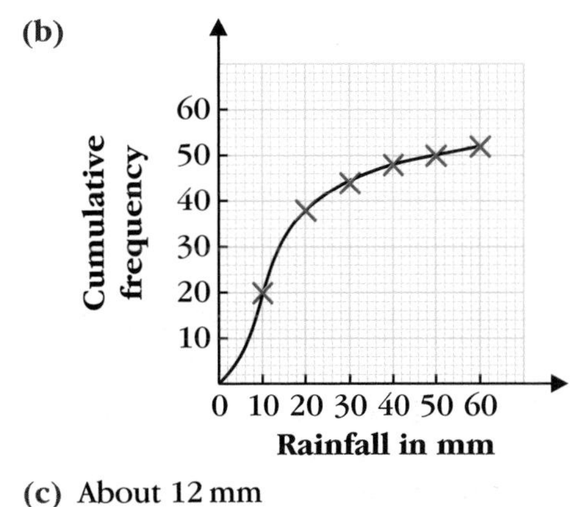

(c) About 12 mm

(d) About 13 mm

(e)

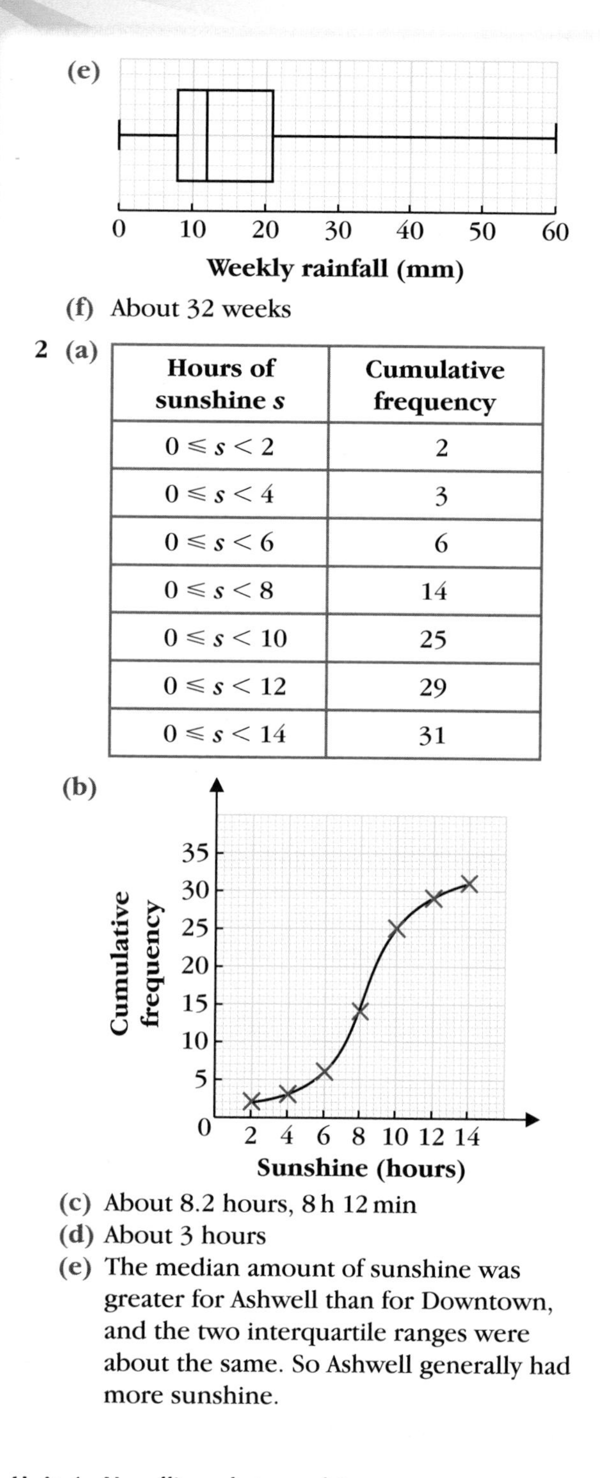

(f) About 32 weeks

2 (a)

Hours of sunshine s	Cumulative frequency
$0 \leqslant s < 2$	2
$0 \leqslant s < 4$	3
$0 \leqslant s < 6$	6
$0 \leqslant s < 8$	14
$0 \leqslant s < 10$	25
$0 \leqslant s < 12$	29
$0 \leqslant s < 14$	31

(b)

(c) About 8.2 hours, 8 h 12 min

(d) About 3 hours

(e) The median amount of sunshine was greater for Ashwell than for Downtown, and the two interquartile ranges were about the same. So Ashwell generally had more sunshine.

Unit 1 Handling data: subject test

1 (a)

2	7, 8
3	3, 6, 7, 9, 9
4	1, 4, 5, 6, 6, 8, 9
5	0, 1, 3
6	0, 2, 5

Key: 2 | 7 means 27 seconds

(b) 45.5 seconds

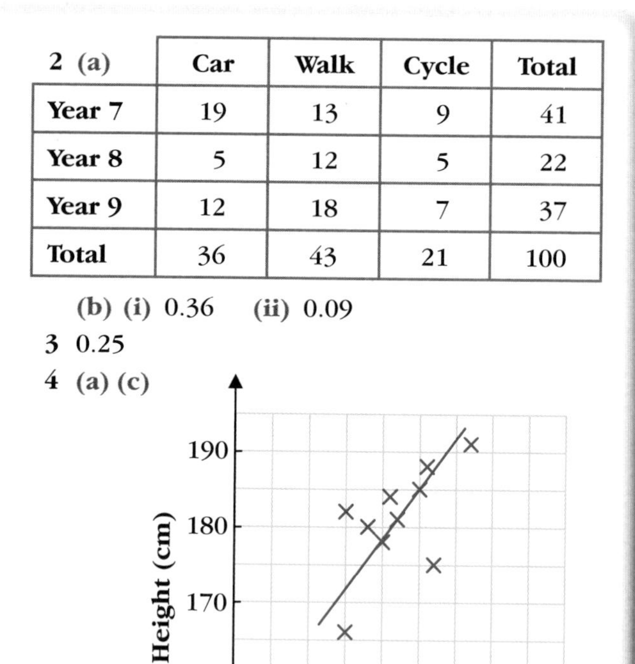

2 (a)

	Car	Walk	Cycle	Total
Year 7	19	13	9	41
Year 8	5	12	5	22
Year 9	12	18	7	37
Total	36	43	21	100

(b) (i) 0.36 (ii) 0.09

3 0.25

4 (a) (c)

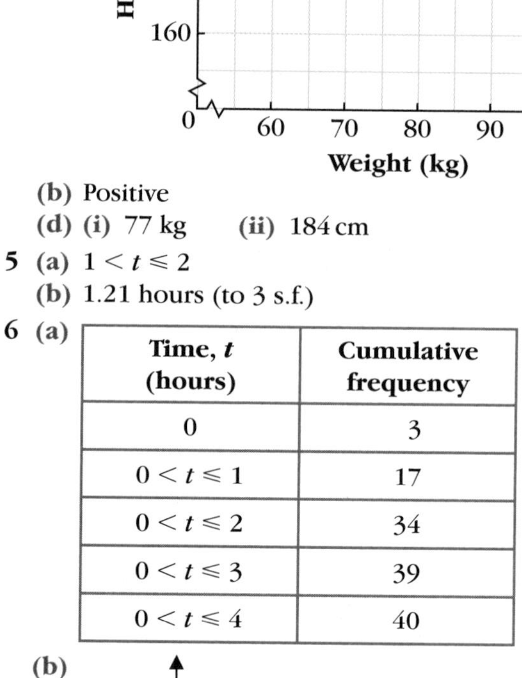

(b) Positive

(d) (i) 77 kg (ii) 184 cm

5 (a) $1 < t \leqslant 2$

(b) 1.21 hours (to 3 s.f.)

6 (a)

Time, t (hours)	Cumulative frequency
0	3
$0 < t \leqslant 1$	17
$0 < t \leqslant 2$	34
$0 < t \leqslant 3$	39
$0 < t \leqslant 4$	40

(b)

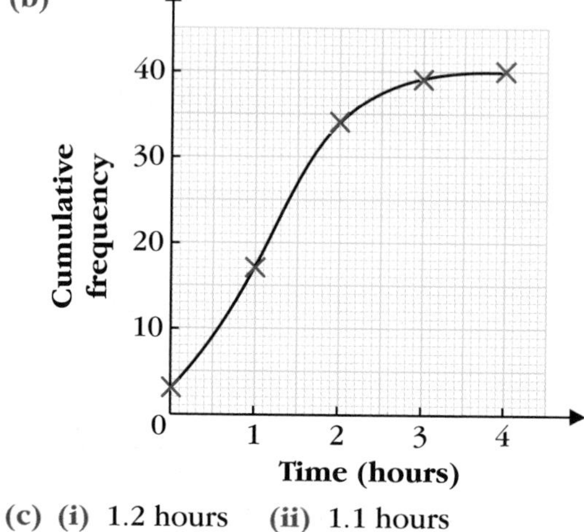

(c) (i) 1.2 hours (ii) 1.1 hours

7 **(a)**

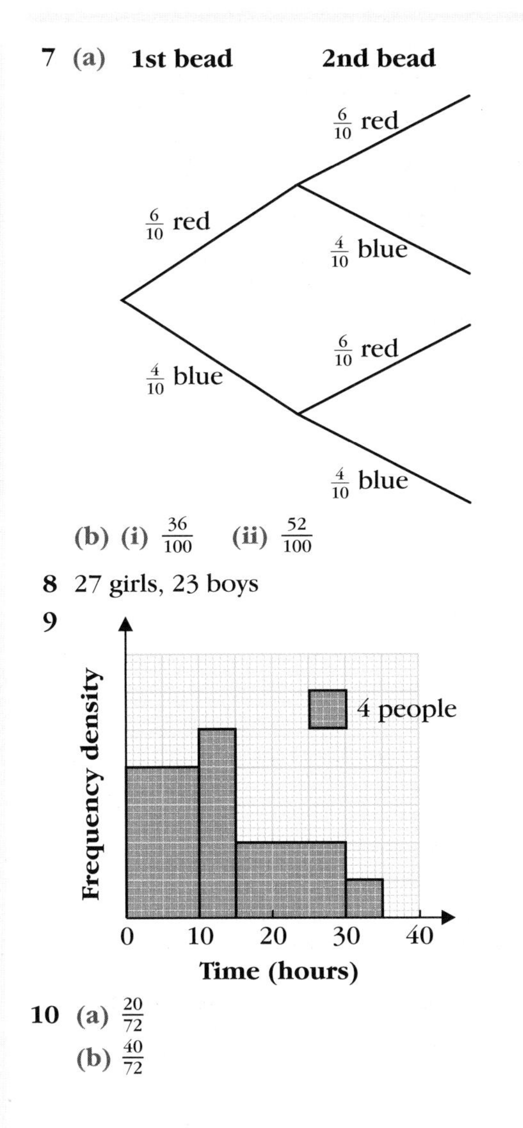

(b) **(i)** $\frac{36}{100}$ **(ii)** $\frac{52}{100}$

8 27 girls, 23 boys

9

10 **(a)** $\frac{20}{72}$
(b) $\frac{40}{72}$

Unit 1 Examination practice paper Calculator

1 **(a)** 0.2 **(b)** 30 seeds

2

1	3 5 7 8
2	4 6 7 9
3	1 2 3 6
4	4 5 8

Key: 1 | 3 means 13

3 **(a)** $20 \leq t \leq 25$ **(b)** 24 minutes

4 5 girls

Unit 1 Examination practice paper Non-calculator

1 **(a)** It is a leading question
(b) How many cakes do you eat each week?

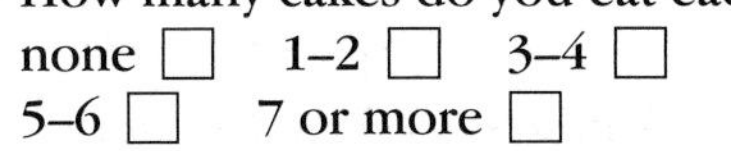

2 **(a)** Positive correlation
(b) Appropriate line of best fit
(c) 41 ± 2

3 **(a)** 73 kg
(b) 15 kg

4 **(a)** 8, 6
(b)

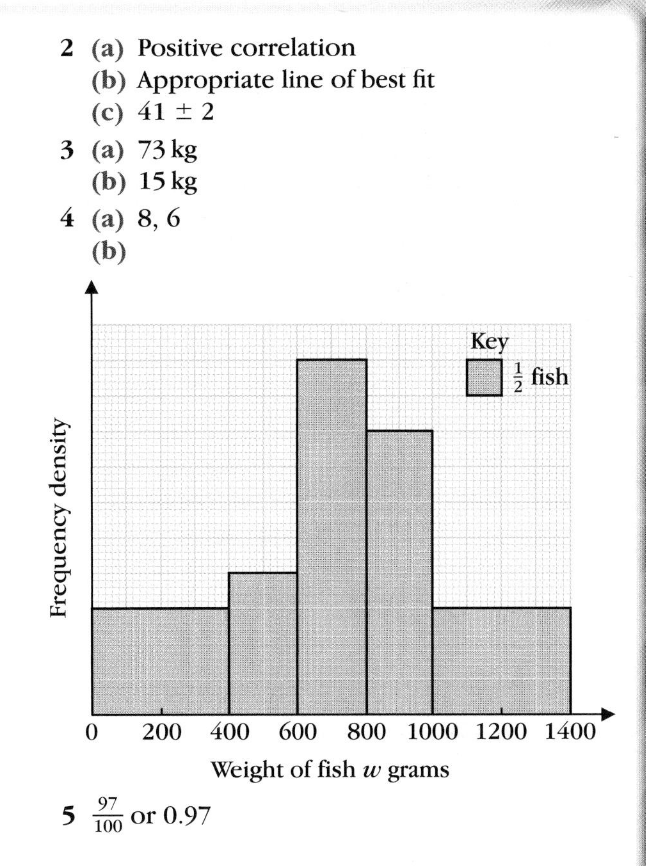

5 $\frac{97}{100}$ or 0.97

Unit 2

Operations

1 **(a)** -4 **(b)** -5 **(c)** -8 **(d)** 8
2 **(a)** -42 **(b)** -6 **(c)** -3 **(d)** 3
3 **(a)** 6970 **(b)** 28 938 **(c)** 13 300
(d) 25 **(e)** 24.5 **(f)** 31
4 **(a)** $16 = 2 \times 2 \times 2 \times 2$
(b) $24 = 2 \times 2 \times 2 \times 3$
5 HFC 8, LCM 48

Indices, powers and roots

1 **(a)** 243 **(b)** 1.3
2 **(a)** 500 **(b)** 375
3 **(a)** 4^5 **(b)** 9^3
4 **(a)** $\frac{1}{9}$ **(b)** $1\frac{1}{5}$ **(c)** $2\frac{1}{4}$
5 **(a)** $\frac{1}{9}$ **(b)** 1
6 **(a)** 4 **(b)** 5

Decimals and rounding

1 (a) (i) 300 000 (ii) 60 000
(iii) 0.3 (iv) 0.0006
(b) (i) 250 000 (ii) 56 900
(iii) 0.347 (iv) 0.000 600

2 (a) 37.8 (b) 66.15 (c) 2.1452

3 (a) 4 (b) 400

4 (a) 9 (b) 3.2 (c) 12

Standard form and bounds

1 (a) 3.56×10^5 (b) 4.567×10^3
(c) 4.5×10^8 (d) 4.5×10^{-4}
(e) 1.5×10^{-7} (f) 1.2×10^{-2}

2 (a) 36 000 (b) 36 000 000
(c) 455 (d) 0.0024
(e) 0.000 002 45 (f) 0.000 675

3 (a) 2×10^8 (b) 4.5×10^9
(c) 5×10^{-2} (d) 4×10^1

4 (a) Upper bound 4.655 m, lower bound 4.645 m
(b) Upper bound 3.45 s, lower bound 3.35 s
(c) Upper bound 16.32005 *l*, lower bound 16.31995 *l*
(d) Upper bound 3.8505 kg, lower bound 3.8495 kg

Fractions

1 (a) $1\frac{1}{4}$ (b) $8\frac{5}{24}$

2 (a) $\frac{13}{24}$ (b) $3\frac{5}{12}$

3 $1\frac{7}{12}$ miles

4 (a) $\frac{7}{16}$ (b) 5

5 (a) $1\frac{7}{18}$ (b) $\frac{2}{3}$

6 (a) $\frac{3}{8}$ (b) $\frac{119}{333}$ (c) $3\frac{8}{33}$

Percentages, fractions and decimals

1 $\frac{3}{5}$, 62.5%, 63%, 0.65, $\frac{2}{3}$, 67%

2 She did equally well (75%) in French and Spanish but worse (67%) in German.

3 £42

4 (a) 22.4 (b) £1.44 (c) £31.50
(d) 0.768 m (e) 96 (f) £2

Unit 2 Number: subject test

1 0.45, $\frac{1}{2}$, 57%, 0.6, 65%, $\frac{3}{4}$

2 (a) 19.6 (b) £1.62 (c) £33.60
(d) 1.536 m (e) 18.4 (f) £3

3 (a) 11 (b) 7 (c) 108 (d) 1.4

4 (a) $\frac{900}{300} = 3$ (b) $\frac{7000 \times 10}{800 - 100} = 100$

5 22

6 (a) 5^9 (b) 3^3 (c) 4^4

7 (a) $2^3 \times 7$ (b) 2^5 (c) $3^2 \times 5$ (d) $2^2 \times 5^2$

8 (a) $20 = 2 \times 2 \times 5$ $30 = 2 \times 3 \times 5$
(b) LCM = 10, HCF = 60

9 (a) 7.66×10^5 (b) 8.65×10^{-4}

10 (a) 760 000 (b) 0.000 054 5

11 (a) 2.94×10^{11} (b) 1.4×10^{-3}
(c) 7.2×10^1 (d) 4×10^{-5}
(e) 1.92×10^{11} (f) 8×10^5

12 (a) $6\frac{3}{20}$ (b) $19\frac{4}{5}$ (c) $2\frac{1}{2}$
(d) $6\frac{1}{4}$ (e) $7\frac{13}{24}$ (f) $2\frac{7}{24}$
(g) $2\frac{7}{12}$ (h) $\frac{11}{16}$ (i) 30
(j) $\frac{1}{4}$ (k) $\frac{33}{40}$ (l) $\frac{25}{32}$

13 (a) 1 (b) 0.04 (c) 216

14 $\frac{4}{11}$

15 (a) Upper bound 5.35 cm, lower bound 5.25 cm
(b) Upper bound 10.65 cm, lower bound 10.55 cm
(c) Upper bound 10.0005 m, lower bound 9.9995 m
(d) Upper bound 1.005 s, lower bound 0.995 s
(e) Upper bound 1.0005 km, lower bound 0.9995 km

Simplifying algebra and expanding brackets

1 (a) $4x - y$ (b) $5x + 8y$

2 (a) t^8 (b) $3p^3$

3 (a) $4a - 8c$ (b) $g^2 - 4g$

4 (a) $x^2 - xy - 2y^2$ (b) $6x^2 - x - 12$

5 (a) p^6q^3 (b) $3y^2$

Factorising and algebraic fractions

1 (a) $7(x - 3y)$ (b) $4t(2t + 1)$

2 (a) $(x - 6)(x + 6)$ (b) $(3x + 5y)(3x - 5y)$

3 (a) $(x + 3y)(x - 4y)$ (b) $(2x - 3)(5x + 7)$

4 (a) $\frac{x}{x + 3}$ (b) $\frac{2y - 1}{y(y - 3)(y + 2)}$

Patterns and sequences

1 $3n + 2$

2 (a) $7n - 4$ (b) 276

3 (a) $9 - 2n$ (b) -91

Coordinates and algebraic line graphs

1 (a) $A(0, 2, 2)$, $D(2, 2, 2)$, $F(2, 0, 0)$
(b) Mid-point is at (1, 1, 1)

2 **(a)**

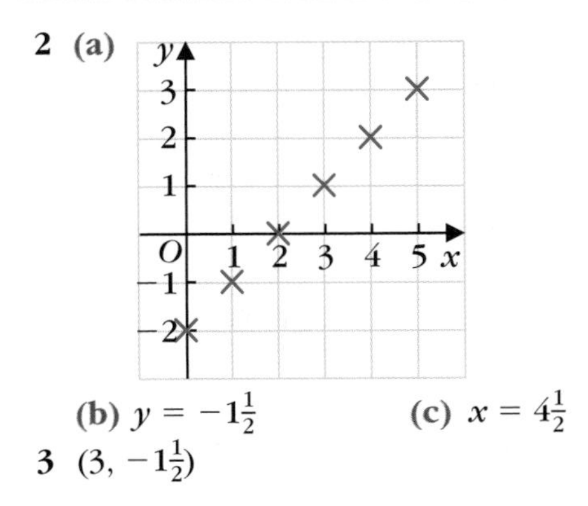

(b) $y = -1\frac{1}{2}$ **(c)** $x = 4\frac{1}{2}$

3 $(3, -1\frac{1}{2})$

Unit 2 Algebra: subject test

1 $12a - 8b$

2 $y(y^2 - 2)$

3 **(a)** **(i)** (3, 2, 0) **(ii)** (0, 2, 2) **(iii)** (3, 2, 2)
(b) (0, 2, 0), (3, 0, 2); mid-point $(1\frac{1}{2}, -1, 1)$

4 $7x + 21$

5 $2t^2 - 3tq$

6 $4x^2$

7 $k(k - 3)$

8 $4n - 1$

9 When $y = 0$, $x = \frac{2}{3}$

10 **(a)** $20 - 3n$ **(b)** -40

11 $2x^2 + 5x - 12$

12 $2x^2 - 5xy - 3y^2$

13 $(x - 3)(x + 2)$

14 $(1\frac{1}{2}, -1)$

15 $(2x - 3y)(2x + 3y)$

16 $8x^3y^6$

17 $(2x + 5)(x - 3)$

18 $2(x - 5)$

19 $\dfrac{4x + 7}{(x + 2)(2x + 5)}$

20 $\dfrac{x - 3}{x + 3}$

Working with angles

1 $a = 47°$ (alternate angles),
$b = 47°$ (corresponding angles),
$c = 84°$ (corresponding angles),
$d = 37°$ (angles in a triangle)

2 **(a)** 35° (corresponding angles)
(b) 70°
Angle *ABE* = 110° (third angle in an isosceles triangle)
Angle *CBE* = 70° (angles on a straight line)
(c) 70° (corresponding angles)

3 Angle *ABC* = 34° (base angle in an isosceles triangle)
Angle *BCD* = 34° (alternate angles)
Angle *CDB* = $\frac{1}{2}(180 - 34) = 73°$ (base angle in an isosceles triangle)

4 Triangle *PXQ* is isosceles.
So angle *PXQ* = angle *PQX*
Angle *PYR* = angle *PXQ* (corresponding angles)
Angle *PRY* = angle *PQX* (corresponding angles)
Hence angle *PYR* = angle *PRY* and triangle *PRY* is isosceles.

Perimeter, area, volume and measures

1 270 000 cm³

2 192 cm³

3 7 cm

4 40 cm²

5 Perimeter 40 cm, area 52 cm²

6 3 hours

7 **(a)** 126 km **(b)** 56 km **(c)** 7 km

8 Volume 1120 cm³, surface area 752 cm²

Shape, space and measure: subject test

1 **(a)** $a = 57°$ (corresponding angles)
$b = 80°$ (angles on a straight line)
$c = 80°$ (angles in a triangle)
(b) $a = 70°$ (corresponding angles)
$b = 50°$ (vertically opposite angles)
$c = 130°$ (angles on a straight line, corresponding angles)

2 9.625 m²

3 **(a)** 700 cm³ **(b)** 9.29 g/cm³

4 44 cm²

5 297.6 km/h

6 160 cartons **7** 50 m/s is faster

8 2690 cubes **9** 4°

Unit 2 Examination practice paper Stage 1

1 B	**2** B	**3** B
4 D	**5** D	**6** A
7 E	**8** C	**9** D
10 B	**11** B	**12** C
13 C	**14** E	**15** D
16 A	**17** C	**18** E
19 C	**20** D	**21** C
22 A	**23** D	**24** B
25 E		

Stage 2

1 (a) $43°$
(b) Angles in a triangle add up to 180°, then corresponding angles

2 9.5 cm

3 $x^2 + 2x - 8$

4 (a) 4^7 or 16384 (b) 5

5 6.25×10^{10}

6 (a) $-3, (-1), 1, 3, (5), 7$
(b) Straight line passing through all the points

7 (a) $(p - 10)(p + 2)$ (b) $\frac{3x^2 - 2}{x(3x - 2)}$

8 (a) $90°$
(b) $140°$
(c) Tangents to a circle from a point are equal so triangle PQT is isosceles. Angle $PTQ = 40°$ so the other two are each 70°.

9 $\frac{14}{99}$

Unit 3

Powers, surds and bounds

1 (a) $\frac{2\sqrt{7}}{7}$ (b) $\frac{1}{2}$
(c) $\sqrt{2}$ (d) $\frac{3\sqrt{5} + 5}{5}$

2 (a) 216 (b) 8 (c) 512
(d) $\frac{1}{4}$ (e) $\frac{1}{9}$ (f) 16

3 Upper bound 5.7148 m/s
Lower bound 5.7138 m/s

4 Upper bound 17.0625 cm²
Lower bound 16.2425 cm²

Percentages

1 (a) £76.50 (b) £63.75

2 £282.00

3 1980–1990

4 20%

5 (a) £120 (b) £132.40

6 £70

Ratio and proportion

1 £4.00

2 (a) 5 eggs (b) 300 g

3 1 km

4 Simon £20, Zoe £25

Unit 3 Number: subject test

1 9

2 £31.50

3 £70.50

4 (a) 30
(b) 8

5 28 g of cheese, 42 g of topping

6 Ron £2.50, Shanna £2.00

7 Amy: £60, Ben: £100, Clive: £140

8 £441

9 £5832

10 (a) £54
(b) £57.30

11 £75

12 (a) 1.813×10^9
(b) 2.5×10^{-3}
(c) 1.536×10^{-8}
(d) 7.22×10^5

13 (a) 27
(b) 64
(c) 0.25
(d) 18

14 (a) $\frac{5\sqrt{17}}{17}$ (b) $\frac{\sqrt{5}}{5}$

15 (a) 10.15 m, 9.05 m
(b) −1.05 m, −2.15 m
(c) 25.425 m, 19.425 m
(d) 0.811 m, 0.619 m

16 Upper bound 8.3341 m/s
Lower bound 8.3326 m/s

Graphs

1

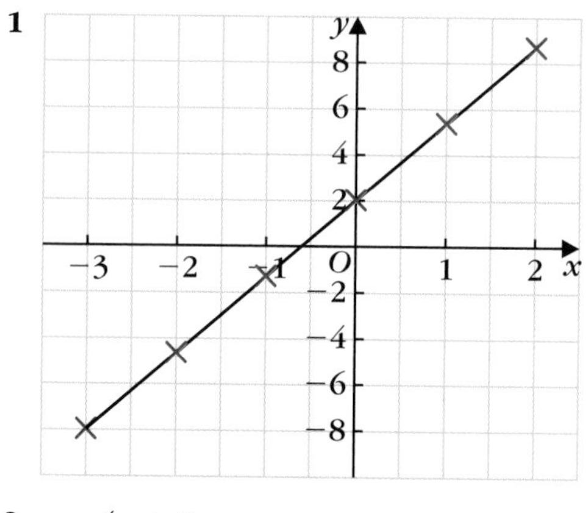

2 $y = 4x + 5$

3 $y = -5\frac{1}{2}x + 15$

4 $y = -\frac{1}{5}x + 4$

More graphs

1 **(a)** 60 km/h

(b)

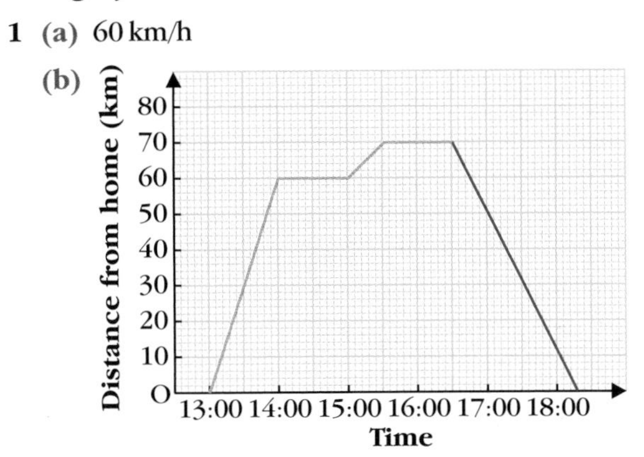

2 $x = 4, y = 1$

3 Between A and B, the car's speed rises at a constant rate; the increase in speed falls off beteen B and C (whilst still getting faster); the speed is then constant for a time (between C and D), before the car comes to an abrupt and sudden halt (at E).

Curved graphs

1 **(a)**

x	0	1	2	3	4
y	0	4	2	0	4

(b)

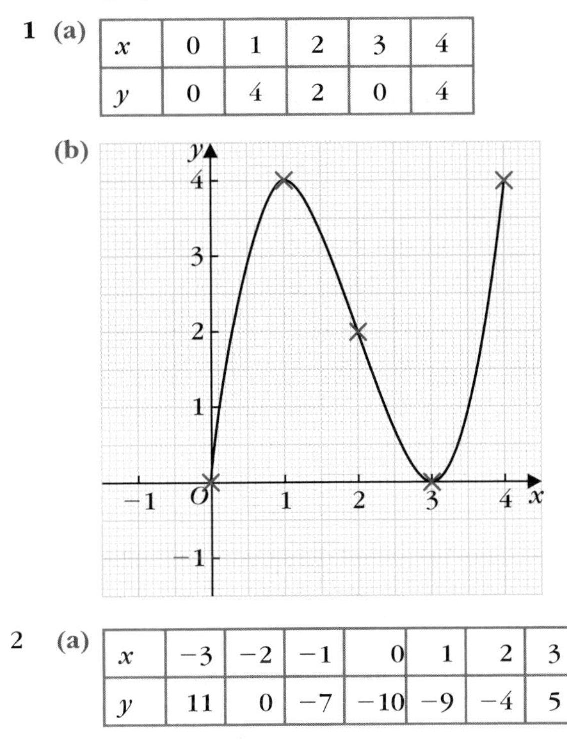

2 **(a)**

x	−3	−2	−1	0	1	2	3
y	11	0	−7	−10	−9	−4	5

(b)

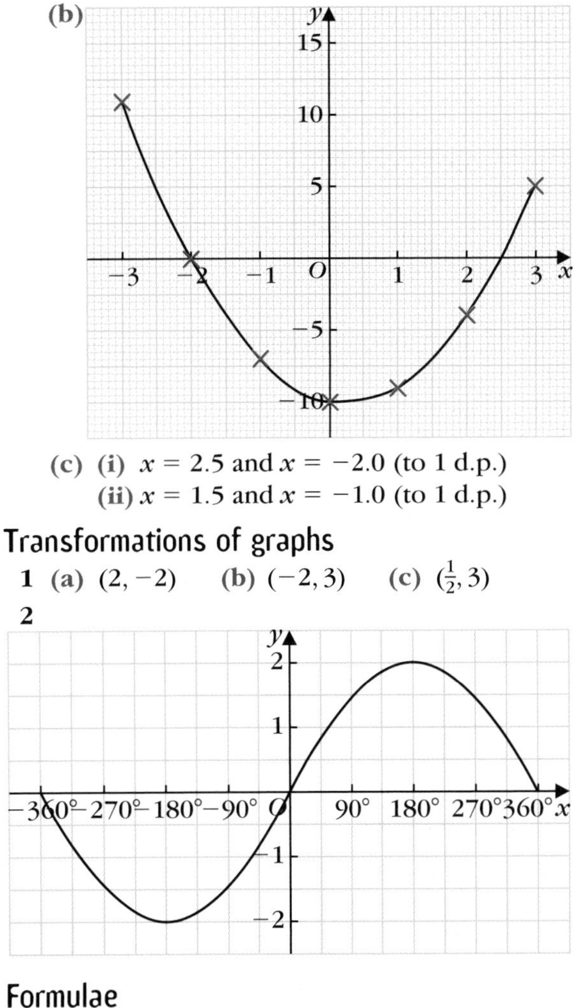

(c) **(i)** $x = 2.5$ and $x = -2.0$ (to 1 d.p.)
(ii) $x = 1.5$ and $x = -1.0$ (to 1 d.p.)

Transformations of graphs

1 **(a)** $(2, -2)$ **(b)** $(-2, 3)$ **(c)** $(\frac{1}{2}, 3)$

2

Formulae

1 $A = 10x + 1$

2 $S = \dfrac{53x + 65y}{100}$

3 $H = 7\frac{1}{2}$

4 $h = 10.2$

Formulae and proof

1 $q = 2(p - 3)$

2 $x = \dfrac{(f + e)^2}{y}$

3 $m = \sqrt{5 - \dfrac{k^2}{9}}$

4 $c = \dfrac{12f + ad}{2a - 3}$

5 Let the odd number be $2n + 1$
$(2n + 1)^3 = 8n^3 + 12n^2 + 6n + 1$
$= 2(4n^3 + 6n^2 + 3n) + 1$
$2(4n^3 + 6n^2 + 3n)$ is even,
so $2(4n^3 + 6n^2 + 3n) + 1$ is odd.

6 Sum of three consecutive numbers $= 3x + 3$.
3 is a factor so the sum of three consecutive numbers is a multiple of 3.

Solving linear equations

1 $b = 5$ **2** $m = 3\frac{1}{2}$ **3** $x = -\frac{2}{3}$

4 $y = 24$ **5** $x = 11$

Solving quadratic and cubic equations

1 $x = 2.4$

2 $x = -5$ and $x = 3$

3 $x = 2.26$ and $x = -0.59$

Solving simultaneous equations

1 $x = 3\frac{1}{2}, y = -2\frac{1}{2}$

2 $x = 6, y = 2$ and $x = -6, y = -4$

Solving inequalities

1 $-3, -2, -1, 0, 1, 2$

2 $-2 < x \leq 2$

3 $x < 8$

4

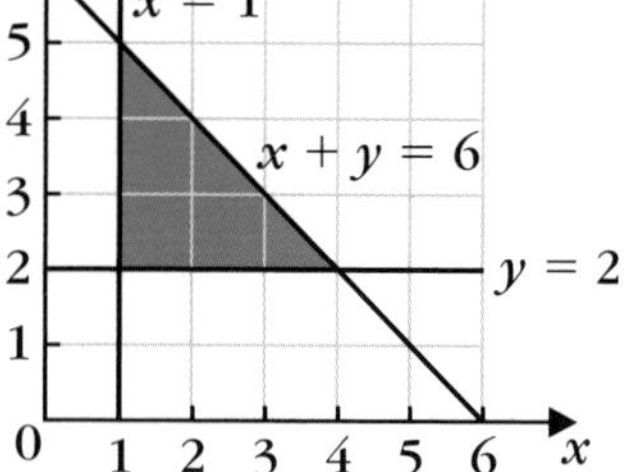

Proportion

1 **(a)** $d = 5t^2$ **(b)** 80 metres **(c)** 2.5 seconds

2 **(a)** $f = \frac{80}{w}$ **(b)** $f = 400$ **(c)** $w = 0.4$

Unit 3 Algebra: subject test

1 **(a)** $8x + 2$ **(b)** $8x + 2 = 180$ **(c)** $x = 22.25$

2 $v = 8$

3

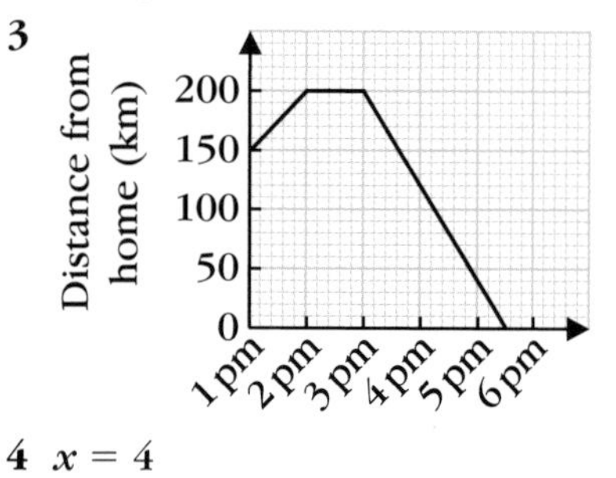

4 $x = 4$

5 $-3, -2, -1, 0, 1, 2, 3$

6 $x = 2.8$

7 **(a)**

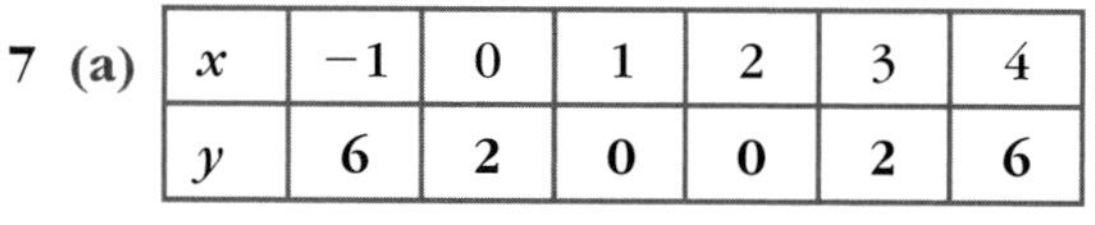

x	-1	0	1	2	3	4
y	6	2	0	0	2	6

(b)

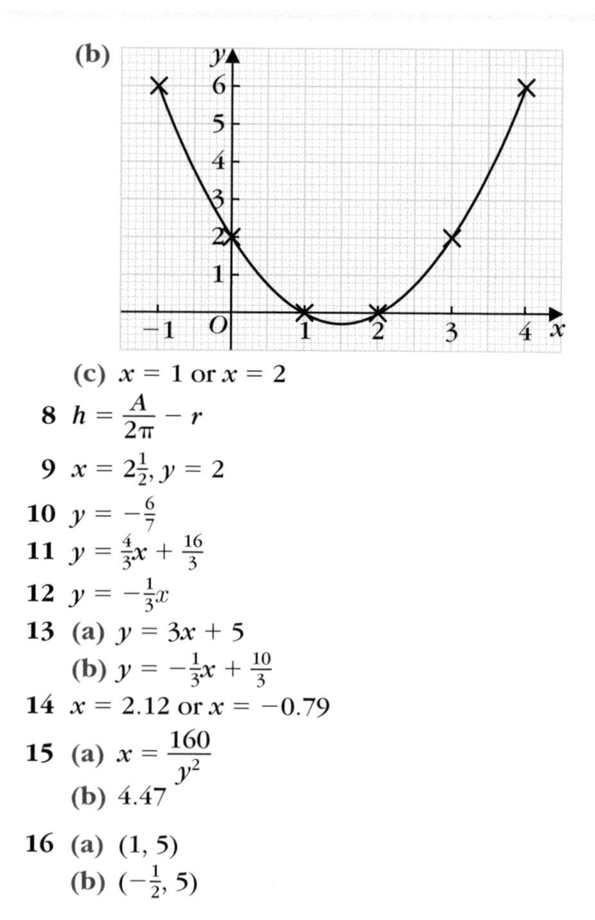

(c) $x = 1$ or $x = 2$

8 $h = \frac{A}{2\pi} - r$

9 $x = 2\frac{1}{2}, y = 2$

10 $y = -\frac{6}{7}$

11 $y = \frac{4}{3}x + \frac{16}{3}$

12 $y = -\frac{1}{3}x$

13 **(a)** $y = 3x + 5$

(b) $y = -\frac{1}{3}x + \frac{10}{3}$

14 $x = 2.12$ or $x = -0.79$

15 **(a)** $x = \frac{160}{y^2}$

(b) 4.47

16 **(a)** (1, 5)

(b) $(-\frac{1}{2}, 5)$

Angles, similarity and congruence

1 **(a)** 105°

(b) $CD = BC = CE$ so triangle BCE is isosceles.
Therefore angle CBE = angle CEB

2 Angle XBC = angle $XCB = 360 \div 5 = 72°$ (exterior angles of regular pentagon)
Angle $BXC = 180 - 72 - 72 = 36°$ (angles in a triangle)

3 **(a)** 20

(b) 3240°

4 **(a)** $BC = ED$ (sides of regular pentagon)
DC is common to both triangles.
Angle DCB = angle CDE (angles of regular pentagon)
Hence triangles DBC and CED are congruent (SAS).

(b) $BC = ED$ (sides of regular pentagon)
$BD = EC$ (corresponding sides of congruent triangles)
EB is common to both triangles.
Hence triangles EBD and BEC are congruent (SSS).

5 162 cm^3

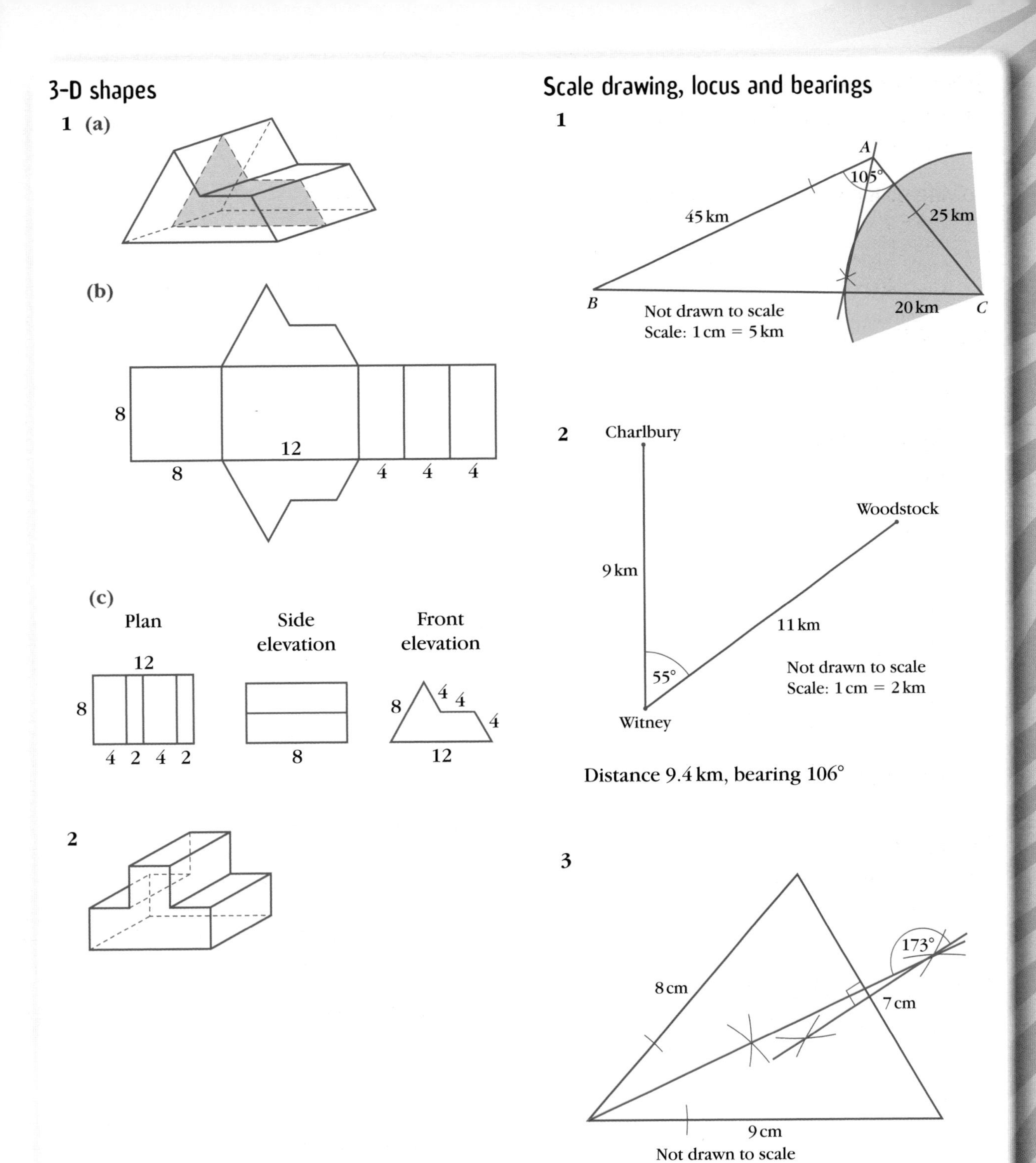
3-D shapes
1 (a)
(b)
8
12
8
4
4
4
(c)
Plan
Side elevation
Front elevation
12
8
4 2 4 2
8
8
4
4
4
12
2
Scale drawing, locus and bearings
1
A
105°
45 km
25 km
B
Not drawn to scale
Scale: 1 cm = 5 km
20 km
C
2
Charlbury
Woodstock
9 km
11 km
55°
Not drawn to scale
Scale: 1 cm = 2 km
Witney
Distance 9.4 km, bearing 106°
3
173°
8 cm
7 cm
9 cm
Not drawn to scale

Circle theorems and proof

1 Angle ECD = angle DEQ = 35°
(alternate segment theorem)
Angle AEC = angle ECD = 35°
(alternate angles)
Angle ABC + angle AEC = 180°
(opposite angles of cyclic quadrilateral)
Angle ABC = 180° − 35° = 145°
Angle BAC = 180° − 145° − 21° = 14°
(angles in a triangle)

2 77°

3 63°

4 Angle PTQ = angle TAQ (alternate segment theorem)
Angle TPA = angle TAP (base angles of isosceles triangle)
So angle PTQ = angle TPQ
So triangle PQT is isosceles

Perimeter, area and volume

1 58 cm

2 Area 44.6 cm^2, perimeter 27.5 cm (to 1 d.p.)

3 Perimeter 488 cm, area 13 100 cm^2 (to 3 s.f.)

4 Volume 12 148 cm^3, surface area 3327 cm^2

Reflections, rotations and enlargements

1

2 (a)

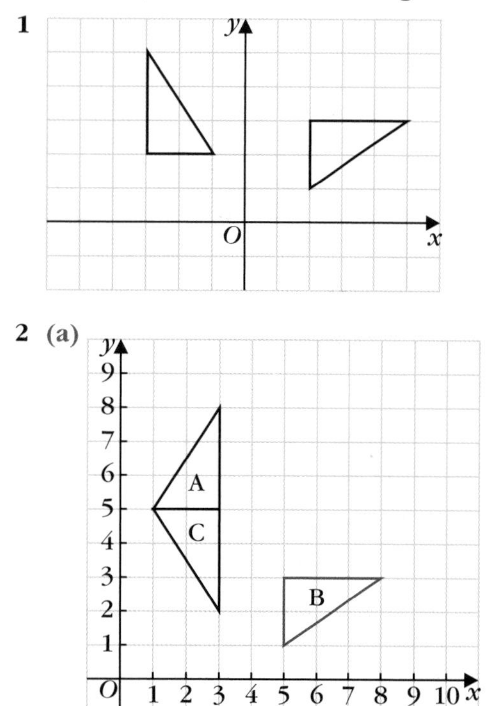

(b) Reflection in the line $y = x$

(c) Rotation 90° clockwise about (5, 5)

3 (a) (b)

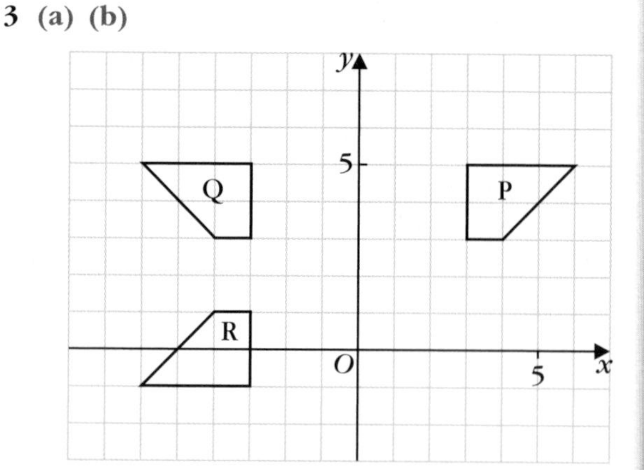

(c) Rotation through 180°, centre (0, 2)

4

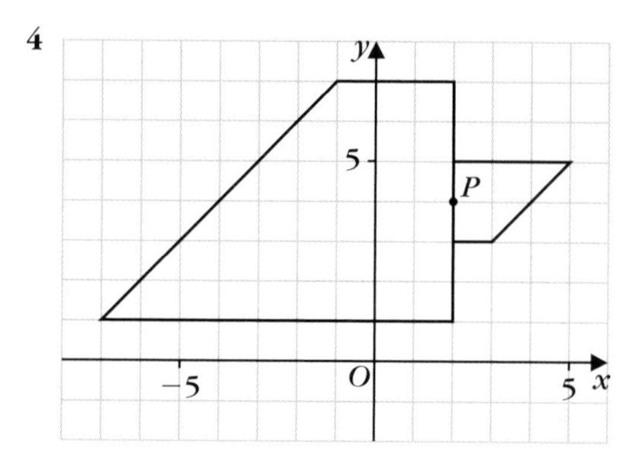

Vectors and translations

1 (a)

(b)

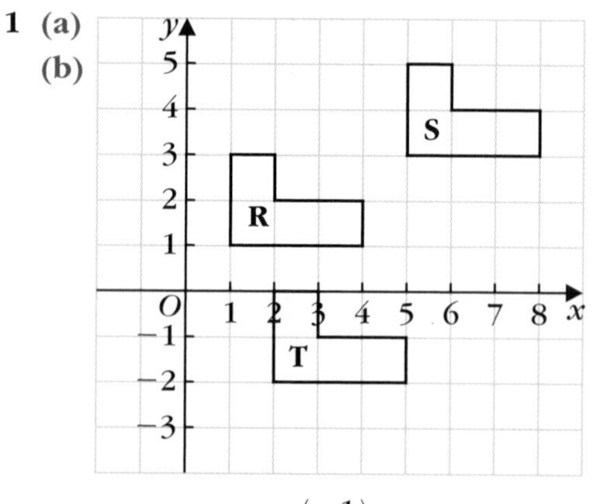

(c) Translation by $\begin{pmatrix} -1 \\ 3 \end{pmatrix}$

2 (a) $\overrightarrow{AP} = 2\mathbf{b}$, $\overrightarrow{AQ} = 4\mathbf{c}$, $\overrightarrow{PQ} = 4\mathbf{c} - 2\mathbf{b}$

(b) $\overrightarrow{AM} = \frac{1}{2}\mathbf{b} + 3\mathbf{c}$

3 (a) $\overrightarrow{AB} = \begin{pmatrix} -1 \\ -7 \end{pmatrix}$, $\overrightarrow{BC} = \begin{pmatrix} 4 \\ 6 \end{pmatrix}$, $\overrightarrow{CA} = \begin{pmatrix} -3 \\ 1 \end{pmatrix}$

(b) $\begin{pmatrix} -1 \\ -1 \end{pmatrix}$

Basic trigonometry

1 **(a)** 16.8 m **(b)** 33.9 m **(c)** 19.3 m

2 **(a)** 66.4° **(b)** 65.8° **(c)** 41.0°

3 **(a)** 12.3° or 167.7°
(b) 127.1° or 307.1°
(c) 43.9° or 316.1°

4 **(a)** 5.46 m **(b)** 27° **(c)** 16.5 m

5

Pythagoras' theorem

1 **(a)** 29 m **(b)** 19.5 m
(c) 15.2 m **(d)** 15 m

2 28.8 cm

3 36 cm²

4 **(a)** 810 cm² **(b)** 20.9 cm (to 1 d.p.)

5 10.6 cm (to 1 d.p.)

6 $4 \pm \sqrt{6}$

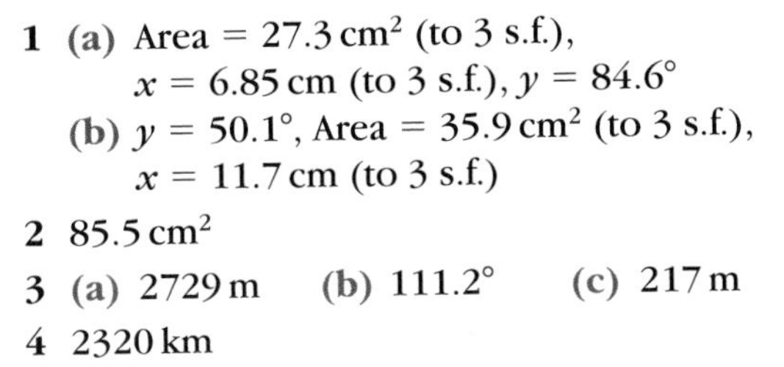

Advanced trigonometry

1 **(a)** Area = 27.3 cm² (to 3 s.f.),
x = 6.85 cm (to 3 s.f.), y = 84.6°
(b) y = 50.1°, Area = 35.9 cm² (to 3 s.f.),
x = 11.7 cm (to 3 s.f.)

2 85.5 cm²

3 **(a)** 2729 m **(b)** 111.2° **(c)** 217 m

4 2320 km

Unit 3 Shape, space and measure: subject test

1

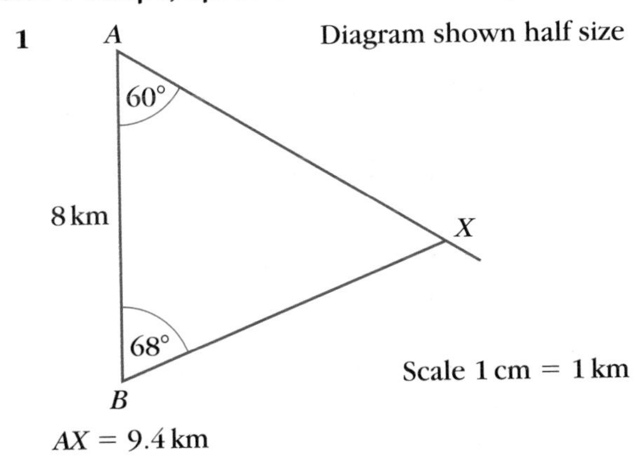

AX = 9.4 km

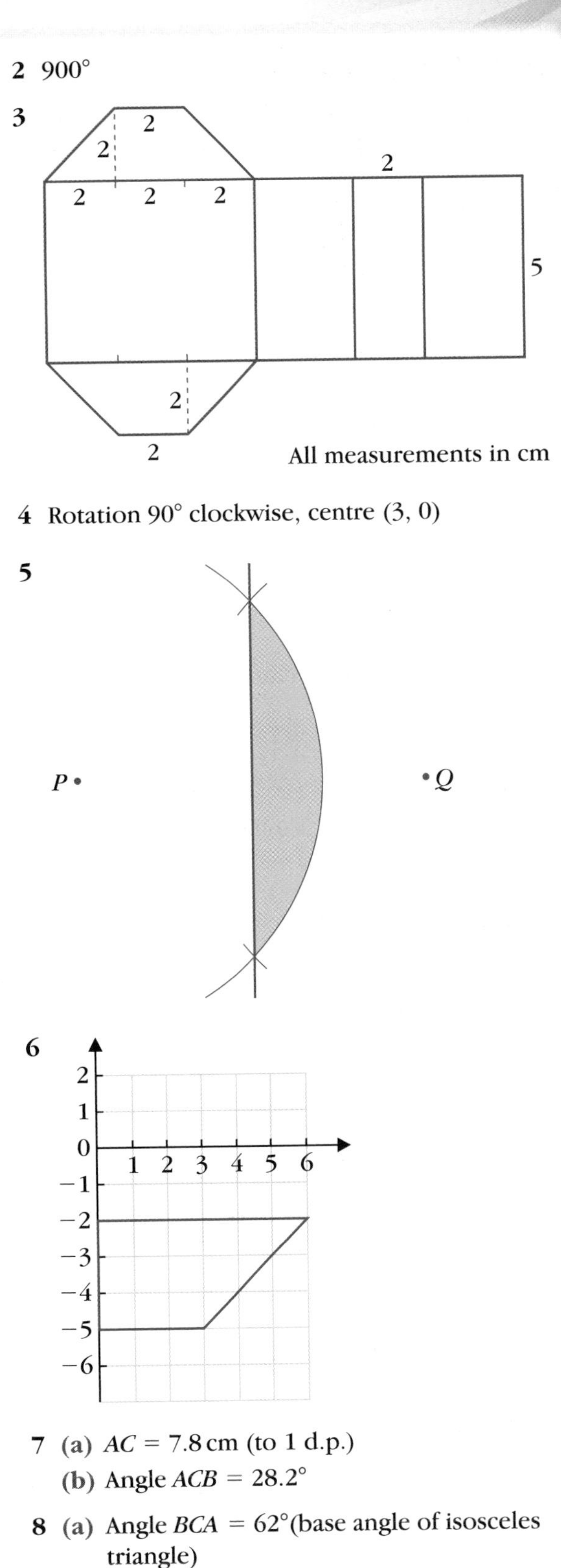

2 900°

3

All measurements in cm

4 Rotation 90° clockwise, centre (3, 0)

5

6

7 **(a)** AC = 7.8 cm (to 1 d.p.)
(b) Angle ACB = 28.2°

8 **(a)** Angle BCA = 62°(base angle of isosceles triangle)
(b) Angle SAB = 62° (alternate segment theorem)
(c) Angle ABC = 56° (angle sum of triangle)
Angle ADC = 124° (opposite angles of cyclic quadrilateral)

9 $BC = DC$
Triangle CBD is isosceles
So angle CBD = angle $CDB = x$
and angle ABC = angle $EDC = 90° + x$
$AB = ED$ (sides of square)
Hence triangles ABC and EDC are congruent (SAS)
$AC = EC$ (corresponding sides of congruent triangles)
So triangle ACE is isosceles.

10 **(a)** **(i)** $249\,mm^2$ (to 3 s.f.)
(ii) 21.7 mm (to 3 s.f.)
(b) $17.1\,cm^2$ (to 3 s.f.)

11 **(a)** $\overrightarrow{PT} = \mathbf{a} + \mathbf{b} + \mathbf{c}$
$\overrightarrow{TQ} = -\mathbf{b} - \mathbf{c}$
$\overrightarrow{UP} = \mathbf{c}$
$\overrightarrow{SU} = -\mathbf{b} - \mathbf{c}$
(b) **(i)** $\overrightarrow{QT}$ and $\overrightarrow{US}$
(ii) $\overrightarrow{ST}$ and $\overrightarrow{UQ}$
(iii) $\overrightarrow{PT}$ and $\overrightarrow{UR}$

12 **(a)** 6.1 cm and 14.8 cm **(b)** 18.6 cm

Unit 3 Examination practice paper Calculator

1 £66.27
2 455 revolutions
3 **(a)** **(i)** a^7 **(ii)** $5xy^{-1}$ **(b)** $n = 6$
4 $x = 2.1$
5 24 cm
6 56
7

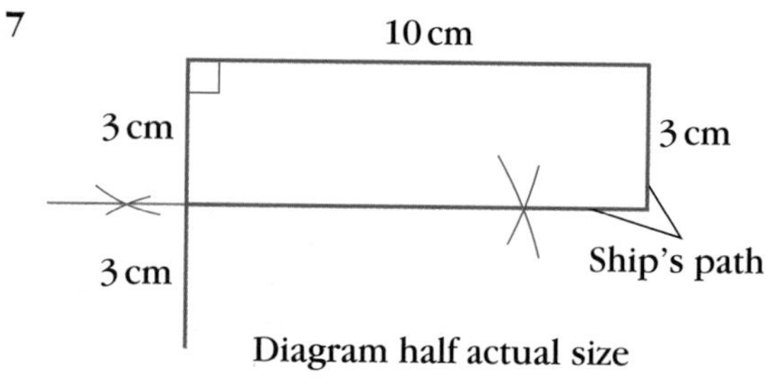

Diagram half actual size

8 $3\,000\,000\,cm^3$ or $3 \times 10^6\,cm^3$
9 £80
10 £249.73
11 0.119 437 862
12 68.9 km/h (to 1 d.p.)
13 10.96 cm (to 2 d.p.)
14 $DA = DG$ so triangle DAG is isosceles
$\angle DAG = \angle DGA$
$\angle BAD = \angle DGF = 90°$ (angles of squares)
So $\angle BAG = \angle FGA$
$BA = FG$ (sides of congruent squares)
So triangles GBA and AFG are congruent (SAS)
15 $279\,cm^2$ (to 3 s.f.)
16 **(a)** $s = 5t^2$ **(b)** 4 seconds
17 $3.46x^2$ (to 2 d.p.)

Unit 3 Examination practice paper Non-calculator

1 8 kg
2 **(a)** $x = 5$ **(b)** $y = 7.5$ **(c)** $x = \frac{1}{2}$
3 $-2, -1, 0, 1$
4 **(a)** $(0, 1)$ **(b)** $y = 3x + 1$
(c) $y = -\frac{1}{3}x + 4\frac{1}{3}$ or $3y + x = 13$
5 **(a)** $\frac{31}{20}$ **(b)** $\frac{18}{5}$
6 12 sides
7

8 $x = 1.5, y = -3$
9 $r = \sqrt{\frac{A}{\pi}}$
10 $a = 29, b = 12$
11 **(a)** Translation 6 right and 2 up or $\begin{pmatrix} 6 \\ 2 \end{pmatrix}$
(b) Rotation 90° clockwise with centre (2, 2)
12 **(a)** $4\mathbf{a}$ **(b)** $-4\mathbf{a} + 2\mathbf{b}$
(c) $2\mathbf{a} + \mathbf{b}$ **(d)** $2\mathbf{a}$
13 $x = 18°$
14 **(a)** 0.0026 **(b)** 2.72×10^{-3}
15 $x = -0.2, y = 1.4$ or $x = -1, y = -1$
16 **(a)** $x = 15$ **(b)** $y = 6.4$
17 **(a)** angle DBC = angle $DCT = 25°$ (alternate segment theorem)
(b) angle ADC + angle $ABC = 180°$ (opposite angles of a cyclic quadrilateral)
angle $ADC = 75°$

Index